D1692671

Stochastic Programming with Multiple Objective Functions

Mathematics and Its Applications *(East European Series)*

I. M. Stancu-Minasian

The Academy of Economic Studies,
Department of Economic Cybernetics, Bucharest, Romania

Stochastic Programming

with
Multiple Objective Functions

Translated from Romanian by
Victor Giurgiuţiu, Ph. D.

The National Institute
for Scientific and Technical Creation,
Bucharest, Romania

EDITURA ACADEMIEI
Bucureşti, Romania

D. Reidel Publishing Company
A MEMBER OF THE KLUWER ACADEMIC PUBLISHERS GROUP
Dordrecht / Boston / Lancaster

Library of Congress Cataloging in Publication Data

Stancu-Minasian, I. M.
Stochastic programming with multiple objective functions.

(Mathematics and its applications. East European series)
Translation of : Programarea stocastică cu mai multe funcţii obiectiv.
Bibliography : p.
Includes index.
1. Stochastic programming. I. Title. II. Title : Stochastic programming. III. Series : Mathematics and its applications (D. Reidel Publishing Company). East European series.
T57.79.S7213 1984 519.7′2 84–4763
ISBN 90–277–1714–1 (Reidel)

Distributors for the U.S.A. and Canada
Kluwer Academic Publishers,
190 Old Derby Street, Hingham, MA 02043, U.S.A.

Distributors for Albania, Chinese People's Republic, Cuba, Czechoslovakia, German Democratic Republic, Hungary, Korean People's Republic, Mongolia, Poland, Romania, the U.S.S.R., Vietnam, and Yugoslavia
Editura Academiei, Bucharest, Romania

Distributors for all remaining countries
Kluwer Academic Publishers Group,
P.O. Box 322, 3300 AH Dordrecht, Holland.

Original title : "*Programarea stocastică cu mai multe funcţii obiectiv*".
First edition published in 1980 by Editura Academiei.
Second edition published in 1984 by Editura Academiei and D. Reidel Publ. Co.

Printed in Romania

To the memory of my mother

Table of Contents

Editor's Preface XI
Preface to the Romanian Edition XIII
Preface to the English Edition XV
Introduction 1

CHAPTER 1 / *Deterministic mathematical programming with several objective functions* 7
1.1. Problem Definition 7
1.2. The Relation between the Various Methods of Solving Mathematical Programming Problems with Multiple Objective Functions 16
1.3. Goal Programming 28
1.4. Maximum Global Utility Method 33
1.5. The Connection with the Theory of Games 43
1.6. POP Method 52
1.7. STEM Method 57
1.8. The Case of Two Objective Functions 62
1.9. A Note on the Present State of the Art 68

CHAPTER 2 / *Stochastic programming with one objective function* 71
2.1. Distribution Problems 72
2.2. Minimum-Risk Problems in Stochastic Programming 92
2.3. Two-Stage Programming under Uncertainty 98
2.4. The Complete Problem 104
2.5. Wets' Algorithm 109
2.6. Chance-constrained Programming 113

CHAPTER 3 / *Approaches to stochastic programming with multiple objective functions* 119
3.1. Chebyshev's Stochastic Problem. The Distribution Problem 121
3.2. Stochastic Fractional Programming. The Distribution Problem 135
3.3. Goal Programming. The Stochastic Case 141
3.4. Group Decision-Making in Stochastic Programming 153
3.5. Efficient Solutions in Stochastic Programming . . 165
3.6. The PROTRADE Method 172
3.7. The Case of Discrete Distributions 183

CHAPTER 4 / *Some generalizations of the minimum-risk problem and Kataoka's problem* 187
4.1. The Minimum-Risk Problem and Kataoka's Problem 187
4.2. Generalizations 201
4.3. The Minimum-Risk Approach to the Chebyshev Problem 223

CHAPTER 5 / *Multiple minimum-risk solutions in stochastic programming* 227
5.1. Problem Definition 227
5.2. The Case of Two Objective Functions 235
5.3. The Case of Three Objective Functions 242
5.4. The Case of r Objective Functions 252
5.5. A Different Approach 256
5.6. An Interactive Approach to Stochastic Programming with Multiple Objective Functions 261

CHAPTER 6 / *The transportation problem with multiple objective functions* 269
6.1. The Deterministic Case 269
6.2. The Stochastic Case 282

CHAPTER 7 / *Applications in economy* 287
7.1. Production Planning with Multiple Efficiency Criteria 287

7.2. Another Example of Production Planning 293
7.3. A Problem of Machinery Loading 297
7.4. A Mixing Problem 298
7.5. Optimisation of Flour Transportation 300
7.6. Efficient Use of Production Capacity in a Metallurgical Workshop 302
7.7. The Mathematical Model for Pig-iron Production . . 304
7.8. Operation Scheduling 305
7.9. The Assignment Problem 307
7.10. The Input-Output Model for Resource Allocation . . 309
References 313
Index 329

Editor's Preface

Growing specialization and diversification have brought a host of monographs and textbooks on increasingly specialized topics. However, the 'tree' of knowledge of mathematics and related fields does not grow only by putting forth new branches. It also happens, quite often in fact, that branches which were thought to be completely disparate are suddenly seen to be related.

Further, the kind and level of sophistication of mathematics applied in various sciences has changed drastically in recent years : measure theory is used (non-trivially) in regional and theoretical economics ; algebraic geometry interacts with physics ; the Minkowsky lemma, coding theory and the structure of water meet one another in packing and covering theory ; quantum fields, crystal defects and mathematical programming profit from homotopy theory ; Lie algebras are relevant to filtering ; and prediction and electrical engineering can use Stein spaces. And in addition to this there are such new emerging subdisciplines as 'completely integrable systems', 'chaos, synergetics and large-scale order', which are almost impossible to fit into the existing classification schemes. They draw upon widely different sections of mathematics.

This program, Mathematics and Its Applications, is devoted to such (new) interrelations as exempli gratia :

— a central concept which plays an important role in several different mathematical and/or scientific specialized areas ;

— new applications of the results and ideas from one area of scientific endeavor into another ;

— influences which the results, problems and concepts of one field of enquiry have and have had on the development of another.

The Mathematics and Its Applications programme tries to make available a careful selection of books which fit the philosophy outlined above. With such books, which are stimulating rather than definitive, intriguing rather than encyclopaedic, we hope to contribute something towards better communication among the practitioners in diversified fields.

Because of the wealth of scholarly research being undertaken in the Soviet Union, Eastern Europe, and Japan, it was decided to devote special attention to the work emanating from these particular regions.

Thus it was decided to start three regional series under the umbrella of the main MIA programme.

The availability of, at first substantial, computing power certainly has been very important in the rise of mathematical programming as a separate mathematical specialization. And, as so often happens, the computing advances have sparked off whole new series of theoretical investigations and advances.

It is a simple observation that for many of the practical problems in management, industry and government to which one would like to apply the techniques and results of mathematical programming, there is not one natural objective function. Also simple examples show (similar to the Endorsed paradox and Arrow impossibility theorems) that there are, in general, no good ways of aggregating several criteria into one objective function. But maybe sometimes there are. Worse, even when there is a natural objective function, but stochastic elements come into play maximizing the expectation will often involve unacceptably large variancces.

Thus, a new area of research arises; a relative newcomer, which deals with programming problems with multiple objectives and stochastic elements. Besides being of practical importance, the area has also generated interesting mathematical programs (and it continues to do so), e.g., the problem of characterizing efficient paths (in the dynamic case) of the set of efficient solutions in the static case, to name one of the oldest and most obvious.

The unreasonable effectiveness of mathematics in science . . .

Eugene Wigner

Well, if you knows of a better 'ole, go to it.

Bruce Bairnsfather

What is now proved was once only imagined.

William Blake

As long as algebra and geometry proceeded along separate paths, their advance was slow and their applications limited.

But when these sciences joined company they drew from each other fresh vitality and thenceforward marched on at a rapid pace towards perfection.

Joseph Louis Lagrange

Amsterdam

Michiel Hazewinkel

Preface to the Romanian Edition

A central problem in today decision theory is how to make a decision when several objectives are followed simultaneously. Such a situation arises in almost all problems of economic and social practice, where a multitude of objectives has become a rule rather than an exception. Such problems cannot be squeezed within the narrow framework of some unique objective, such as profit maximization or cost minimization. Thus, it is often difficult to identify the programme objectives in many public departments — e.g. education, welfare, urbanism — and express them in terms compatible with the classical methods of decision-theory. Additionally, these objectives may be even contradictory.

Recently, the management science underwent a far-reaching change in that the main emphasis was placed on a comprehensive multiple criteria decision theory. But in this enterprise, one is often confronted with problems of unprecedented complexity. Thus, the new methods entail consideration of psychological, humanitarian, juridical and political aspects of decision-making and should include the existing methods for the classical case of simple decisions. In this way, a new interdisciplinary science is about to be born.

This book outlines the state of the art in an important chapter of multiple criteria decision theory — the stochastic programming with several objective functions. The birth and subsequent development of this chapter of operation research came off as an answer to several legitimate objections to the classical deterministic programming: the random character of the actual data and the ambiguous, "non-mathematical" definition of the notion of "objective" which requires consideration of several functions to be optimized.

The author made commanding efforts in systematizing the comprehensive literature on the subject. Add to this the great number of original contributions, some of which have already secured international recognition. Mention should be also made that the theoretical results given herein may be readily used in applications as this has been successfully verified in practice.

The present book should be considered as a valuable Romanian contribution to the present efforts in establishing a mathematical theory to deal with the complex problems raised by nature and society.

MARIUS IOSIFESCU, D. Sc.

Preface to the English Edition

Roughly speaking, this book preserves the main ideas presented in the Romanian edition. However, numerous changes and additions have been performed. Thus, a note on the present state of the art in deterministic mathematical programming with multiple objective functions has been appended to Chapter 1 (Section 1.9). Chapter 4 has been extended to a minimum-risk approach to the Chebyshev problem (Section 4.3.). Chapter 5 contains additionally an interactive approach to stochastic programming with multiple objective functions (Section 5.6). Numerous examples were added in Chapter 7 to illustrate applications in presentday economy (Sections 7.6 – 7.10). These examples deal with the efficient use of a production capacity in a metallurgical workshop, the mathematical model for pig-iron production, operation scheduling, the assignment problem and the input-output model for resource allocation. Additionally, the bibliography of the English edition has been substantially extended.

I would like to express my gratitude to Mr. Victor Giurgiuțiu for his well-cast translation. Thanks are also due to Editura Academiei and D. Reidel Publishing Company for their conjoined efforts in publishing this book. Last but not least, I am under obligation to Mrss. Sorana Gorjan and Liana Petrescu, and Mr. Petre Mocanu for their commanding work in editing this book.

Dr. I. M. STANCU-MINASIAN

Bucharest, January 1984

Introduction

The last decades witnessed a deeper thrust of science into the production of material goods. Thus, the successful management of research in such domains and the basic functions of management and scientific organization can be expressed almost entirely through quantitative methods. This explains the increasing use of mathematical methods in applied work, as well as the present efforts to improve them.

In the present condition of a complex economy under continuing progress, the large-scale use of modern computational methods has become peremptory. With the advent of electronic computers, mathematical methods have become increasingly involved in a variety of domains.

The importance of these research domains is likewise highlighted by the hosts of monographs published of late, the studies on the efficiency of enterprises and the numerous approaches to the use of mathematical methods in economy.

Special heed is paid to mathematical programming, the methods of which can lead to an economical use of resources in attaining an economic objective. Basically, the principal merit in using mathematical programming techniques as objective methods for developing decisions is that by considering a large number of factors and the relationships between them, we can calculate the effect of the variation of these factors with sufficient accuracy.

The earliest approach dealt with the linear programming case with one objective function in terms of post-optimization, parametrization and transportation. Practical requirements called for a mathematical programming with several objective functions.

2 — 583

Two objections have been so far raised to linear programming, and these hold even when the linearity assumption is found to be satisfactory. The first states that in applications at least some of the coefficients of the linear programming problems must be assumed to be random. The answer to this objection is furnished by stochastic linear programming, which underwent a rapid development over the last two decades.

The second objection raised to mathematical programming as a whole and to the classical decision theory states that in practical problems in economics, engineering or sociology, as well as in the language used by the expert, the terms "optimization" and "optimum" are not as strict in meaning as they are in contemporary mathematics, but rather vague and cannot be expressed by means of only one criterion or function to be optimized. The answer furnished by mathematical programming to these objections is the multiple objective function programming, which soon became the source of a comprehensive literature.

This book shows how mathematical programming is toying to answer both objections using a model that should be as faithful as possible to real problems. This orientation, called the stochastic programming with multiple objective functions, is of undeniable service in practice but is still confronted with intractable theoretical difficulties.

This modern topic is coupled with numerous practical applications concerning the simultaneous optimization of several functions with constrained independent variables.

The book presents the results obtained by the author in a department of mathematical programming where there exist several optimum criteria of stochastic type. It should be noted that such problems fall in the province of operational research, decision theory, cybernetics and modern management as well.

Mathematical programming with multiple objective functions — also called vector programming — is a recent development of mathematical programming, and emerged from an attempt to tackle the problems raised by the present development of science, engineering, industry, economy, etc. Due to the complexity of these problems, several objectives had to be incorporated in the optimization process. Basically, the problem consists in optimizing several objective functions (some of which must be maximized, some minimized) provided the variables satisfy the linear and nonlinear constraints.

The main difficulty in solving vector programming problems is that it is generally impossible to obtain a feasible solution which should simultaneously optimize all the efficiency functions. As a rule, not all the objective functions will reach their optimum (maximum or minimum) values at the same point x of the feasible domain. The optimal solution for one function is not ordinarily optimal for the others, and sometimes it may be even disadvantageous. That is why various approaches are adopted to find the best compromise solution (or solutions).

Many mathematical programming problems are stochastic because some of the data involved are random variables. This may be due to

— errors and variations in the problem parameters, which are often associated with probabilities;

— risk and uncertainty, which may sometimes allow a significant numerical representation of the utility function of the decision-maker (e.g. the von Neumann axiomatic system);

— the need for optimum decision rules connected with some statistical decision functions, etc.

Hence, one has to study vector programming under the assumption that some of the problem parameters are random variables.

We should note that even in the case of one objective function with random coefficients of given distribution, we must consider two deterministic objective functions: the maximization of the mean value of the objective function and the minimization of the objective function variance. It is not sufficient to find a solution which maximizes the mean value, since this may become unacceptable when the resulting variance is large.

Several problems of stochastic mathematical programming with multiple objective functions are examined in this book. The text divides into seven chapters. Each chapter brings together the author's results, some of which have already been published in several papers and in his doctoral thesis, and some contributions by other writers. Results already known are given a unified treatment.

Chapter 1 is an introduction to the notions and methods of deterministic programming with several objective functions.

Stress is laid on the maximum global utility technique (which is partly due to the author) for multiple criteria problems, which is extensively applied in economic problems.

Chapter 2 deals with the principal problems arising from stochastic programming with one objective function. Among these, let us mention the distribution of the optimum, two-stage programming and chance-constrained programming. Most of the results already known are given original interpretations.

Chapter 3 gives solutions to several difficult problems of stochastic mathematical programming with several objective functions. Section 3.1 deals with the stochastic Chebyshev problem and the distribution shape of the optimum (Theorem 3.2). Section 3.2 discusses stochastic fractional programming and develops the distribution of the optimum (Theorem 3.3). Theorems 3.2 and 3.3 generalize the results of stochastic linear programming concerning the explicit formulae for the distribution of the optimum in linear programs with cost coefficients, affine functions of random variable. Section 3.3 discusses four variants of goal programming in stochastic terms and derives the deterministic mathematical programming problem. The problem of group-decision in stochastic programming with multiple objective functions is treated in Section 3.4. Section 3.5 deals with the stochastic programming problem with one objective function to which we can associate two deterministic objective functions (the mean and the variance). This case is framed against the general theory presented in Chapter 1, and thus the results obtained by Bereanu, Charnes and Cooper are here derived in a distinct approach. Theorems defining the efficient solutions for this particular case are also given. Section 3.6 presents the PROTRADE method for solving stochastic programming problems with multiple objective functions, as was developed by Goicoechea, Duckstein and Bulfin. This method can be considered to be a stochastic analog of the POP and STEM methods presented in Chapter 1. Section 3.7 examines the discrete distribution case, which is a particular case of the problem presented in Section 3.1.

Chapter 4 presents Kataoka's problem, i.e. the determination of the smallest (or largest) value of the objective function such that the probability of obtaining such a result exceeds a certain threshold. The subsequent minimum-risk problem consists in the maximization of the probability that the value of the objective function exceeds a certain threshold. Some of the results obtained by Bereanu, Cooper, Dragomirescu, Geoffrion and Kataoka concerning the minimum risk problem and Kataoka's problem are sketched in Section 4.1.

Section 4.2 deals with more general cases of the above two problems : the case when the objective function appears as a product or ratio of linear functions. Deterministic equivalents are given for the two problems (Theorems 4.7, 4.7′, 4.8, 4.9, 4.9′ and 4.10). Sub-section 4.2.5 discusses Kataoka's problem for a ratio of linear functions. It is shown that under certain assumptions the equivalent deterministic problem has a finite optimum solution (Theorem 4.11) and that the objective function is explicitly quasi-concave (Theorem 4.12). Next, the solution methods used for the equivalent deterministic problem (Theorems 4.14 and 4.15) employ the transformation method and the duality theorems. Section 4.3. is a minimum-risk approach to the Chebyshev problem.

Chapter 5 deals with a more general case of the minimum risk problem, i.e. the maximization of the probability that the value of r objective functions exceeds some given thresholds. In order to solve this problem under the ordinary normality assumptions for the random coefficients, the author suggests a method which consists in the sequential solution of some minimum-risk problems (containing also quadratic constraints). The presentation of this method is an opportunity for the author to introduce the notion of "multiple minimum-risk solution", and a relaxation-type method for obtaining such solutions in the case of two or three objective functions (Theorems 5.1 and 5.2) and in the general case of r objective functions (Theorem 5.3). In addition to some quantitative results, numerical solution methods are suggested. An alternate solution to the minimum-risk problem for r objective functions is shown in Section 5.5. Section 5.6 gives an interactive approach to stochastic programming with multiple objective functions.

The transportation problem with several objective functions is the object of Chapter 6. The first section refers to the deterministic case and describes two solution methods using a synthesis function. The first method employs a utility function and the second resorts to a fractional objective function. A local minimum criterion is given here for the second problem (Theorem 6.2), which is then solved by use of a change of variables. The second part of Chapter 6 deals with the stochastic case and applies the results obtained by Szwarc and Corban to the problem with a fractional objective function.

Besides their mathematical relevance, the methods presented have large applicability. This is illustrated in Chapter 7 which gives several concrete applications in economics.

Most of the theoretical results presented herein derive from the practical problems tackled by the author at the Laboratories of the Department of Economic Cybernetics, the Academy of Economic Studies. This reveals by itself the manner in which mathematics is entangled with economic practice.

The list of references appended at the end of the book gives only a glimpse of the huge number of works published on this topic.

The problems treated in this book fall into three chapters of mathematical programming: stochastic programming, multiple objective function programming and fractional programming. The author has long dwelled on these topics and this work came off as five comprehensive bibliographies [137, 139, 144, 146, 235] which cover more than 2000 titles. Because of the limited space of this book, reference was only made to those works whose results are used in the text.

Acknowledgements. Many of the results presented here were obtained as I was preparing my doctoral thesis under the supervision of M. Iosifescu, D. Sc., whose competent guidance and kind understanding have been invaluable to me. I take this opportunity to express my gratitude to him.

I would also like to thank B. Bereanu, D. Sc., and dr. C. Bergthaller for their interesting suggestions to many of the problems treated in this book. Useful remarks are also acknowledged to Professor dr. I. Marușciac and dr. T. Postelnicu.

My warmest thanks are also due to Staff of the Laboratories of the Chair of Cybernetics, the Academy of Economic Studies, and to Academician Manea Mănescu in particular, for the facilities they provided in carrying out my work.

Chapter 1

Deterministic mathematical programming with several objective functions

1.1 PROBLEM DEFINITION

Consider the linear programming problem with several objective functions

$$(1.1) \quad \text{(optimize) } F = Cx$$

subject to

$$(1.2) \quad Ax \leqslant b$$

$$(1.3) \quad x \geqslant 0,$$

where

$A = ((a_{ij})), i = 1, \ldots, m; j = 1, \ldots, n$ is an $m \times n$ matrix;

$b = (b_1, \ldots, b_m)' \in R^m$ is an m-dimensional column vector;

$x = (x_1, \ldots, x_n)' \in R^n$ is an n-dimensional unknown column vector;

$F = (F_1, \ldots, F_r)'$ is an r-dimensional column vector of objective functions;

and

$C = ((c_{ij})), i = 1, \ldots, r; j = 1, \ldots, n$ is an $r \times n$ matrix, and the prime (′) denotes the transposed of a vector.

Denote by $D = \{x = (x_1, \ldots, x_n)' \mid Ax \leqslant b, x \geqslant 0\}$ the set of feasible solutions to problem (1.1)—(1.3).

Note that "optimize" means here either "maximize" or "minimize". With no loss of generality one can assume in what follows that for optimum all the objective functions must be maximized. If not so from the beginning, this can be achieved through the usual transformation $\min F(x) = -\max[-F(x)]$.

The mathematical programming problem with several objective functions consists in finding the vector (or vectors) $x^* = (x_1^*, \ldots, x_n^*)' \in D$ which is *best* for the assembly of objective functions F_h, $(h = 1, \ldots, r)$.

Since the definition of x^* is somewhat ambiguous (*the best* is not quantifiable and insufficiently clear), several views have already been expressed with respect to the definition and solution of a multiple criteria problem.

Since the vector space V of the vector objective functions $F(x)$, defined as

$$V = \{F(x) \mid F(x) = (F_1(x), \ldots, F_r(x))' ; x \in D\}$$

is not totally ordered, it is not usually possible to find a feasible solution to optimize all objective functions simultaneously.

Hence, for more than one objective function, an optimal solution for one function is not necessarily optimum for others. Thus, one has to introduce the notion of the *best compromise solution*, also known as *nondominated solution*, *efficient solution*, *noninferior solution*, *Pareto's optimum solution*, *etc.*

The various ways of defining x^* can be grouped as follows:
1° Let x^* be the vector which optimizes (maximizes or minimizes) a synthesis-function of the r efficiency functions:

$$h(F) = h[F_1, \ldots, F_r].$$

Several expressions are possible for h [74]:

a) $h[F_1, \ldots, F_r] = \underset{i=1,\ldots,r}{\text{optimum}} \{F_i(x)\}$.

For instance, if F_i are *maximum* functions, one takes

$$h[F_1, \ldots, F_r] = \min_{i=1,\ldots,r} \{F_i(x)\}$$

and then maximizes it, i.e.

$$\max_{x \in D} \min_{1 \leq i \leq r} \{F_i(x)\}.$$

If F_i are *minimum* functions, one considers

$$h[F_1, \ldots, F_r] = \max_{i=1,\ldots,r} \{F_i(x)\}$$

and then minimizes it, i.e.

$$\min_{x \in D} \max_{1 \leqslant i \leqslant r} \{F_i(x)\}.$$

Thus, one obtains the so-called *Chebyshev problem*, the solution of which can be obtained through several algorithms available in the literature [27].

b) For $r = 2$, one can consider that

$$h[F_1, F_2] = \frac{F_1(x)}{F_2(x)},$$

which entails an ordinary fractional programming problem

c) $h[F_1, \ldots, F_r] = \sum_{i=1}^{r} \alpha_i [F_i(x)]^{\beta_i}, \quad \alpha_i \geqslant 0, \; \beta_i > 0.$

d) $h[F_1, \ldots, F_r] = - \sum_{i=1}^{r} \alpha_i \exp[-F_i(x)], \quad \alpha_i \geqslant 0.$

e) $h[F_1, \ldots, F_r] = \prod_{i=1}^{r} F_i.$

2° Let x^* be the vector which minimizes the function [167]

$$h(x) = h(\Psi_1(x - X_1), \ldots, \Psi_r(x - X_r)), \tag{1.4}$$

where $X_j = (x_{1j}, \ldots, x_{nj})'$ $(j = 1, \ldots, r)$ is the optimal solution of the objective function F_j, and Ψ_k is the distance function between the vector $x \in D$ and the optimum solution X_k of the function F_k. Certain expressions for h and Ψ_k can lead to particular forms of (1.4), e.g.

a) $h(x) = \sum_{k=1}^{r} \alpha_k \sum_{j=1}^{n} (x_j - x_{jk})^2, \ \alpha_k \geqslant 0$;

b) $h(x) = \sum_{k=1}^{r} \alpha_k \sum_{j=1}^{n} |x_j - x_{jk}|, \ \alpha_k \geqslant 0$;

c) $h(x) = \max_{k=1,\ldots,r} \Psi_k(x - X_k)$;

d) $h(x) = \sum_{k=1}^{r} \Psi_k(x - X_k)$;

e) $h(x) = \prod_{k=1}^{r} \Psi_k(x - X_k)$.

If one takes

$$\Psi_k(x - X_k) = \alpha_k |F_k(X_k) - F_k(x)|, \ \alpha_k \geqslant 0$$

and the function $h(x)$ conformal to c), then the best compromise x^* is that for which

$$\Phi(x^*) = \min_{x \in D} \max_{k=1,\ldots,r} \alpha_k |F_k(X_k) - F_k(x)|.$$

If one takes

$$\Psi_k(x - X_k) = \alpha_k |F_k(X_k) - F_k(x)|, \ \alpha_k \geqslant 0$$

and the function $h(x)$ according to d), then the vector x^* is determined such that

$$\Phi(x^*) = \min_{x \in D} \sum_{k=1}^{r} \alpha_k |F_k(X_k) - F_k(x)|.$$

In economics or physics, the interpretation of such synthesis functions as in 1° and 2° is not always straightforward, and hence its particular choice depends very much on the problem under consideration.

3° Let x^* belong to a set of *efficient points* [28], [29], [35], [90], [95] defined as follows : $x^* \in D$ is *efficient point (solution)* if and only if there exists no $x \in D$ such that $F_h(x) \geqslant F_h(x^*)$ for $h = 1, \ldots, r$ and such that for at least one h_0 one has $F_{h_0}(x) > F_{h_0}(x^*)$ (assuming that maximum is sought for all objective functions). That is, x^* is an efficient point if it has the property that there exists no other point x to improve at least one objective function while the others remain unchanged.

The notion of efficient solution plays an important role in economics, game theory, statistical decision-making, and generally in any decision problem with several uncomparable criteria.

As will be seen in Section 1.2., the determination of efficient points is the most general method among the optimisation methods of mathematical programming with multiple objective functions.

4°. Let x^* be an optimum solution obtained by ordering the criteria in the following way [97] : solve r mathematical programming problems restraining each time the domain D by changing into constraints the optimum solutions obtained by solving a certain problem having only one objective function. That is, consider the following sequence of domains :

$$D_0^* = D$$

$$D_1^* = \{x \mid F_1(x) = \underset{y \in D_0^*}{\text{optimum}}\, F_1(y)\,;\ x \in D_0^*\},$$

$$D_2^* = \{x \mid F_2(x) = \underset{y \in D_1^*}{\text{optimum}}\, F_2(y)\,;\ x \in D_1^*\},$$

$$\cdots\cdots\cdots\cdots\cdots\cdots$$

$$D_k^* = \{x \mid F_k(x) = \underset{y \in D_{k-1}^*}{\text{optimum}}\, F_k(y)\,;\ x \in D_{k-1}^*\},$$

$$\cdots\cdots\cdots\cdots\cdots\cdots$$

$$D_r^* = \{x \mid F_r(x) = \underset{y \in D_{r-1}^*}{\text{optimum}}\, F_r(y)\,;\ x \in D_{r-1}^*\}.$$

According to Lebedev et al. [97], solving the mathematical programming problem with several objective functions means to find one or more points of the set D_r^*.

Obviously, the set D_r^* depends on the way in which functions are ordered. Usually, to two different ordering schemes there correspond different sets of points.

5° Let x^* be a point in the domain of feasible solutions which can be obtained through a search procedure based on various criteria. For instance, one can use the POP (Progressive Orientation Procedure) [12] or STEM (Step Method) methods [11].

6° Let x^* belong to a set of *properly efficient solutions* defined as follows (Geoffrion [75]) : x^* is said to be a properly efficient solution of problem (1.1)—(1.3) if it is efficient and if there exists a scalar $M > 0$ such that for each i one has

$$\frac{f_i(x^*) - f_i(x)}{f_j(x) - f_j(x^*)} \leqslant M$$

for some j such that $f_j(x) < f_j(x^*)$ whenever $x \in D$ and $f_i(x) > f_i(x^*)$.

From this definition we derive that the set of properly efficient points is included in that of efficient points. For the linear case, the two sets coincide (Izerman [202]).

Saska [122] suggests that the solution of a linear programming problem with multiple objective function can be obtained by a method of type 1°a), i.e. by finding the vector $x^* \in D$ which minimizes the function

$$(1.5) \quad h(x) = \max_{1 \leqslant h \leqslant r} \frac{|\mathbf{X}_h - F_h(x)|}{\sqrt{\sum_{j=1}^{n} c_{hj}^2}},$$

where $\mathbf{X}_h = \underset{x \in D}{(\text{optimum})} F_h(x)$.

Thus, the vector $x^* = (x_1^*, \ldots, x_n^*)' \in D$ has the property of being at the least distance from the hyperplanes determined by the r objective functions; i.e.

$$\max_{1\leqslant h\leqslant r}\frac{|\mathbf{X}_h - F_h(x^*)|}{\sqrt{\sum_{j=1}^{n} c_{hj}^2}} = \min_{x\in D}\left[\max_{h=1,\ldots,r}\frac{|\mathbf{X}_h - F_h(x)|}{\sqrt{\sum_{j=1}^{n} c_{hj}^2}}\right] \equiv \delta. \tag{1.6}$$

The function $h(x)$ is convex * (i.e. the existence of a minimum is ensured) and nonlinear and so the effective determination of the optimum solution through numerical means raises intractable difficulties.

In the same paper [122], Saska reduces the nonlinear problem to an equivalent linear problem of the form :

$$\min \Psi(x) = x_{n+1}, \tag{1.7}$$

subject to

$$Ax \leqslant b;\ x \geqslant 0,$$

$$V^{(k)}(x) + x_{n+1} \geqslant \widetilde{\mathbf{X}}_k,$$

$$V^{(k)}x - x_{n+1} \leqslant \widetilde{\mathbf{X}}_k,$$

where

$$V^{(k)} = (V_{ki}) = \frac{c_{ki}}{\sqrt{\sum_{j=1}^{n} c_{kj}^2}},\ (k = 1, \ldots, r;\ i = 1, \ldots, n)$$

$$\widetilde{\mathbf{X}}_k = \frac{\mathbf{X}_k}{\sqrt{\sum_{j=1}^{n} c_{kj}^2}},\quad (k = 1, \ldots, r)$$

and x_{n+1} is a complementary variable such that

$$\frac{|\mathbf{X}_k - F_k(x)|}{\sqrt{\sum_{j=1}^{n} c_{kj}^2}} \leqslant x_{n+1}\ (k = 1, \ldots, r). \tag{1.8}$$

---

* See Definition 1.4.

The link between the two problems is given by

THEOREM 1.1. *The optimum solution* $(\hat{x}, \hat{x}_{n+1})$ *of the mathematical programming problem (1.7) is equal to the optimum solution* (x^*, δ) *of the nonlinear programming problem*

$$\min_{x \in D} \left[\max_{h=1,\ldots,r} \frac{|\mathbf{X}_h - F_h(x)|}{\sqrt{\sum_{j=1}^{n} c_{hj}^2}} \right]$$

subject to

$$Ax \leqslant b;\ x \geqslant 0.$$

Proof. One writes:

$$(1.9)\quad \min_x \Psi(x) = \hat{x}_{n+1} \geqslant \max_{1 \leqslant k \leqslant r} \frac{|\mathbf{X}_k - F_k(\hat{x})|}{\sqrt{\sum_{j=1}^{n} c_{kj}^2}} =$$

$$= \max_{1 \leqslant k \leqslant r} |\widetilde{\mathbf{X}}_k - V^k \hat{x}| \geqslant \min_x \{ \max_{1 \leqslant k \leqslant r} |\widetilde{\mathbf{X}}_k - V^k x| \} =$$

$$= \min_x \left\{ \max_{1 \leqslant k \leqslant r} \frac{|\mathbf{X}_k - F_k(x)|}{\sqrt{\sum_{j=1}^{n} c_{kj}^2}} \right\} =$$

$$= \max_{1 \leqslant k \leqslant r} \frac{|\mathbf{X}_k - F_k(x^*)|}{\sqrt{\sum_{j=1}^{n} c_{kj}^2}} = \delta.$$

Relation (1.9) means that

$$(1.9')\quad \hat{x}_{n+1} \geqslant \delta.$$

Since $\min x_{n+1} = \hat{x}_{n+1}$, it follows that

$$(1.10)\quad \hat{x}_{n+1} \leqslant \delta.$$

From (1.9′) and (1.10) it results that

$$\hat{x}_{n+1} = \max_{1 \leqslant k \leqslant r} |\tilde{\mathbf{X}}_k - V^k \hat{x}| = \max_{1 \leqslant k \leqslant r} |\hat{\mathbf{X}}_k - V^k x^*| = \delta.$$

The proof is complete.

Saska also gives another method for solving a multiple objective function problem where function (1.5) is replaced by the minimizing function

$$h_1(x) = \max_{1 \leqslant k \leqslant r} \frac{|\mathbf{X}_k - F_k(x)|}{\mathbf{X}_k}.$$

Thus, the solution of the multiple objective functions problem reduces to solving

$$\min_{x \in D} \max_{1 \leqslant k \leqslant r} \frac{|\mathbf{X}_k - F_k(x)|}{\mathbf{X}_k}.$$

The same idea is also found in the works of Tamm [153] and Nykowski [113], [114]. For r given maximizing functions $F_1, \ldots, F_r$ on a domain D, Tamm suggests a best compromise solution by solving the mathematical programming problem:

$$\min x_{n+1},$$

subject to

$$\frac{\mathbf{X}_i - c_i' x}{|\mathbf{X}_i|} \leqslant x_{n+1}, \quad (i = 1, \ldots, r)$$

$$x \in D$$

where

$$\mathbf{X}_i = \max_{x \in D} F_i(x).$$

We shall give now the following definitions:

DEFINITION 1.1. The function $f: D \to R$, where D is a convex set of R^n and R is the real axis, is concave (strictly concave) if for any x and y of D and any $\lambda \in (0, 1)$, one has

$$f(\lambda x + (1 - \lambda)\, y) \geqslant (>)\, \lambda f(x) + (1 - \lambda)\, f(y).$$

When f is differentiable, the analogous definition becomes

$$f(x) - f(y) \leqslant (<)\, (x - y)' \,\nabla f(y) \qquad \forall\, x, y \in D,$$

where $\nabla f(x)$ denotes the gradient of f.

DEFINITION 1.2. The function $f : D \to R$ is quasi-concave (explicitly quasi-concave) if for any x and y of D, and any $\lambda \in (0,1)$, one has $f(\lambda\, x + (1 - \lambda)y) \geqslant (>) \min (f(x), f(y))$.

When $f(x)$ is a differentiable function, the alternative definition of the quasi-concave (explicitly quasi-concave) function is as follows :

$$f(x) \geqslant (>) f(y) \Rightarrow (x - y)' \,\nabla f(y) \geqslant (>)\, 0.$$

DEFINITION 1.3. The function $f : D \to R$, differentiable on D, is pseudoconcave if for any x and y of D such that $(y - x)'\nabla$ $\nabla f(x) \leqslant 0$, one has $f(y) \leqslant f(x)$.

DEFINITION 1.4. The function $f : D \to R$, is convex, strictly convex, quasi-convex, explicitly quasi-convex and pseudoconvex if the function $-f$ is concave, strictly concave, quasiconcave, explicitly quasi-concave and pseudoconcave, respectively.

1.2. THE RELATION BETWEEN THE VARIOUS METHODS OF SOLVING MATHEMATICAL PROGRAMMING PROBLEMS WITH MULTIPLE OBJECTIVE FUNCTIONS

Denote by $D^{**} \subset D$ the set of efficient solutions and by D^*_r the set of optimum solutions obtained through ordering the criteria. The link between these two sets is shown by the following theorem.

THEOREM 1.2. *Any optimal solution obtained by ordering the criteria is an efficient point of these criteria, i.e.*

$$D_r^* \subseteq D^{**}.$$

Proof. Assume that $D_r^* \nsubseteq D^{**}$ i.e. there exists a point $x' \in D_r^*$ such that $x' \notin D^{**}$. Since the point x' is not efficient, there exists another point $x'' \in D$ such that

$$F_k(x'') \geqslant F_k(x') \qquad (k = 1, \ldots, r)$$

and for at least one value l of the index k, the inequality is strict, i e.

$$(1.11) \quad F_l(x'') > F_l(x').$$

Let $k = 1$. Since $x' \in D_r^*$ and $D_r^* \subset D_1^*$, it follows that $x' \in D_1^*$ and hence x' is a maximum point for F_1. Since also $F_1(x'') \geqslant F_1(x')$, only the equality is possible, i.e. $F_1(x'') = F_1(x')$

For $k = 2$, one also gets $F_2(x') = F_2(x'')$ since $x'' \in D_2^*$. On similar grounds one can prove that $F_k(x'') = F_k(x')$, for any $k = 1, \ldots, r$. But this contradicts (1.11) and hence the proof is complete.

Let us now assume that in 2°, Section 1.1,

$$\Psi_i(x - X_i) = F_i(X_i) - F_i(x)$$

and

$$h(\Psi_1(x - X_1), \ldots, \Psi_r(x - X_r)) = \sum_{i=1}^{r} [F_i(X_i) -$$

$$- F_i(x)]^2 = \Phi(x).$$

Huang [80] considers the problem of determining the vector x^* which minimizes the function $\Phi(x)$ (the mean-square strategy) and defines it as the vector which solves the mathematical programming problem with multiple objective functions. Thus x^* minimizes the sum of squared deviations from the absolute optimum values.

One can prove that x^* is an efficient point.

THEOREM 1.3. *The optimal solution x^* of the problem*

$$\min \left\{ \Phi(x) = \sum_{i=1}^{r} [F_i(X_i) - F_i(x)]^2 \right\}$$

subject to

$$x \in D$$

is an efficient solution of problem (1.1)–(1.3).

Proof. Assume that x^* minimizes $\Phi(x)$, i.e. for any other point $x \in D$,

$$\Phi(x^*) = \sum_{i=1}^{r} [F_i(X_i) - F_i(x^*)]^2 \leqslant \sum_{i=1}^{r} [F_i(X_i) - F_i(x)]^2 = \Phi(x). \tag{1.12}$$

If x^* is not an efficient solution of problem (1.1)–(1.3), then there exists a point $\bar{x} \in D$ such that

$$F_i(x^*) \leqslant F_i(\bar{x}), \quad (i = 1, \ldots, k), \tag{1.13'}$$

$$F_i(x^*) < F_i(\bar{x}), \quad (i = k + 1, \ldots, r). \tag{1.13''}$$

On the other hand, since X_i is the optimal solution of F_i, one can write

$$F_i(X_i) - F_i(x^*) \geqslant 0, \quad (i = 1, \ldots, r), \tag{1.14}$$

$$F_i(X_i) - F_i(\bar{x}) \geqslant 0, \quad (i = 1, \ldots, r). \tag{1.15}$$

Subtracting (1.13) from $F_i(X_i)$ and using (1.14)–(1.15), one gets

$$\begin{aligned} &F_i(X_i) - F_i(x^*) \geqslant F_i(X_i) - F_i(\bar{x}) \geqslant 0, \ (i = 1, \ldots, k), \\ &F_i(X_i) - F_i(x^*) > F_i(X_i) - F_i(\bar{x}) \geqslant 0, \\ &\qquad\qquad (i = k + 1, \ldots, r). \end{aligned} \tag{1.16}$$

Squaring and summing up for $i = 1, \ldots, r$ both sides of (1.16) yields

$$\Phi(x^*) = \sum_{i=1}^{r} [F_i(X_i) - F_i(x^*)]^2 > \sum_{i=1}^{r} F_i(X_i) -$$

$$- F_i(\bar{x})]^2 = \Phi(x).$$

But this contradicts (1.12) and hence x^* is an efficient solution of (1.1)–(1.3). The proof is complete.

Let now

$$\Psi_i(x - X_i) = F_i(X_i) - F_i(x)$$

and

$$h(\Psi_1(x - X_1), \ldots, \Psi_r(x - X_r)) = \max_{i=1,\ldots,r} [F_i(X_i) -$$

$$- F_i(x)] = \Psi(x).$$

The vector x^* which minimizes $\Psi(x)$ can be taken to represent the solution of the multiple objective functions problem. The point x^* minimizes the maximum deviation from the optimum values of each objective function.

THEOREM 1.4. *The optimal solution x^* of the problem*

$$\min \{\Psi(x) = \max_{i=1,\ldots,r} [F_i(X_i) - F_i(x)]\},$$

subject to

$$x \in D,$$

is an efficient solution of problem (1.1)–(1.3).

Proof. If x^* is not an efficient solution, then there exists an x' such that $F_i(x') \geqslant F_i(x^*)$, and at least one index i_0 for which the inequality is strict. Hence,

$$F_i(X_i) - F_i(x') \leqslant F_i(X_i) - F_i(x^*)$$

which for $i = i_0$ is a strict inequality. Then

$$\max_i [F_i(X_i) - F_i(x')] \leqslant \max_i [F_i(X_i) - F_i(x^*)],$$

i.e. $\Psi(x') \leqslant \Psi(x^*)$, which contradicts the assumption that x^* minimizes the function $\Psi(x)$. Hence, the proof is complete.

To find the entire set of efficient solutions raises intractable difficulties in computation, even when $F_i(x)$ are linear and D is a convex polyhedron. The set of efficient solutions contains very many optimum solutions of the multiple objective functions problem, which can be derived by use of the methods already stated.

Let us examine now a method for obtaining the efficient solutions. Since efficient solutions can be characterized in more than one way, we shall confine our discussion to only one of them and show the link between this problem and that of parametric programming (Dinkelbach [52], [54], Geoffrion [72], [75], Karlin [90], etc.).

For a better characterization of the efficient solutions, we have

LEMMA 1.1. ([202]). *The point $x^0 \in D$ is an efficient solution of the linear programming problem with multiple objective functions (1.1)–(1.3) if and only if*

Problem P

$$\max \sum_{j=1}^{r} d_j y_j, \quad (d_j > 0),$$

subject to

$$\sum_{j=1}^{n} a_{ij} x_j = b_i, \quad (i = 1, \ldots, m)$$

$$-\sum_{j=1}^{n} c_{sj} x_j + y_s = -\sum_{j=1}^{n} c_{sj} x_j^0, \quad (s = 1, \ldots, r)$$

$$x_j \geqslant 0;\ y_s \geqslant 0, \quad (j = 1, \ldots, n;\ s = 1, \ldots, r)$$

has an optimal solution $(\hat{x}, \hat{y})$ with $\hat{y} = 0$.

The proof is straightforward.

LEMMA 1.2. [202]. *The point $x^0 \in D$ is an efficient solution of (1.1)–(1.3) if and only if*

Problem D

$$\min(u'b - w'Cx^0),$$

subject to

$$u'A - w'C \geqslant 0,$$

$$w \geqslant d > 0,$$

has an optimal solution $(\hat{u}, \hat{w})$ *with* $\hat{u}'b - \hat{w}'Cx^0 = 0$.

Proof. Problems P and D are dual to each other. Hence, by virtue of the duality theory, $(\hat{x}, \hat{y})$ is an optimal solution for Problem P if and only if Problem D has an optimal solution $(\hat{u}, \hat{w})$ with $d'\hat{y} = \hat{u}'b - \hat{w}'Cx^0$. Hence, according to lemma 1.1, x^0 is an efficient solution of (1.1)–(1.3) if and only if $(\hat{u}, \hat{w})$ is optimal for Problem D and $\hat{u}'b - \hat{w}'Cx^0 = 0 = d'\hat{y}$.

A procedure for obtaining efficient solutions by use of parametric programming is given by the following

THEOREM 1.5. *The feasible solution* x^* *is efficient for problem (1.1)–(1.3) if and only if it is optimal for the problem*

$$\max_{x \in D} \sum_{i=1}^{r} \lambda_i F_i(x)$$

with a given $\lambda = (\lambda_1, \ldots, \lambda_r)$, *where* $\lambda_i > 0$, $(i = 1, \ldots, r)$ *and* $\sum_{i=1}^{r} \lambda_i = 1$.

Proof. Sufficiency. Let x^* be the optimal solution for

$$\mathbf{F}(x) = \sum_{i=1}^{r} \lambda_i F_i(x) \text{ ; i.e.}$$

$$(*) \qquad \sum_{i=1}^{r} \lambda_i F_i(x^*) \geqslant \sum_{i=1}^{r} \lambda_i F_i(x), \quad \forall\, x \in D$$

and assume that x^* is not efficient.

Hence, there exists an $\bar{x} \in D$ such that

$$F_i(\bar{x}) \geqslant F_i(x^*), \quad (i = 1, \ldots, r)$$

the strict inequality taking place for at least one index i. Multiplication by $\lambda_i > 0$ and summation for all i's yields

$$\sum_{i=1}^{r} \lambda_i F_i(\bar{x}) > \sum_{i=1}^{r} \lambda_i F_i(x^*)$$

which contradicts the inequation (*). Hence x^* is an efficient solution.

Necessity. Let x^* be an efficient solution and consider the vector function

$$F^*(x) = F(x) - F(x^*) : D \subset R^n \to Z \subset R^r,$$

where $F(x) = (F_1(x) \ldots F_r(x))$. Then

$$F^*(x^*) = 0 \in Z.$$

Since x^* is an efficient solution, it follows that there exists no other $x \in D$ such that

$$F^*(x) = F(x) - F(x^*) \geqslant 0 \text{ and } F(x) \neq F(x^*).$$

Denoting $z = F^*(x)$, one can say that there exists no $z \in Z$ such that $z \geqslant 0$, $z \neq 0$. This is tantamount to saying that $0 \in Z$ is a boundary point for Z.

The boundary hyperplanes through $0 \in Z$ together with the corresponding half-planes form a convex polyhedral cone $C \supset Z$:

$$C = \{z \in R^r \mid C^* z \leqslant 0\}.$$

Let $\mathbf{1} = (1, 1, \ldots, 1)$ be the r-dimensional vector whose components are all equal to unity. Hence there exists no $z \in C$ such that $z \geqslant 0$ and $\mathbf{1}'z > 0$, i.e. for any $z \in C$, $z \geqslant 0$ implies $\mathbf{1}'z \leqslant 0$.

Taking into account the manner in which C was defined, one may conclude that for any $z \in R^r$ with $\begin{pmatrix} C^* \\ -E \end{pmatrix} z \leqslant 0$ we have $\mathbf{1}'z \leqslant 0$. According to Farcaş' Lemma, there exists $[u',v'] \geqslant \geqslant 0'$ such that $\mathbf{1}' = (u', v')\begin{pmatrix} C^* \\ -E \end{pmatrix} = u'C^* - v'$. Hence, $u'C^* = = w' > 0$. This implies that

$$w'z = u'C^*z \leqslant 0, \quad z \in C.$$

Let us take $\lambda = \frac{w}{\sum_{i=1}^{r} w_i}$. Obviously, $\lambda > 0$ and $\sum_{i=1}^{r} \lambda_i = 1$. Since $w'z \leqslant 0$, one gets

$$\lambda'z \leqslant 0\,;\; z \in Z \subset C,$$

i.e

$$\lambda'[F(x) - F(x^*)] \leqslant 0, \quad x \in D,$$

or

$$\sum_{i=1}^{r} \lambda_i F_i(x) \leqslant \sum_{i=1}^{r} \lambda_i F_i(x^*), \quad x \in D.$$

Thus, the proof is complete.

Let the parameters space, $\Lambda = \left\{(\lambda_1, \ldots, \lambda_r) \mid 0 \leqslant \lambda_i \leqslant 1, i = 1, \ldots, r, \sum_{i=1}^{r} \lambda_i = 1\right\}$, be the subset of the r-dimensional Euclidean space E^r.

The existence of a vector $\lambda = (\lambda_1, \ldots, \lambda_r)\left(\lambda_i > 0, \sum_{i=1}^{r} \lambda_i = 1\right)$ associated with an efficient point may be interpreted as the existence of a probability distribution so that

$$F^*(x) = \sum_{i=1}^{r} \lambda_i F_i(x)$$

may appear as a weighted mean of the individual objectives.

As $\sum_{i=1}^{r} \lambda_i = 1$, one of the parameters λ_i can be expressed as a function of the other $r-1$, and thus the dimension of the parameters space Λ reduces by one . This observation can be useful in some practical applications.

Let

$$F_\lambda^* = \sum_{i=1}^{r} \lambda_i F_i = \sum_{i=1}^{r} \lambda_i \sum_{j=1}^{n} c_{ij} x_j = \sum_{j=1}^{n} \left(\sum_{i=1}^{r} \lambda_i c_{ij} \right) x_j.$$

Express λ_r as

$$\lambda_r = 1 - \sum_{i=1}^{r-1} \lambda_i,$$

and hence

$$F_\lambda^* = \sum_{j=1}^{n} \left[\sum_{i=1}^{r-1} \lambda_i c_{ij} + \left(1 - \sum_{i=1}^{r-1} \lambda_i \right) c_{ij} \right] x_j =$$

$$= \sum_{j=1}^{n} \left[c_{rj} + \sum_{i=1}^{r-1} (c_{ij} - c_{rj}) \lambda_i \right] x_j = \sum_{j=1}^{n} [c_{rj} + c_j(\lambda)] x_j,$$

where $c_j(\lambda) = \sum_{i=1}^{r-1} (c_{ij} - c_{rj}) \lambda_i$. As $\lambda_r \geqslant 0$, one writes $\sum_{i=1}^{r-1} \lambda_i \leqslant 1$. Consequently the parameters space Λ can be expressed as

$$\Lambda = \left\{ \lambda | \lambda \in R^{r-1}, \sum_{i=1}^{r-1} \lambda_i \leqslant 1, \lambda_i \geqslant 0, \ i = 1, \ldots, r-1 \right\}.$$

The interior of this domain is defined as

$$\text{int}\, \Lambda = \left\{ \lambda | \lambda \in R^{r-1}, \ \sum_{i=1}^{r-1} \lambda_i \leqslant 1, \lambda_i > 0, i=1,\ldots, r-1 \right\}.$$

The structure of the set Λ is described by the following

THEOREM 1.6 [125]. *The set of λ values which render basic feasible solutions optimal form a convex polyhedral set in the space E^r. Such sets are separated by common boundary hyperplanes.*

Proof. For any basic solution x_B let us denote by $\Lambda(x_B)$ the part of Λ for which x_B remains optimal. If B is a base for the parametric problem, then

$$x_B = B^{-1}b.$$

Consider that the functions are written as

$$F_i = \sum_{j=1}^{n} c_{ij}x_j \qquad (i = 1, \ldots, r)$$

and denote by $I \subset \{1,2, \ldots, n\}$ the indices corresponding to the vectors belonging to the base and by J those from the outside. Define

$$c_j(\lambda) = \sum_{i=1}^{r} \lambda_i c_{ij},$$

$$c_B(\lambda) = \{c_k(\lambda)\}, \qquad k \in I.$$

$$z_j = c_B(\lambda)\, B^{-1}a_j.$$

The optimum conditions for x_B are $z_j - c_j(\lambda) \geqslant 0$, $j \in J$; $c(\lambda) = \{c_j(\lambda)\}$. Considering that the left-hand side of the inequations is linear in $\lambda_1, \ldots, \lambda_r$, one gets

$$\sum_{i=1}^{r} q_{ij}\lambda_i \geqslant 0, \qquad j \in J.$$

Hence, the set $\Lambda(x_B)$ assumes the following form :

$$\Lambda(x_B) = \left\{\lambda \mid \sum_{i=1}^{r} q_{ij}\lambda_i \geqslant 0\,;\, 0 < \lambda_i < 1,\, \sum_{i=1}^{r} \lambda_i = 1,\, j \in J\right\}.$$

Since each linear inequality from the definition of $\Lambda(x_B)$ defines a half-space in Λ, it follows that $\Lambda(x_B)$ is the intersection of $n - m$ such half-spaces, and hence it is a convex polyhedral set. This proves the first part of the theorem.

Next let B' be a new base obtained from B by substituting the vector a_r belonging to the base with the vector a_k from its outside. Since a_k comes into the base,

$$z_k - c_k = \min_j \{z_j - c_j \mid z_j - c_j < 0\} =$$

$$= \min_j \{c_B(\lambda)\, B^{-1}a_j - c_j(\lambda) \mid c_B(\lambda) \cdot B^{-1}a_j - c_j(\lambda) < 0\}.$$

Since a_r comes out of the base

$$\frac{x_{Br}}{y_{rj}} = \min_j \left\{\frac{x_{Bi}}{y_{ij}} \mid y_{ij} < 0\right\},$$

where $y_j = \{y_{kj}\} = B^{-1}a_j(k \in I)$. The other differences $z_j - c_j(j \in J)$ become

$$z_j' - c_j = (z_j - c_j) - \frac{y_{rj}}{y_{rk}}(z_k - c_k), \quad j \in J,$$

and they should be non-negative for x_B' to be optimal. The set of values λ for which

$$z_k - c_k = c_B(\lambda) B^{-1} a_k - c_k(\lambda) = 0$$

makes up a boundary hyperplane for the convex polyhedral $\Lambda(x_B)$. For λ belonging to this hyperplane

$$z_j' - c_j = z_j - c_j, \quad j \in J,$$

$$z_j - c_j \geqslant 0, \qquad j \in J,$$

$$z_r - c_r = 0 = z_r' - c_r.$$

One can check that

$$\Lambda(x_{B'}) = \left\{\lambda \mid z_j' - c_j \geqslant 0, j \in J ; 0 < \lambda_i < 1 ; \sum_{i=1}^{r} \lambda_i = 1\right\}.$$

Hence, the boundary hyperplane $\{\lambda \mid z_k - c_k = 0\}$ of $\Lambda(x_B)$ is included into $\Lambda(x'_B)$ and it is also a boundary hyperplane, namely, $\{\lambda \mid z'_r - c_r = 0\}$ of $\Lambda(x'_B)$. Hence $\Lambda(x_B)$ and $\Lambda(x'_B)$ are connected along a common boundary hyperplane.

The domains $\Lambda(x_B)$ are disjointed and their reunion for all λ is the space Λ [168].

Observation 1. Let $x^i = (x_1, \ldots, x_m, 0, \ldots, 0)$ be a nondegenerated solution of the parametric problem. According to the previous theorem, this solution is optimal, hence efficient, if the following optimality conditions are satisfied:

$$z_j^*(\lambda) = z_j(\lambda) - c_j(\lambda) \geqslant 0, \quad j \in J.$$

These conditions (constraints) define the domain $\Lambda(x^i)$. Part of these constraints may be redundant and in particular applications their elimination is advisable and even necessary since otherwise they can lead to dominated solutions that are apparently non-dominated.

Observation 2. The separation hyperplane between two adjacent domains $\Lambda(x^i)$ and $\Lambda(x^j)$ denoted by

$$H_k = \{\lambda \mid \lambda \in R^{r-1}, z_k^*(\lambda) = 0\}$$

has the following properties:

1) $\Lambda(x^i) \subset \{\lambda \mid \lambda \in R^{r-1}, z_k(\lambda) \geqslant 0\}$,

2) $\Lambda(x^j) \subset \{\lambda \mid \lambda \in R^{r-1}, z_k(\lambda) \leqslant 0\}$,

3) $H_k \cap \Lambda(x^i) = H_k \cap \Lambda(x^j) = \Lambda(x^i) \cap \Lambda(x^j)$.

The hyperplane H_k is useful in eliminating nonefficient solution. If $H_k \cap \Lambda(x^i) \cap \operatorname{Int} \Lambda \neq \emptyset$, then x^j obtained from x^i by introducing the k-th column into the base is a nondominated (efficient) solution.

1.3. GOAL PROGRAMMING

The multiple objective function problem has so far been approached through a synthesis function for the r objective functions (case 1°) or by first solving r mathematical programming problems and then constructing a synthesis function for the optimum values or for the optimal solutions (case 2°). Here we shall consider another type of solution methods which is currently known as "goal programming". This method was introduced by Charnes and Cooper [35] and subsequently developed by Ijiri [81], Lee [98], Contini [44], etc. In particular, Lee [98] made an extensive use of goal programming in solving problems of production planning, marketing, financial decisions, academic planning and medical care. An excellent survey of goal programming methods is due to Kornbluth [207].

Consider the goal vector $\bar{F} = (\bar{F}_1, \bar{F}_2, \ldots, \bar{F}_r)$, where the $\bar{F}_i$'s are the levels to be reached by the objective functions. Obviously, it is usually nearly impossible to find a point x^* in the feasible solutions domain such that the objective functions are of the required levels,

$$F_1(x^*) = \bar{F}_1, \; F_2(x^*) = \bar{F}_2, \ldots, F_r(x^*) = \bar{F}_r,$$

i.e.

$$\bar{F} = F = Cx.$$

For a given $x \in D$, certain differences (plus or minus) will exist between $F_i(x)$ and $\bar{F}_i$ and hence the problem consists in minimizing the distance between the vector $\bar{F}$ and the vector containing the possible values of the objective functions. Thus, one obtains the problem

$$(1.17) \quad \min_{x \in D} d(\bar{F}, F(x)).$$

Consider now the n-dimensional vector space R^n endowed with the classical norm $\|\cdot\|$. For any two points $q = (q_1, \ldots, q_n)$ and $r = (r_1, \ldots, r_n)$ of R^n, one can express the distance $d(q, r)$

as the norm $\|q - r\|$, $(d(q, r) = \|q - r\|)$ and hence the closiness of F and $\bar{F}$ can be measured by $\|F - \bar{F}\|$.

Hölder's norm, L_p,

$$\|x\|_p = \left(\sum_{i=1}^{n} |x_i^p|\right)^{1/p}, \qquad p \geqslant 1.$$

is widely used, and for $p = 2$ one gets the Euclidean distance

$$\|x\|_2 = \left(\sum_{i=1}^{n} |x_i|^2\right)^{1/2}.$$

Other norms also used in R^n are

$$\|x\|_1 = \sum_{i=1}^{n} |x_i| \quad \text{or} \quad \|x\|_\infty = \max_{1 \leqslant i \leqslant n} \{|x_i|\}.$$

Applying these norms to $F - \bar{F}$ leads to the following goal programming models:

$$\textit{Model 1}: \min_x \left\{\|F - \bar{F}\|_p = \left[\sum_{i=1}^{r} |F_i - \bar{F}_i|^p\right]^{1/p} \,\middle|\, Ax = b,\ x \geqslant 0\right\};$$

$$\textit{Model 2}: \min_x \left\{\|F - \bar{F}\|_1 = \sum_{i=1}^{r} |F_i - \bar{F}_i| \,\middle|\, Ax = b,\ x \geqslant 0\right\};$$

$$\textit{Model 3}: \min_x \left\{\|F - \bar{F}\|_2 = \left[\sum_{i=1}^{r} |F_i - F_i|^2\right]^{1/2} \,\middle|\, Ax = b,\ x \geqslant 0\right\};$$

$$\textit{Model 4}: \min_x \{\|F - \bar{F}\|_\infty = \max_{1 \leqslant i \leqslant r} |F_i - \bar{F}_i| \mid Ax = b,\ x \geqslant 0\}.$$

The existence of a minimum value for these models is ensured by Hölder's inequality which indicates $\|\cdot\|_p$ to be convex for any p.

Model 3 has two alternative forms:

$$\textit{Model } 3': \min_x \left\{ \|F - \bar{F}\|_2 = \sum_{i=1}^{r} (F_i - \bar{F}_i)^2 \,|\, Ax = b,\, x \geqslant 0 \right\}$$

$$\textit{Model } 3'': \min_x \left\{ \|F - \bar{F}\|_2 = \sum_{i=1}^{r} \alpha_i (F_i - \bar{F}_i)^2 \,|\, Ax = b,\, x \geqslant 0 \right\},$$

where α_i are weighting coefficients for F_i.

Minimization of $\|F - \bar{F}\|_p$ for $p \geqslant 2$ leads to nonlinear programming, whereas the minimization of $\|F - \bar{F}\|_1$ and $\|F - \bar{F}\|_\infty$ can be done by the Simplex method of linear programming.

Indeed, let us consider Model 2, studied by Charnes and Cooper [35] and Ijiri [81], which minimizes the sum of absolute values of the errors.

Denote, for each F_k and each $x \in D$, the plus and minus differences of $F_k(x)$ from $\bar{F}_k$ by $d_k^+(x)$ and $d_k^-(x)$ (or briefly by d_k^+, d_k^-). Now one has to minimize the sum of these differences.

Note that if $d_k^+ > 0$, then $d_k^- = 0$ and *vice-versa*, if $d_k^- > 0$, then $d_k^+ = 0$ since, in any given point, the difference between F_k and $\bar{F}_k$ must be either positive or negative, and never both.

Hence, $|F_k - \bar{F}_k| = |d_k^+ - d_k^-| = d_k^+ + d_k^-$ since $d_k^+ d_k^- = 0$.

Model 2 becomes

$$\text{(1.18)} \quad \min \sum_{k=1}^{r} (d_k^-(x) + d_k^+(x)),$$

subject to

$$F_k(x) + d_k^- - d_k^+ = \bar{F}_k, \quad (k = 1, \ldots, r),$$

$$d_k^+(x),\ d_k^-(x) \geqslant 0,$$

$$x \in D.$$

Denote by e a row vector of r unity elements, I the unity matrix of order r and d^+, d^- the vectors of components d_i^+ and d_i^-. The model (1.18) is then written as

$$\min \, [ed^+ + ed^-],$$

subject to

$$F(x) - Id^+ + Id^- = \bar{F},$$

$$Ax = b,$$

$$x, d^+, d^- \geqslant 0.$$

If $\bar{F}_i \geqslant \max_{x \in D} F_i(x)$ $(i = 1, \ldots, r)$, then the solution of problem (1.18) is efficient. In order to have efficient solutions for any $\bar{F} \in R^r$, we substitute (1.18) by (1.18′), where the objective function reads as $\sum_{k=1}^{r} (d_k^-(x) - d_k^+(x))$.

THEOREM 1.7 *Any optimal solution $\tilde{x}$ of problem (1.18′) is an efficient solution.*

Proof. Assume that $\tilde{x}$ is not efficient, and hence there exists a point $x \in D$ such that

$$(1.19') \quad F_k(x) \geqslant F_k(\tilde{x}), \qquad (k = 1, \ldots, r)$$

and for at least one k

$$(1.19'') \quad F_k(x) > F_k(\tilde{x}).$$

Denote by $\tilde{d}_k^+$ and $\tilde{d}_k^-$ the deviations of the objective functions F_k from the values $\bar{F}_k$ at a point $\tilde{x}$. However, both x and $\tilde{x}$ satisfy the constraints of problem (1.18′), i.e.

$$F_k(x) + d_k^- - d_k^+ = \bar{F}_k$$

$$F_k(\tilde{x}) + \tilde{d}_k^- - \tilde{d}_k^+ = \bar{F}_k$$

or

$$d_k^- - d_k^+ = \bar{F}_k - F_k(x)$$

$$\tilde{d}_k^- - \tilde{d}_k^+ = \bar{F}_k - F_k(\tilde{x}).$$

Hence, according to (1.19),

$$\tilde{d}_k^- - \tilde{d}_k^+ \geqslant d_k^- - d_k^+ \text{ for all } k = 1, \ldots, r$$

and

$$\tilde{d}_k^- - \tilde{d}_k^+ > d_k^- - d_k^+ \text{ for at least one } k.$$

Summing up one gets

$$\sum_{k=1}^{r} (\tilde{d}_k^- - \tilde{d}_k^+) > \sum_{k=1}^{r} (d_k^- - d_k^+)$$

which contradicts the fact that $\tilde{x}$ is an optimal solution for (1.18′). The proof is complete.

Model 4 (Zuhovitskii and Avdeeva [172]) can be reduced to the following linear model:

$$\min \lambda,$$

subject to

$$\bar{F}_i - F_i \leqslant \lambda,$$

$$-\bar{F}_i + F_i \leqslant \lambda,$$

$$Ax = b,$$

$$x, \lambda \geqslant 0.$$

Model 3 can be approached with the generalized inverse method [81]. A solution of $\bar{F} = Cx$, $x \geqslant 0$ can assume the form

$$x = C^+ \bar{F} + C^0 \beta,$$

where C^+ is the generalized inverse (of order $n \times r$) of C (rank r), C^0 is an nx$(n - r)$-matrix of a null space, with the property that $CC^0 = 0$, and β is an arbitrary $(n - r)$-vector.

The matrix C^0 and vector β are determined from the other constraints imposed on x. For example, from $x \geqslant 0$, i.e.

$$C^+\bar{F} + C^0\beta \geqslant 0,$$

one determines C^0, whereas from $Ax = b$, i.e.

$$AC^{+}\bar{F} + AC^{0}\beta = b,$$

one gets β.

1.4. MAXIMUM GLOBAL UTILITY METHOD

One of the solution methods mentioned in Section 1.1 consists in weighting the objective functions and then adding them up to obtain an efficiency function such that the problem reduces to an ordinary linear programming problem.

In what follows, we shall present the author's Maximum Global Utility Method [30], [31], [130]. Observing that the simple weighted summation of the objective functions does not bear an economic justification since more often than not incomensurable objectives are stated together (minimum cost and maximum production output), we have replaced the actual, economically meaningful objective functions by utility functions of von Neumann-Morgenstern [100] type such that the summation becomes by all means possible. Throughout the development of this method, it was considered that the linear programming problem with multiple criteria can be reduced to a multidimensional decision problem.

Consider a decision-requiring situation with m strategies and n criteria (Table 1.1).

Table 1.1

Strategies \ Criteria	C_1	C_2	$\dots$	C_n
S_1	x_{11}	x_{12}	$\dots$	x_{1n}
S_2	x_{21}	x_{22}	$\dots$	x_{2n}
.	.	.		.
.	.	.		.
.	.	.		.
S_m	x_{m1}	x_{m2}	$\dots$	x_{mn}

In order to define the notion of utility in the von Neuman-Morgenstern sense, one considers that the decision-maker compares two consequences x_{ik} and x_{jk} from the standpoint of criterion k and selects one of the following options :

— prefers x_{ik} to $x_{jk}(x_{ik} \succ x_{jk})$;

— prefers x_{jk} to $x_{ik}(x_{jk} \succ x_{ik})$;

— prefers neither x_{ik} nor $x_{jk}(x_{ik} \sim x_{jk})$.

We shall introduce another type of consequences, called the probability mixture of two basic consequences x_{ik} and x_{jk}, i.e. $x' = [px_{ik}, (1 - p)x_{jk}]$, where p is the probability of occurrence of consequence x_{ik} and $1-p$ is the probability of occurrence of consequence x_{jk}.

The utility $u(x_{ik})$ of consequence x_{ik} can be determined assuming that the utilities $u(x_{hk})$ and $u(x_{jk})$ are known. If $x_{hk} \succ x_{jk}$, then $u(x_{hk}) = 1$ and $u(x_{jk}) = 0$, and one of the following situations can be obtained :

a) $x_{hk} \succ x_{ik} \succ x_{jk}$; then one estimates the probability p for which

$$x_{ik} \sim [px_{hk}, (1 - p) x_{jk}]$$

and assumes that $u(x_{ik}) = p$;

b) $x_{ik} \succ x_{hk} \succ x_{jk}$; then one estimates the probability q for which

$$x_{hk} \sim [qx_{ik}, (1 - q) x_{jk}]$$

and assumes that $u(x_{ik}) = \dfrac{1}{q}$;

c) $x_{hk} \succ x_{jk} \succ x_{ik}$; one estimates the probability r for which

$$x_{jk} \sim [rx_{hk}, (1 - r) x_{ik}]$$

and assumes that $u(x_{ik}) = -\dfrac{r}{1-r}$.

The theory of von Neumann-Morgenstern is based on a system of axioms which ensures the existence of real-valued function, *the utility function* u, such that :

a) if A and B are the consequences of two distinct ways of action, then $A \succ B$ (A is preferred to B) if and only if $u(A) > u(B)$;

b) if C is a probability mixture of two consequences A and B,

$$C = [pA, (1 - p)B], \; 0 \leqslant p \leqslant 1, \quad \text{then}$$

$$u(C) = pu(A) + (1 - p)u(B);$$

c) if u has the properties a) and b), then it can undergo a linear positive transformation

$$u'(A) = au(A) + b, \quad a > 0.$$

In order to use the notion of "utility" in solving the multi-dimensional decision-making problem, one has to examine the manner in which the utilities can be summated. In this respect, Fishburn [65], [66] showed that the utilities are additive only if the associated criteria are independent in the sense of the utility theory. This is defined by Fishburn as follows.

To each strategy given in Table 1.1, one can associate an n-tuple of consequences $(x_1, x_2, \ldots, x_n)$, where $x_j \in X_j = \{x_{1j}, x_{2j}, \ldots, x_{mj}\}$. In addition to the mn-tuples, one can also consider the $r = m^n$ n-tuples belonging to the Cartesian product $X = X_1 \times X_2 \times \ldots \times X_n$. Let us denote them by $x^1, x^2, \ldots, x^r$.

One can define a set G of mixture pairs (ω_1, ω_2), where

$$\omega_1 = (p_1 x^1, \; p_2 x^2, \; \ldots, p_r x^r), \; \sum_{j=1}^{r} p_j = 1, \; x^j \in X \text{ for any } j$$

$$\omega_2 = (q_1 x^1, \; q_2 x^2, \; \ldots, q_r x^r), \; \sum_{k=1}^{r} q_k = 1, \; x^k \in X \text{ for any } k$$

such that $\omega_1 \neq \omega_2$ and the total probability of an $x_j \in X_j (j = 1, \ldots, n)$ is the same in both mixtures.

According to Fishburn, n given criteria $X_1, X_2, \ldots, X_n$ are mutually independent in the sense of the utility theory if and only if $\omega_1 \sim \omega_2$ for any $(\omega_1, \omega_2) \in G$.

The maximum global utility method consists in the following steps [30], [31], [130]:

1) For each objective function one determines the optimum and the pessimistic values $X_j = \underset{x \in D}{\text{optimum}}\ F_j(x)$, $Y_j = \underset{x \in D}{\text{pessimum}}\ F_j(X)$.

2) Using von Neumann-Morgenstern's method or other methods one determines the utilities u_j of these values on the set of all optimum and pessimistic values already determined. In ref. [194] Fishburn gives twenty-four methods for estimating additive utilities:

$$\begin{pmatrix} X_1 \ldots X_r & Y_1 & \ldots & Y_r \\ u_1 \ldots u_r & u_{r+1} & \ldots & u_{2r} \end{pmatrix}.$$

3) The objective functions $F_1, \ldots, F_r$ having concrete economic significance are transformed into utility functions $F'_1, \ldots, F'_r$. This is achieved by solving r linear systems

$$\begin{aligned} \alpha_j X_j + \beta_j &= u_j \\ & \qquad (j = 1, \ldots, r). \\ \alpha_j Y_j + \beta_j &= u_{j+r} \end{aligned}$$

With the coefficients α_j, β_j thus obtained, one makes transformations of the form

$$F'_j = \alpha_j F_j + \beta_j = \sum_{k=1}^{n} \alpha_j c_{jk} x_k + \beta_j = F''_j + \beta_j, \qquad (j = 1, \ldots, r).$$

4) The linear programming problem

$$\max \left\{ F^* = \sum_{j=1}^{r} F''_j = \sum_{j=1}^{r} \sum_{k=1}^{n} \alpha_j c_{jk} x_k \right\}$$

is solved to obtain the solution of maximum utility $X^* = (x_1^*, \ldots, x_n^*)$. The maximum global utility will then be:

$$U(X^*) = F^*(X^*) + \sum_{j=1}^{r} \beta_j.$$

Numerical example. Consider a dietitian's problem with several efficiency functions. The nutriments S_1, S_2 and S_3 come into the composition of the aliments A_1, A_2, A_3, A_4, A_5 and A_6 as is shown in Table 1.2. This table also shows the calories content and unit prices in conventional monetary units.

Table 1.2

Nutriments	Measure	Aliments					
		A_1	A_2	A_3	A_4	A_5	A_6
S_1	kg/kg	0.1	—	0.05	—	—	1
S_2	kg/kg	—	0.7	0.02	0.3	1	—
S_3	kg/kg	0.8	0.05	0.06	0.3	—	—
Calories	Calories/kg	3500	2000	700	3000	4000	1000
Cost	Monetary units/kg	25	2	1.6	6	10	12

The problem is to determine a diet which contains
- 0.080 Kg of nutriment S_1
- 0.250 Kg of nutriment S_2
- 0.220 Kg of nutriment S_3

subject to the following criteria :
- minimum cost
- maximum calories
- minimum total weight of the diet.

The constraints system for this problem is

$$0.1x_1 + 0.05x_3 + x_6 = 0.08$$

$$0.7x_2 + 0.02x_3 + 0.3x_4 + x_5 = 0.25$$

$$0.8x_1 + 0.05x_2 + 0.06\, x_3 + 0.3x_4 = 0.22$$

$$x_i \geqslant 0, \qquad (i = 1, \ldots, 6)$$

and the three efficiency functions are

$$F_1 = 25\, x_1 + 2x_2 + 1.6x_3 + 6x_4 + 10x_5 + 12x_6 : \text{(minim)},$$

$$F_2 = 3.5\, x_1 + 2x_2 + 0.7x_3 + 3x_4 + 4x_5 + 8x_6 : \text{(maxim)},$$

and

$$F_3 = x_1 + x_2 + x_3 + x_4 + x_5 + x_6 : \text{(minim)}.$$

One solves 6 linear programming problems.

– 3 problems defined by the constraints set and the "optimum" functions

– 3 problems defined by the constraints set and the "pesimum" functions

The solutions are :

$$S_1 = S_{F_1(\min)} = (x_2 = 0.1446\,;\ x_3 = 1.6\,;\ x_4 = 0.389\,;$$
$$x_1 = x_5 = x_6 = 0),$$

$$S_2 = S_{F_2(\max)} = (x_4 = 0.734\,;\ x_5 = 0.03\,;\ x_6 = 0.08\,;$$
$$x_1 = x_2 = x_3 = 0),$$

$$S_3 = S_{F_3(\min)} = (x_1 = 0.275\,;\ x_5 = 0.75\,;\ x_6 = 0.0525\,;$$
$$x_2 = x_3 = x_4 = 0),$$

$$S_4 = S_{F_1(\max)} = (x_1 = 0.275\,;\ x_5 = 0.25\,;\ x_6 = 0.0525\,;$$
$$x_2 = x_3 = x_4 = 0),$$

$$S_5 = S_{F_2(\min)} = (x_1 = 0.2535\,;x_2 = 0.375\,;\ x_6 = 0.0547\,;$$
$$x_3 = x_4 = x_5 = 0)$$

$$S_6 = S_{F_3(\max)} = (x_2 = 0.1446\,;\ x_3 = 1.6\,;\ x_4 = 0.389\,;$$
$$x_1 = x_5 = x_6 = 0).$$

To these values there correspond the following values of the efficiency functions :

– Optimum values : $X_1 = 5.182$; $X_2 = 2.960$; $X_3 = 0.578$;

– Pessimistic values : $Y_1 = 10.005$; $Y_2 = 2.039$; $Y_3 = 2.134$.

The estimated utilities ($u_{\max} = 1$, $u_{\min} = 0$) are given in table 1.3.

Table 1.3

The value of the efficiency function	5.182	2.960	0.578	10.005	2.039	2.134
The utility	0.9	1	0.9	0	0.5	0.2

Next, one sets up the following linear sets of equations:

$$\begin{cases} 10.005\alpha_1 + \beta_1 = 0 \\ 5.182\alpha_1 + \beta_1 = 0.9 \end{cases}$$

$$\begin{cases} 2.960\alpha_2 + \beta_2 = 1 \\ 2.039\alpha_2 + \beta_2 = 0.5 \end{cases}$$

$$\begin{cases} 0.578\alpha_3 + \beta_3 = 0.9 \\ 2.134\alpha_3 + \beta_3 = 0.2. \end{cases}$$

The solution is

$$\alpha_1 = -0.186\,;\ \beta_1 = 1.87\,;$$

$$\alpha_2 = 0.54\,;\ \beta_2 = -0.60\,;$$

$$\alpha_3 = -0.45\,;\ \beta_3 = 1.16.$$

Using these values, transform the functions F_1, F_2 and F_3 in utility functions and by summation we obtain the summated-utility function F^* to be maximized:

$$F^* = -0.186(25x_1 + 2x_2 + 1.6x_3 + 6x_4 + 10x_5 + \\ + 12x_6) + 0.54\,(3.5x_1 + 2x_2 + 0.7x_3 + 3x_4 + \\ + 4x_5 + 8x_6) - 0.45(x_1 + x_2 + x_3 + x_4 + x_5 + \\ + x_6) = -3.21x_1 + 0.258x_2 - 0.369x_3 + 0.046x_4 + \\ - 0.15\,x_5 + 1.638\,x_6.$$

Solving the linear programming problem defined by the function F^* and the set of constraints yields the maximum utility solution :

$$S^* = (x_2 = 0.0452\,;\ x_4 = 0.7243\,;\ x_6 = 0.08\,;$$

$$x_1 = x_3 = x_5 = 0)$$

and the corresponding utility

$$u(S^*) = 0.18 + \beta_1 + \beta_2 + \beta_3 = 0.18 + 1.86 -$$

$$- 0.60 + 1.16 = 2.61.$$

For the sake of comparison, the same diet problem will be next solved through other methods :

a) *Method* 2°a), Section 1.1. Choose

$$\Psi_k(x - S_k) = \sum_{j=1}^{n} (x_j - x_{jk})^2, \ (k = 1,2,3)$$

and

$$h(x) = \sum_{k=1}^{3} \alpha_k \sum_{j=1}^{n} (x_j - x_{jk})^2$$

with $\alpha_1 = 0.25$, $\alpha_2 = 0.01$, $\alpha_3 = 0.16$.

The following quadratic programming problem with linear constrains is obtained :

$$\min \{0.25[(x_1^2 + (x_2 - 0.144)^2 + (x_3 - 1.6)^2 +$$

$$+ (x_4 - 0.389)^2 + x_5^2 + x_6^2] + 0.01\,[x_1^2 + x_2^2 +$$

$$+ x_3^2 + (x_4 - 0.734)^2 + (x_5 - 0.03)^2 + (x_6 -$$

$$- 0.08)^2] + 0.16[(x_1 - 0.275)^2 + x_2^2 + x_3^2 +$$

$$+ x_4^2 + (x_5 - 0.75)^2 + (x_6 - 0.52)^2]\}.$$

subject to

$$0.1x_1 + 0.05x_3 + x_6 = 0.08,$$

$$0.7x_2 + 0.02x_3 + 0.3x_4 + x_5 = 0.25,$$

$$0.8x_1 + 0.05x_2 + 0.06x_3 + 0.3x_4 = 0.22,$$

$$x_i \geqslant 0, \qquad (i = 1, \ldots, 6).$$

After some calculations the efficiency function becomes

$$\min \; [0.42(x_1^2 + x_2^2 + x_3^2 + x_4^2 + x_5^2 + x_6^2) -$$

$$-(0.088x_1 + 0.072\, x_2 + 0.8x_3 + 0.2091\, x_4 +$$

$$+ 0.0806x_5 + 0.168\, x_6) + 0.753837].$$

Wolfe's algorithm yields the solution

$$x_1 = 0.267527\,;\; x_2 = 0.119559\,;\; x_3 = 0\,;\; x_4 = 0\,;$$

$$x_5 = 0.166308\,;\; x_6 = 0.053247.$$

b) *Method* 4°, Section 1.1, based on criteria ordering. Denote by D the initial domain of feasible solutions, and order the criteria by their importance as F_1, F_2, F_3. The results thus obtained are given in table 1.4 below.

Table 1.4

Problem to be solved	Solution (Value of efficiency function)
$\min\limits_{x \in D} F_1 = \min(25x_1 + 2x_2 + 16x_3 + 6x_4 + 10x_5 + 12x_6)$	$x_1 = 0$ $x_2 = 0.1446$ $x_3 = 1.6$ $x_4 = 0.389$ $x_5 = x_6 = 0$ $\min F_1 = 5.182\ (= X_1)$

Table 1.4 continued

Problem to be solved	Solution (Value of efficiency function)
$x \in D$ $25x_1 + 2x_2 + 1.6x_3 + 6x_4 + 10x_5 + 12x_6 \leqslant 5.182 (= X_1)$ $\max F_2 = \max(3.5x_1 + 2x_2 + 0.7x_3 + 3x_4 + 4x_5 + 8x_6)$	$x_1 = 0$ $x_2 = 0.1444$ $x_3 = 0.5972$ $x_4 = 0.38981$ $x_5 = 0$ $x_6 = 0.000138$ $\max F_2 = 2.57749 (= Y_1)$
$x \in D$ $25x_1 + 2x_2 + 1.6x_3 + 6x_4 + 10x_5 + 12x_6 \leqslant 5.182 (= X_1)$ $3.5x_1 + 2x_2 + 0.7x_3 + 3x_4 + 4x_5 + 8x_6 \geqslant 2.57749 (= Y_1)$ $\min F_3 = \min(x_1 + x_2 + x_3 + x_4 + x_5 + x_6)$	$x_1 = 0$ $x_2 = 0.14444$ $x_3 = 1.5972$ $x_4 = 0.38989$ $x_5 = 0$ $x_6 = 0.000138$ $\min F_3 = 2.13162 (= Z_1)$
$x \in D$ $25x_1 + 2x_2 + 1.6x_3 + 6x_4 + 10x_5 + 12x_6 \leqslant 5.7002$ $(= X_1 + 10\%)$ $\max F_2 = \max(3.5x_1 + 2x_2 + 0.7x_3 + 3x_4 + 4x_5 + 8x_6)$	$x_1 = x_2 = x_3 = 0$ $x_4 = 0.7333$ $x_5 = 0.029999$ $x_6 = 0.07999$ $\max F_2 = 2.95999$
$x \in D$ $25x_1 + 2x_2 + 1.6x_3 + 6x_4 + 10x_5 + 12x_6 \leqslant 5.7002$ $(= X_1 + 10\%)$ $\min F_3 = \min(x_1 + x_2 + x_3 + x_4 + x_5 + x_6)$	$x_1 = 0.032759$ $x_2 = 0.08647$ $x_3 = 0$ $x_4 = 0.63156$ $x_6 = 0.07672$ $\min F_3 = 0.82751$
$x \in D$ $25x_1 + 2x_2 + 1.6x_3 + 6x_4 + 10x_5 + 12x_6 \leqslant 5.7002$ $(= X_1 + 10\%)$ $3.5x_1 + 2x_2 + 0.7x_3 + 3x_4 + 4x_5 + 8x_6 \geqslant 2.959$ $\min F_3 = \min(x_1 + x_2 + x_3 + x_4 + x_5 + x_6)$	$x_1 = 0.0307597$ $x_2 = 0.086473$ $x_3 = 0$ $x_4 = 0.631561$ $x_5 = 0$ $x_6 = 0.219083$ $\min F_3 = 0.827519$
$x \in D$ $25x_1 + 2x_2 + 1.6x_3 + 6x_4 + 10x_5 + 12x_6 \leqslant 5.7002$ $(= X_1 + 10\%)$ $3.5x_1 + 2x_2 + 0.7x_3 + 3x_4 + 4x_5 + 8x_6 \geqslant 2.664$ $(= X_2 - 10\%)$ $\min F_3 = \min(x_1 + x_2 + x_3 + x_4 + x_5 + x_6)$	$x_1 = 0$ $x_2 = 0.11561$ $x_3 = 1.02055$ $x_4 = 0.495537$ $x_5 = 0$ $x_6 = 0.028972$ $\min F_3 = 1.660675$

1.5. THE CONNECTION WITH THE THEORY OF GAMES

Consider again problem (1.1)—(1.3) of maximizing r objective functions on a domain D.

In compliance with Theorem 1.5., Section 1.2, an efficient solution x^* may be found if we determine a vector $\lambda^* = (\lambda_1^*, \ldots, \lambda_r^*)$ and solve a maximization problem

$$\max_{x \in D} \sum_{i=1}^{r} \lambda_i^* F_i(x).$$

Determination of all the efficient solutions is frequently superfluous as one solution alone may be sufficient. Hence, the problem consists in finding only one vector λ^*.

In several cases, λ_i^* are the weights attached to the objective functions and reflect the relative importance of the latter. Hence, their determination is mostly subjective. However, Belenson and Kapur [10] suggest that $\lambda^* = (\lambda_1^* \ldots, \lambda_r^*)$ can be determined by solving a zero-sum two-person game with mixed strategies. The algorithm suggested there can be extended to also take into account the preferrences for a certain solution. In such a case, a maximum of r other solutions can be investigated.

Before presenting the algorithm, we shall recall some elements of the theory of games.

A zero-sum two-person game is represented by an $(r \times r)$ pay-off matrix

$$M = ((f_{ij})), \qquad (i, j = 1, \ldots, r),$$

where f_{ij} is the win of player I adopting the strategy i while the second player is adopting the strategy j. If λ_i is the probability (frequency) of player I using the strategy i and μ_j is the probability of the second player using the strategy j, then the expected win per play is

$$P = \sum_{i=1}^{r} \sum_{j=1}^{r} f_{ij} \lambda_i \mu_j,$$

where, obviously,

$$\sum_{i=1}^{r} \lambda_i = 1, \quad \lambda_i \geqslant 0, \qquad (i = 1, \ldots, r),$$

$$\sum_{j=1}^{r} \mu_j = 1, \quad \mu_j \geqslant 0, \qquad (j = 1, \ldots, r).$$

The solution of the game can be obtained by solving the following linear programming problems:

$$\min \left(\frac{1}{P_0} \right) = \sum_{i=1}^{r} r_i$$

subject to

$$\sum_{i=1}^{r} f_{ij} r_i \geqslant 1, r_i \geqslant 0, \quad (j = 1, \ldots, r),$$

or

$$\max \left(\frac{1}{P^0} \right) = \sum_{j=1}^{r} s_j,$$

subject to

$$\sum_{j=1}^{r} f_{ij} s_j \leqslant 1, s_j \geqslant 0, \qquad (i = 1, \ldots, r),$$

where P_0 and P^0 denote the minimum and maximum values of P, respectively. For the optimum win of game P^* one should have $P_0^* = P^{0*} = P^*$ and $\lambda_i^* = r_i^* P^*, \mu_j^* = s_j^* P^*$.

We revert now to the problem of maximizing r objective functions and choose the pay-off matrix

$$f_{ij} = F_i(X_j) = c_i' X_j, \qquad (i, j = 1, \ldots, r),$$

where X_j is the optimal solution of function F_j.

Solving the zero-sum two-person game one finds the weights $\lambda^* = (\lambda_1^*, \ldots, \lambda_r^*)$ to be used in the summation of the objective functions in order to obtain an efficient solution.

Since the objective functions F_i may have different economical relevance, the pay-off matrix should be normalized. Let

$$M_i = F_i(X_i)$$

and

$$M' = ((f'_{ij})) = \left(\left(\frac{f_{ij}}{M_i}\right)\right) = \left(\left(\frac{c'_i X_j}{M_i}\right)\right)$$

be the normalized pay-off matrix. Solving the zero-sum two-person game with the pay-off matrix M' yields the solution $\lambda' = (\lambda'_1, \ldots, \lambda'_r)$.

The solution of the initial game is

$$\lambda^*_k = \frac{n_k}{\sum_{k=1}^{r} n_k}, \qquad (k = 1, \ldots, r),$$

where $n_k = \dfrac{\lambda'_k}{M_k}$

If there is a line i such that $f_{ij} \leqslant 0$ $(j = 1, \ldots, r)$, then one chooses

$$K = -\min_{i,j} f_{ij}$$

$$M_i = \max_j f_{ij} + K, \quad (i = 1, \ldots, r)$$

and

$$f'_{ij} = \frac{f_{ij} + K}{M_i}, \quad (i, j = 1, \ldots, r).$$

We give below the algorithm for solving the multiple objective functions programming problem using the theory of games.

The following steps are to be performed :

I. For a given mathematical programming problem with multiple objective functions, one solves r problems with one objective function

$$\max_{x \in D} F_i(x)$$

and finds the solutions X_i.

II. One constructs the $(r \times r)$ pay-off matrix, its elements being

$$f_{ij} = F_i(X_j) = c_i' X_j.$$

(a) Is a given solution X_k satisfactory from the standpoint of the decision-maker?

If YES, then the problem is solved and the given solution is optimal.

If NO, then go to step III.

III. Determine the constant K such that:

If for at least one i, $f_{ij} \leqslant 0$ for any j, then

$$K = -\min_{i,\, j} f_{ij},$$

otherwise $K = 0$.

IV. Determine

$$M_i = \max_j f_{ij} + K, \quad (i = 1, \ldots, r)$$

and construct the normalized pay-off matrix with entries

$$f'_{ij} = \frac{f_{ij} + K}{M_i}.$$

V. Solve the zero-sum two-person normalized game through linear programming and find the solution $\lambda' = (\lambda_1', \ldots, \lambda_r')$.

VI. Determine the solution of the initial (non-normalized) game

$$\lambda_k^* = \frac{n_k}{\sum_{k=1}^{r} n_k}, \quad (k = 1, \ldots, r),$$

with $n_k = \dfrac{\lambda_k'}{M_k}$.

VII. Determine the optimal solution of the linear programming problem

$$\max_{x \in D} F_{r+1}(x) = \sum_{k=1}^{r} \lambda_k^* F_k(x).$$

Let this solution be X_{r+1}.

a) Has solution X_{r+1} been previously considered?

If YES, then the problem cannot have a satisfactory solution.
If NO, then go to step VII b.

b) Is solution X_{r+1} satisfactory?

If YES, then the problem is solved.

If NO, then go to step VIII.

VIII. Substitute the solution X_α by X_{r+1}, where X_α is the optimal solution of the function $F_\alpha(x)$ which is the least preferred by the decision-maker. A new game is set up and Steps III—VII are followed in order to find a new efficient solution X_{r+2}.

a) Has solution X_{r+2} been previously considered?

If YES, then substitute X_β by X_{r+1}, where X_β is the optimal solution for the next least-preferred function and go to step IX.

If NO, then go to step VIII b.

b) Is solution X_{r+2} satisfactory?

If YES, then the problem is solved.

If NO, then substitute X_β by X_{r+1}, where X_β is already defined and go to step IX.

IX. Reiterate step VIII until all the least-preferred functions have been considered.

Finally, either one finds a satisfactory solution, or such a solution does not exist.

Example 1. Consider the following linear programming problem with two objective functions:

$$\max F_1(x) = 3.5x_1 + 2x_2 + 0.7x_3 + 3x_4 + 4x_5 + 8x_6,$$

$$\max F_2(x) = x_1 + x_2 + x_3 + x_4 + x_5 + x_6,$$

subject to

$$0.1x_1 \qquad\qquad + 0.05x_3 \qquad\qquad + x_6 = 0.08,$$

$$0.7x_2 + 0.02x_3 + 0.3x_4 + x_5 = 0.25,$$

$$0.8x_1 + 0.05x_2 + 0.06x_3 + 0.3x_4 \qquad = 0.22,$$

$$x_1 \ldots, x_6 \geqslant 0.$$

Stage I. Solving the two independent optimisation problems yields :

$$X_1 = (0\,;\ 0\,;\ 0\,;\ 0.734\,;\ 0.03\,;\ 0.08) \text{ and } F_1(X_1) = 2.96$$

and

$$X_2 = (0\,; 0.14\,; 1.6\,; 0.38\,; 0\,; 0) \text{ and } F_2(X_2) = 2.13.$$

Stage II. Consider the zero-sum two-person game with the pay-off matrix :

$\|f_{ij}\|$	X_1	X_2
F_1	2.96	2.54
F_2	0.84	2.13 .

Stage III. Determine the constant K. As $f_{ij} > 0$ for all i; $j = 1,2$, then $K = 0$.

Stage IV. Determine

$$M_1 = 2.96 \text{ și } M_2 = 2.13.$$

Normalize the pay-off matrix and get

$\|f'_{ij}\|$	X_1	X_2
F_1	1	0.85
F_2	0.39	1

Stage V. Solve the game defined by the matrix of Stage IV using linear programming, and obtain

$$\lambda_1' = 0.9 \text{ and } \lambda_2' = 0.04.$$

Stage VI. Calculate

$$n_1 = \frac{0.9}{2.96} = 0.3, \quad n_2 = \frac{0.04}{2.13} = 0.02,$$

hence, $\lambda_1^* = 0.94$, $\lambda_2^* = 0.06$.

Stage VII. Construct a new function :

$$F_3(x) = 0.94\, F_1(x) + 0.06\, F_2(x) = 3.35x_1 + 1.94x_2 +$$

$$+ 0.71x_3 + 2.88x_4 + 3.82x_5 + 7.58x_6.$$

Maximizing $F_3(x)$ on the given domain yields $X_3 = X_1$, i.e., according to this algorithm, the problem does not assume a satisfactory solution.

Exemple 2 [10]

$$\max F_1(x) = -2x_1 + x_2,$$

$$\max F_2(x) = 3x_1 - x_2,$$

subject to

$$-x_1 + x_2 \leqslant 1$$

$$x_1 + x_2 \leqslant 7$$

$$x_1 \leqslant 5$$

$$x_2 \leqslant 3$$

$$x_1, x_2 \geqslant 0.$$

Stage I. Solving the two problems separately yields

$$X_1 = (0, 1) \text{ and } X_2 = (5, 0) \text{ with } F_1(X_1) = 1\,;\ F_2(X_2) = 15.$$

Stage II. The pay-off matrix is :

$\|f_{ij}\|$	X_1	X_2
F_1	1	−10
F_2	−1	15 .

Stages III—IV. $K^0 = 0$, $M_1^0 = 1$, $M_2^0 = 15$, whereas the normalized pay-off matrix is

$\|f_{ij}'^0\|$	X_1	X_2
F_1'	1	−10
F_2'	$-\frac{1}{15}$	1 .

Stages V—VI. Find

$$\lambda_1'^0 = \frac{16}{181}\,;\ \lambda_2'^0 = \frac{165}{181},$$

$$n_1^0 = \frac{16}{181}\,;\ n_2^0 = \frac{11}{181}\,;\ \sum_k n_k^0 = \frac{27}{181},$$

$$\lambda_1^{*0} = \frac{16}{27}\,;\ \lambda_2^{*0} = \frac{11}{27}.$$

Stage VII. Solve

$$\max_{x \in D} F_3(x) = \frac{16}{27} F_1(x) + \frac{11}{27} F_2(x) = \frac{1}{27} x_1 + \frac{5}{27} x_2$$

and get $X_3 = (4,3)$ such that $F_1(X_3) = -5$, $F_2(X_3) = 9$. Notice that X_3 does not coincide with the previous solutions.

Let us assume that X_3 is not a satisfactory solution.

Stage VIII. Consider that the function $F_\alpha(x) = F_2(x)$ is least-preferred by the decision-maker. Substitute X_2 by X_3 and estart from stage II.

Stage II. The pay-off matrix is

$\|f_{ij}^{\prime 1}\|$	X_1	X_3
F_1	1	-5
F_2	-1	9

Stages III—IV. $K^1 = 0$, $M_1^1 = 1$, $M_2^1 = 9$ and the normalized pay-off matrix is

$\|f_{ij}^{\prime 1}\|$	X_1	X_3
F_1'	1	-5
F_2'	$-\frac{1}{9}$	1

Stages V—VI. One finds

$$\lambda_1^{\prime 1} = \frac{5}{32}, \quad \lambda_2^{\prime 1} = \frac{27}{32},$$

$$n_1' = \frac{5}{32}, \quad n_2^1 = \frac{3}{32}, \quad \sum_k n_k^1 = \frac{1}{4},$$

$$\lambda_1^{*1} = \frac{5}{8}, \quad \lambda_2^{*1} = \frac{3}{8}.$$

Stage VII. Solve

$$\max_{x \in D} F_4(x) = -\frac{1}{8} x_1 + \frac{1}{2} x_2$$

and get $X_4=(2, 3)$ such that $F_1(X_4) = -1$ and $F_2(X_4) = 3$. The solution X_4 is considered to be satisfactory and hence we choose it as the solution of the initial problem.

1.6. POP METHOD

Recent investigation due to French researchers – mainly those working at SEMA – give solutions to the linear programming problem with multiple objective functions. One of these solutions can be obtained by use of the POP method, as described by Benayoun and Tergny [12]. Like Saska's method, the POP method aims at obtaining a "best compromise" vector X.

The set of feasible solutions D can be partially ordered using the following

DEFINITION. Given two feasible solutions $X_1 = (x_{11}, x_{12}, \ldots, x_{1n})$ and $X_2 = (x_{21}, x_{22}, \ldots, x_{2n})$, one says that solution X_1 *is preferred to* solution X_2 (or X_1 *uniformly dominates* X_2) and write $X_1 \succcurlyeq X_2$ if

$$F_j(X_1) = \sum_{i=1}^{n} c_{ji} x_{1i} \geqslant F_j(X_2) = \sum_{i=1}^{n} c_{ji} x_{2i};$$

$$j = 1, 2, \ldots, j_0 - 1, j_0 + 1, \ldots, r,$$

and for at least one index j_0

$$F_{j_0}(X_1) = \sum_{i=1}^{n} c_{j_0 i} X_{1i} > F_{j_0}(X_2) = \sum_{i=1}^{n} c_{j_0 i} x_{2i}.$$

Definition. An *efficient solution* is any maximal solution in the sense of the above defined order, i.e. a solution X such that if any other solution Y is such that $Y \succcurlyeq X$, then $X \equiv Y$.

One can notice that the efficient solution defined by the French researchers is no other than the efficient solution already defined in Section 1.1.

Definition. The vector $S_j = (x_{j1}, x_{j2}, \ldots, x_{jn})$ is a *marginal solution* relative to the criterion F_j if it maximizes the function $F_j(X) = \sum_{i=1}^{n} c_{ji}x_i$ on the domain D.

For each of the r efficiency functions F_j, one can find a marginal solution S_j. Hence, one can construct an r by r *table of consequences* (the win table):

Table 1.5

Marginal Solutions \ Criteria	F_1	F_2	. . .	F_r
S_1	y_{11}	y_{12}	. . .	y_{1r}
S_2	y_{21}	y_{22}	. . .	y_{2r}
.	.	.		.
.	.	.		.
.	.	.		.
S_r	y_{r1}	y_{r2}	. . .	y_{rr}

where $y_{ij} = F_j(S_i) = \sum_{k=1}^{n} c_{jk}x_{ik}$.

Notice that this table contains the values assumed by the efficiency function F_i for the marginal solutions S_j.

Definition. *The mixt solution* associated to the feasible solutions $S_1, S_2, \ldots, S_r$ is any solution of the form

$$X = \sum_{k=1}^{r} \lambda_k S_k; \; \sum_{k=1}^{r} \lambda_k = 1, \; 0 \leqslant \lambda_k \leqslant 1, 1 \leqslant k \leqslant r.$$

The mixt solutions associated to solutions $S_1, S_2 \ldots, S_r$ are points (other than the corners) of the least convex polyhedron $\mathscr{P} = \mathscr{P}(S_1, \ldots, S_r)$ containing the points $S_1, \ldots, S_r$.

Obviously, if the solutions $S_1, \ldots, S_r$ are feasible, then the associated mixed solutions are also feasible.

If $X = \sum_{k=1}^{r} \lambda_k S_k$ is a mixed solution, then the table of consequences can be used to derive the relative values of each criterion :

$$F_j(X) = \sum_{k=1}^{r} \lambda_k y_{kj}, \qquad (j = 1, \ldots, r).$$

One can assume that all the marginal solutions from the table of consequences are efficient. If, for example, the marginal solution S_{j_0} is not efficient, then it can be replaced by the solution obtained by solving the problem :

$$\max F_{j_0} = F_{j_0}(X) + \varepsilon \sum_{j \neq j_0} F_j(X), \quad (\varepsilon > 0, \text{sufficiently small})$$

subject to

$$AX \leqslant b$$

$$X \geqslant 0.$$

It can be easily shown [12] that the solution of this problem is efficient.

The POP method consists in the sequential exploration of the efficient solutions. The decision-maker is allowed to interfere with the solution process and guide the search for the most efficient feasible solutions.

The method is iterative, each iteration consisting of two stages :

Stage A. One starts from the table of consequences associated to the efficient solutions $S_1, S_2, \ldots, S_r$. Consider the polyhedron $\mathscr{P} = \mathscr{P}(S_1, S_2, \ldots, S_r)$ generated by these points.

In order to lessen the number of efficient solutions under consideration, one can imagine various methods for reducing the domain of feasible solutions D, namely by eliminating some of the corners S_j and introducing a smaller number of other corners. Thus, one constructs the sequence of decreasing polyhedra :

$$\mathscr{P} \supset \mathscr{P}_1 \supset \ldots \supset \mathscr{P}_j \supset \mathscr{P}_{j+1} \supset \ldots \supset \mathscr{P}_s = \mathscr{P}'.$$

The decision-maker stops at polyhedron $\mathscr{P}_s$ and proceeds to stage B provided that he does not wish to eliminate or introduce further corners or $\mathscr{P}_s$ reduces to a single point S. In the latter case, if S is an efficient solution, then this is the sought for solution; if not, one proceeds to stage B to find a better one. If at the end of stage A, one finds a solution which, in the opinion of the decision-maker, satisfies all the efficiency functions, then the algorithm is stopped.

One of the methods by which one can pass from polyhedron $\mathscr{P}_j$ to polyhedron $\mathscr{P}_{j+1}$ is that of *comparing with the average value* (Benayoun and Tergny [12]). This consists in adding the average solution to the marginal solutions

$$\bar{S} = \frac{1}{r} \sum_{i=1}^{r} S_i$$

and comparing each corner S_i with the average $\bar{S}$ in order to decide whether it can be eliminated or not.

To this purpose, one computes two values :

p_i^* = the percentage of criteria F_j for which the solution S_i has

$$y_{ij} > \frac{1}{r} \sum_{i=1}^{r} y_{ij};$$

and

$$q_i^* = \max\left(-\min_j \frac{y_{ij} - \frac{1}{r}\sum_{i=1}^{r} y_{ij}}{M_j - m_j}, 0\right)$$

with

$$M_j = \max y_{ij} \text{ şi } m_j = \min y_{ij}.$$

These are compared with the threshold values P and Q. The solution where $p_i^* < P$ and $q_i^* > Q$ are eliminated.

Stage B. In this stage, the decision-maker performs a new optimisation on the restricted feasible domain, obtained by adding to the initial system of constraints a new constraint obtained by assigning certain values l_j^k to some of the objective functions. This actually means to identify at each iteration k a subdomain D^k, in the domain D, containing new efficient solutions. The subdomain D^k is defined by the intersection between D and the set of inequalities $F_j(X) \geqslant l_j^k$, $(j = 1, \ldots, m)$.

Let $\mathscr{P}^k$ and $\mathscr{P}'^k$ be the solution polyhedra obtained at iteration k. The values prescribed for some of the objective functions depend on the values assumed by various corners of the polyhedra $\mathscr{P}^k$ and $\mathscr{P}'^k$. If

$$y_j'^k = \min_{X \in \mathscr{P}'^k} F_j(X), \qquad (j = 1, \ldots, m)$$

and

$$y_j^k = \min_{X \in \mathscr{P}^k} F_j(X),$$

then one takes

$$l_i^k = y_j^k.$$

Since $y_j'^k \geqslant l_j^k$, any solution contained in D^k will assume, for all the criteria, a value at least equal to the least values assumed by these criteria over the set of points which had been eliminated.

For the feasibility domain D^k thus found one calculates the new extreme solutions, using as the former plus a "mixt" criterion defined as the weighted sum of criteria

$$F^k = \sum_j \pi_j^k F_j(X).$$

The weights π_j^k can be calculated as follows:

$$\pi_j^k = 1/(y_j^k - F_j(\bar{S}'^k)),$$

where $\bar{S}'^k$ is the average solution on $\mathscr{P}'^k$.

At the end of this stage, the decision-maker may retain one of the solutions found to be optimal (if this is of some interest), or proceed again to stage A. The flowchart for this algorithm is given in fig. 1.1.

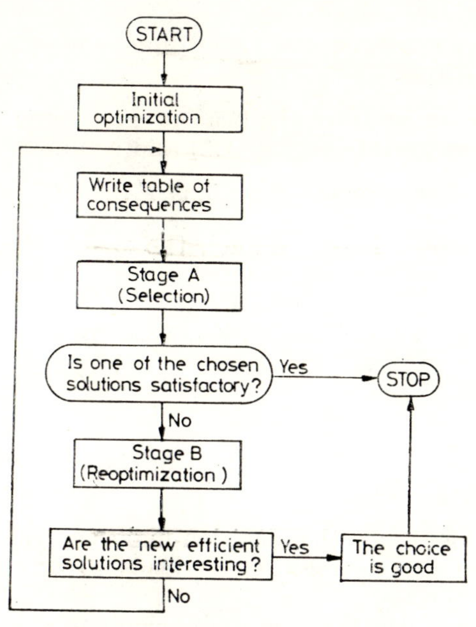

Fig. 1.1.

1.7. STEM METHOD

In emulation of the POP method, the French researchers developed a more systematic method — STEM — which is more serviceable in practice. Thus, the STEM method has the advantage that it requires only a minimum information from the decision-maker.

Thus, through an alternate use of computational and decision phases, the decision-maker can better manage the search for convenient solutions.

The steps to be followed in the STEM method are the following.

Stage 1. Solve r linear programming problems as defined by the constraints (1.2)–(1.3) and each of the functions F_k.
Let

$$Z_1 = (\max_{x \in D} F_1(X), \ldots, \max_{x \in D} F_r(X)) = (F_{11}, F_{12}, \ldots, F_{1r})$$

be the vector containing the maximum values taken by the efficiency functions.

Stage 2. Solve an additional linear programming problem defined by the constraints (1.2)–(1.3) and a function F^* defined as the weighted sum of the functions $F_1, \ldots F_r : F^* = \sum_{j=1}^{r} \pi_j F_j$.

As will be shown in the sequel, the weights π_j are chosen depending on the available information about the relative importance of the criteria employed.

Let S^* be the solution maximizing the function F^* and let

$$Z(S^*) = (F_1(S^*), F_2(S^*), \ldots, F_r(S^*))$$

be the vector containing the values of the efficiency functions $F_1, \ldots, F_r$ for the solutions S^*.

Stage 3. In this stage the decision-maker is first challenged to decide whether the solution S^* leads to acceptable values for $F_1, \ldots, F_r$ or not.

To this purpose, the decision-maker compares the components of vector $Z(S^*)$ with those of vector Z_1.

If all the functions take acceptable values for solution S^*, the problem is solved and S^* becomes the optimal solution. If not, then the decision-maker has to indicate for that function F_k, which is not yet acceptable, the threshold F^*_{k1} over which their values will become acceptable.

Stage 4. Restart the problem from Stage 1 considering the initial set of constrains plus the addition constraint introduced at Stage 3.

If $D_1 \subset D$ is the new domain of feasible solutions defined by the constraints

$$AX \leqslant b; \; X \geqslant 0$$

$$F_k \geqslant F^*_k,$$

then Stage 1 leads to a new vector :

$$Z_2 = (\max_{x \in D_1} F_1(X), \ldots, \max_{x \in D_1} F_r(X)) = (F_{21}, F_{22}, \ldots, F_{2r})$$

containing the maximum values of the efficiency functions on the restricted domain D_1.

Stage 5. This is the second stage in which the decision-maker has to intervene. In order to do this, he has to answer the question : "Will decreasing D lead to maximum inadmissible values for certain functions ?". If the answer is "Yes", the process is resumed from Stage 1 with a smaller threshold $F_k^{**} < F_k^*$. If "No", proceed to Stage 2 for the domain D_1.

Three situations are distinguished in ref. [11] for selecting the weights π_j used in Step 2 :

1° The importance of each function can be expressed through a weighting number ;

2° The importance relationships between any two criteria are known ;

3° No information is given regarding the importance of function F_h ;

Since in 1°, the weight p_j of each criterion F_j is assumed to be known $\pi_j = p_j$. Then, throughout the computation, the values of π_j remain unchanged.

In 2°, one compares any two criteria F_i and F_j and state whether they are equally important or one is more important than the other. Then associate to functions $F_1, \ldots, F_r$ the square matrix $A = ((a_{ij}))$, $(i, j = 1, 2, \ldots, r)$ defined as :

$$a_{ij} = \begin{cases} 1 \text{ if } F_i \text{ and } F_j \text{ are equally important;} \\ 2 \text{ if } F_i \text{ is more important than } F_j; \\ 4 \text{ if } F_i \text{ is much more important than } F_j; \\ 0 \text{ otherwise.} \end{cases}$$

Let $p_i = \sum_{j=1}^{r} a_{ij}$. The values of p_i give some information on whether and how F_i exceeds the other criteria. The weights π_i and π_j are then taken such that

$$\frac{\pi_i}{p_i} = \frac{\pi_j}{p_j}, \quad (i, j = 1, 2, \ldots, r).$$

As in the previous situation, once the weights π_i are determined, they remain unchanged throughout the remainder of the algorithm.

In 3° one has to consider the table of consequences associated to solutions $X_1, \ldots, X_r$. Let $C_s = (Z_{1s}, Z_{2s}, \ldots, Z_{rs})$ be a column of the table of consequences. The difference between the maximum value in this column and the remainder of values has to be considered. If these differences are small, then this particular criterion is of small importance. Conversely, if they are large, the criterion is very important and hence it is given a larger weight. Since the amplitude of the differences is generally relative, it is more useful to consider some normalized quantities, such as the utilities, instead of the consequences.

Thus, a criterion F_s assumes utility 1 for the maximum value and utility 0 for the minimum value. Then the values associated to a criterion F_h will increase as the values in the corresponding column approach zero.

Let m_k be the smallest utility value associated to the function F_k and let $\alpha_k = 1 - m_k$. Then the weights π_i and π_j are taken such that

$$\frac{\pi_i}{\alpha_i} = \frac{\pi_j}{\alpha_j}, \qquad \sum_{i=1}^{r} \pi_i = 1.$$

As distinct from the other two situations, the values of the weights π_i will now change from one iteration to the next.

In order to improve the convergence of the algorithm and to lessen the need for session with the decision-maker, three acceleration methods are suggested [11]:

1. The parametrization of the thresholds;
2. The minimal degradation method;
3. The marginal conditioning method.

Using acceleration methods 1 and 2, Benayoun et al. [11] proposed a modified, quickly convergent algorithm which can be used when there exists no information about the functions F_h. This algorithm, named STEM 2, consists of the following stages:

Stage 1. Start with the table of consequences associated to the maximal solutions of the functions F_h. Denote the maximum and minimum values of the function F_j by M_j and m_j, respectively, and let $\overline{m}_j$ be the least value in the column corresponding to F_j.

Stage 2. Let $\gamma_j = \dfrac{M_j - \overline{m}_j}{M_j - m_j}$, $(j = 1, 2, \ldots, r)$ and choose the weights of F_j's such that

$$\frac{\pi_j}{\gamma_j} = \alpha, \quad \sum_{j=1}^{r} \pi_j = 1,$$

where α is constant.

Stage 3. Solve the linear programming problem with parameter λ (L.P.(λ)) :

$$\min \lambda$$

subject to

$$\lambda \geqslant (M_j - F_j(X))\pi_i, \quad (j = 1, \ldots, r).$$
$$x \in D$$

Its solution S^* leads to the vector $Z(S^*)$.

Stage 4. Here one proceeds as in Stage 3 of the general STEM algorithm. The decision-maker compares the components of the vectors $Z(S^*)$ and Z_1, and hence may be faced with one of the following situations :

a) The value taken by each function when $S = S^*$ is acceptable. The S^* is the required solution and the search is stopped.

b) No function takes acceptable values when $S = S^*$ and hence there exists no solution.

c) There exist functions which though assuming unacceptable values may be improved since there exists a function F_{j^*} whose maximum value can be decreased.

Stage 5. Let the decrease of F_{j^*} be ΔF_{j^*}. The problem is resumed from Step 2 assuming that $\pi_{j^*} = 0$ and adding the constraints

$$F_{j^*}(X) \geqslant F_{j^*}(S_j^*) - \Delta F_{j^*},$$

$$F_j(X) \geqslant F_j(S^*), \; j \neq j^*.$$

This method of improving the solution leads within at most r iterations to a solution which is "closest" to Z_1. The flowchart of STEM 2 method is given in fig. 1.2.

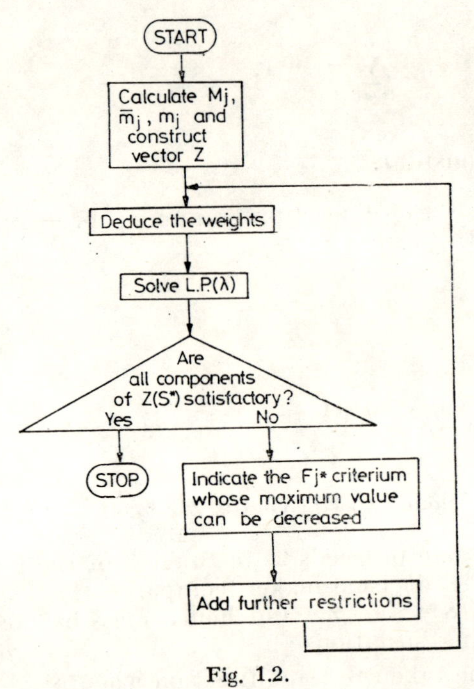

Fig. 1.2.

1.8. THE CASE OF TWO OBJECTIVE FUNCTIONS

Since stochastic programming entangles the existence of two objective functions (the mean values and the variance), we shall examine the case of two objective functions separately. One should mention that this problem has been approached by Swarup, Bector, Dinkelbach, Zionts, Geoffrion, Martos, Charnes and Cooper, a.o.

Consider the problem :

$$(1.20)\quad \max_{x \in D} h(f_1(x), f_2(x)),$$

where f_1 and f_2 are concave real-valued continuous functions, h is a nondecreasing real-valued continuous function and D is a nonempty convex compact set.

Let us first examine the case when the synthesis function assumes the form:

$$h(f(x), g(x)) = z = \frac{f(x)}{g(x)},$$

where $f(x)$ and $g(x)$ are not necessarily linear and $g(x)$ preserves its sign (say, $+$) on D.

The link between fractional programming and parametric programming problems is established through the following

THEOREM 1.8. *The vector x_0 is an optimal solution of the fractional programming problem*

$$(1.21)\quad \max_{x \in D} \frac{f(x)}{g(x)} = \frac{f(x_0)}{g(x_0)} = z_0$$

if and only if

$$(1.22)\quad F(z_0) = \max_{x \in D} [f(x) - z_0 g(x)] = 0.$$

Proof. If x_0 is an optimal solution for the fractional programming problem (1.21), then

$$z_0 = \frac{f(x_0)}{g(x_0)} \geqslant \frac{f(x)}{g(x)}, \qquad x, x_0 \in D,$$

i.e.

$$f(x) - z_0 g(x) \leqslant 0, \qquad x \in D,$$

since

$$g(x) > 0, \qquad x \in D.$$

But for x_0 the equality sign is established and, hence,

$$F(z_0) = \max_{x \in D} [f(x) - z_0 g(x)] = 0.$$

Conversely, if for x_0 and $z_0 = \frac{f(x_0)}{g(x_0)}$,

$$F(z_0) = \max_{x \in D} [f(x) - z_0 g(x)] = 0,$$

then

$$f(x) - z_0 g(x) \leqslant f(x_0) - z_0 g(x_0) = 0, \quad x, x_0 \in D,$$

i.e.

$$\frac{f(x)}{g(x)} \leqslant z_0 = \frac{f(x_0)}{g(x_0)} \quad \forall\, x \in D.$$

Hence x_0 is a solution of the fractional programming problem (1.21). The proof is complete now.

This theorem also furnishes a method for computing the optimal solution for the fractional programming problem.

Let $F(z) = \max_x [f(x) - zg(x)]$, $x \in D$, $z \in R$. If one chooses two numbers z_1 and z_2 such that $z_2 > z_1$ and if for z_i $(i = 1, 2)$ fixed, x_i maximizes $F(z_i)$, then

$$F(z_2) = \max_x [f(x) - z_2 g(x)] = f(x_2) - z_2 g(x_2) <$$

$$< f(x_2) - z_1 g(x_2) \leqslant \max_x [f(x) - z_1 g(x)] = F(z_1).$$

This means that $F(z)$ is strictly decreasing in z and consequently $F(z) = 0$ has a unique solution.

The algorithm for solving the fractional programming problem consists in the following steps:

Step 1. Consider $z = z_1 = 0$ and determine X_1 as the optimal solution of $F(z_1) = F(0) = \max_x f(x)$.

Let $z_2 = \frac{f(x_1)}{g(x_1)}$.

Step 2. Determine X_2 as the optimal solution of the mathematical programming problem:

$$F(z_2) = \max_x [f(x) - z_2 g(x)].$$

Let $z_3 = \frac{f(x_2)}{g(x_2)}$

The algorithm develops iteratively.

Step 3. Stop at iteration k if $F(z_k)$ is zero (according to theorem 1.8) or less then a quantity $\delta > 0$ infinitesimally small.
Numerical Example. Maximize

$$z = \frac{x_1 - 3x_2 + 3x_3}{x_1 + 2x_2 + x_3 + 1} = \frac{f(x)}{g(x)},$$

subject to

$$\begin{aligned} x_1 + x_2 \quad - x_3 &\leqslant 6 \\ -2x_1 - x_2 \quad + 2x_3 &\leqslant 2 \\ 2x_1 - 3x_2 + x_3 &\leqslant 3 \\ x_1, x_2, x_3 &\geqslant 0. \end{aligned}$$

By adding the new variables x_4, x_5, x_6 the constraints become :

$$\begin{aligned} x_1 + x_2 - x_3 + x_4 &= 6 \\ -2x_1 - x_2 + 2x_3 + x_5 &= 2 \\ 2x_1 - 3x_2 + x_3 + x_6 &= 3 \\ x_1, x_2, x_3, x_4, x_5, x_6 &\geqslant 0. \end{aligned}$$

Consider $z_1 = 0$ and solve the linear programming problem $F(z_1) = F(0) = \max\limits_{x \in D} f(x) = \max\limits_{x \in D} (x_1 - 3x_2 + 3x_3)$. The optimal solution $X_1\left(\frac{37}{3};\ 14;\ \frac{61}{3};\ 0;\ 0;\ 0\right)$ yields $F(z_1) = F_1 = \frac{94}{3} =$ $= 33.13$. Let $z_2 = \frac{f(X_1)}{g(X_1)} = \frac{94}{185} \approx 0.5$. Solve the linear programming problem :

$$F_2 = F(z_2) = F(0.5) = \max_{x \in D} (f(x) - 0.5g(x)) =$$

$$= \max_{x \in D} \left(\frac{91}{185} x_1 - \frac{743}{185} x_2 + \frac{461}{185} x_3 - \frac{94}{185}\right).$$

The optimal solution

$$X_2\left(\frac{2}{3}; 0; \frac{5}{3}; 7; 0; 0\right) \text{ yields } F(z_2) = F_2 =$$

$$= \frac{2205}{555} \approx 3.98.$$

Let $z_3 = \dfrac{f(X_2)}{g(X_2)} = \dfrac{17}{10} = 1.7.$
Solve the problem:

$$F_3 = F(z_3) = F(1.7) = \max_{x \in D} (f(x) - 1.7g(x)) =$$

$$= \max_{x \in D} \left(-\frac{7}{10} x_1 - \frac{64}{10} x_2 + \frac{13}{10} x_3 - \frac{17}{10} \right).$$

The optimal solution $X_3\left(\frac{2}{3}; 0; \frac{5}{3}; 7; 0; 0; 0\right)$ yields $F_3 = 0$.

According to the algorithm under discussion, X_3 is the optimal solution of the fractional programming problem.

Let P_λ be the following problem:

$$\max_{x \in D} [\lambda f_1(x) + (1 - \lambda) f_2(x)]$$

and let $D(\lambda)$ be the set of optimal solutions of P_λ for $\lambda \in [0, 1]$. Denote

$$H(\lambda) = \max_{x \in D(\lambda)} h(f_1(x), f_2(x)), \qquad \lambda \in [0, 1],$$

$$H(\lambda^*) = \max_{\lambda \in [0, 1]} H(\lambda).$$

The following theorem establishes the link between problem (1.20) and P_λ. This theorem is a restriction of Theorem 1.5 to the case of only two objective functions.

THEOREM 1.9. *Any optimal solution for the problem*

$$\max_{x \in D} h[f_1(x), f_2(x)]$$

lies among the optimal solutions of P_λ for a given $\lambda \in [0, 1]$.
This means that if

$$H(\lambda^*) = \max_{\lambda \in [0,1]} H(\lambda),$$

then the optimal solution of problem (1.20) is that $x \in D(\lambda^*)$ such that

$$h[f(x)] = H(\lambda^*).$$

In the following chapters a theorem more general than Theorem 1.9 will be required. Whence,

THEOREM 1.10 (Geoffrion [74]). *Let D be a convex compact set in $D \subset R^n$; $f_1, f_2, u_1(f_1), u_2(f_2)$ concave functions on D, uibeing strictly increasing on the image of D under f_i $(i = 1, 2)$, and φ non-decreasing and quasi-concave on the convex hull of the image of D under $(u_1(f_1), u_2(f_2))$, all functions being continuous. Assume that a parametric programming algorithm is available for solving:*

$$\min_{x \in D} [\lambda f_1(x) + (1 - \lambda) f_2(x)]$$

for each value of the parameter λ in the unit interval and that the resulting optimal solution function $x(\lambda)$ is continuous on $[0, 1]$. Then the function

$$\Phi(\lambda) = \psi(x(\lambda)) = \varphi(u_1(f_1(x(\lambda))), u_2(f_2(x(\lambda))))$$

is continuous and unimodal on $[0, 1]$ and, if λ^* maximizes Φ on $[0, 1]$, then $x(\lambda^*)$ is optimal for the problem

$$\min_{x \in D} \psi(x) = \varphi(u_1(f_1(x), u_2(f_2(x))).$$

1.9. A NOTE ON THE PRESENT STATE OF THE ART

Without aiming at an exhaustive picture of mathematical programming with multiple objective functions, we shall however give bellow a brief view of the present "state of the art". A considerable amount of research-work has been under progress regarding the multiple criteria problem and reference to these can be found in the works of Zeleny [253], Stancu-Minasian [137], Nehse [220], Achilles, Elster and Nehse [173].

A complete survey of multicriteria optimization was given by Stadler [239].

More than forty Ph. D. theses have been written on multicriteria decision-taking. A number of books, of which many Conference Proceedings, have been published ([43, [193], [237], [244], [252], [257]).

Using the first formulation of an optimum problem with vector criteria which is due to Pareto [225], Koopmans introduced the concept of "Pareto optimality" [206].

The polyhedral and polar cones are important in the study of multicriteria decision problems. Their properties were used in the study of the solution to such problems by Tamura [234] and Yu [250].

An important class of methods used for solving the multiple objective programming problems is that of interactive methods. In recent years, the use of these methods has increased considerably. This arises from the fact that these methods allow a direct session between the computer and the decision-maker during the solution process. Among these, the POP and the STEM methods have been discussed in Sections 1.6 and 1.7. Other important interactive methods are : the Geoffrion-Dyer-Feinberg algorithm of man-machine interaction [195], the interactive rectangle elimination method for bi-objective decision-making [227]; and the interactive multi-objective decision-making technique for solving large-scale multiobjective optimisation problems of the block angular structures [229]. Ecker suggested [192] an interactive method for obtaining a subset of the nondominated set to a multiple objective linear program. This subset is named "the trade-off compromise set". Another interactive method has recently been applied in solving a multidimensional location model, by Nijkamp and Spronk [222].

Though in classical mathematical programming problems the coefficients are assumed to be known with accuracy, this is seldom so in practice because of the errors of measurement or variation with market conditions. Bitran [182] approaches linear multiple objective programs with the criteria coefficients given on intervals and presents an algorithm to obtain a set of efficient extreme points.

The techniques of mathematical programming with multiple objective functions were extended to special classes of mathematical programming problems as integer and mixed integer programming, fractional programming [233] and programming in complex space [190], [205].

Zionts [258] reviewed the techniques for solving multiple objective integer linear programming. He contended that "a method for solving multiple criteria integer linear programming problems is not an obvious extension of the methods that solve multiple criteria linear programming problems". Hence he [256] gave two ways of adapting the Zionts-Wallenius algorithm [255] for solving multicriteria mixed integer programming : the cutting plane approach and the branch-and-bound approach.

Willareal and Karwan [249] suggested a general procedure for solving multicriteria discrete mathematical programming problems using a recursive dynamic programming approach. Bowman, Jr. [185] proposed a generalized Tchebycheff norm to generate all efficient integer solutions.

Lupşa [216] established a link between the multiple objective function mathematical programming and a certain problem of parametric programming and developed an algorithm for solving multiple criteria integer programming problems.

Lee [214] developed a goal programming approach to multicriteria integer programming. He examined three integer programming algorithms (the dual cutting plane method, the branch-and-bound method and the implicit enumeration method) against the goal programming framework.

Bitran [183] presents an algorithm for obtaining all the efficient points of the linear multi-objective problem with zero-one variables (for the multi-objective integer programming see also [177]). A branch-and-bound algorithm was also developed [184] and then applied to facility location problems.

Slowinski [230] studies the allocation of different categories of resources in a network project under multiple conflicting criteria and describes two possible approaches for solving the problem (see also [224]). Huckert, Rhode, Roglin and Weber [201] consider a mixed integer linear vector minimization problem (MILVMP) for solving the scheduling problem. In doing this, they take into account the following criteria:

— the minimizing Make-Span
— the minimizing total flow time
— the minimizing total duration and idle time
— the minimizing tardiness
— the minimizing maximum tardiness
— the minimizing of the number of tardy jobs.

The main difficulty encountered in solving MILVMP is that the feasible region is not convex (as for problem (1.1)—(1.3)) and hence the necessity condition in Theorem 1.5 (Section 1.1) does not hold. Consequently, the interactive method based on weighting objectives cannot be used since the parametrization is not sufficient to generate all the efficient integer solutions. In order to overcome this shortcoming, the above mentioned authors define the notion of *essential efficiency* (i.e. an efficiency relative to the convex hull of the feasible set) which generalizes the notion given by Brucker [32] in the case of finite alternative. An interactive method based on the Tchebycheff-approximation is given. For large problems, heuristically efficient solutions are defined.

Recently, the fuzzy approach was used for solving linear programming problems with multiple objective functions (Zimmermann [254]).

Takeda and Nishida [238] fuzzify the concepts of domination structures and nondominated solutions, which were introduced by Yu [250], to tackle general situations in which there exists some information concerning the decision-maker's preferences. Introducing the concept of fuzzy convex cones and fuzzy polar cones, they showed how some of the main results obtained by Yu [250] can be extended.

One should also mention the algorithms given by Philip [116], [117] for defining and constructing efficient solutions, and those developed by Maruşciac and Rădulescu [106], [107], [108], for obtaining efficient solutions in quadratic programming.

Chapter 2

Stochastic programming with one objective function

Consider the linear programming problem

(2.1) $\max(\min) Z = c'x,$

subject to

(2.2) $Ax \leqslant b$

(2.3) $x \geqslant 0,$

where A is an $m \times n$ matrix, c and x are n-vectors, and b is an m-vector.

Such a problem is called a stochastic programming problem if one or more of the coefficients from the set (A, b, c) are random variables on the probability space $\{\Omega, K, P\}$.

Obviously, when the coefficients are random variables, the ordinary problem of linear programming applies no longer since (for a minimum problem, for example) one cannot find a feasible vector x_0 such that $c'(\omega)x_0 \leqslant c'(\omega)x$ for every $\omega \in \Omega$ and for every feasible x.

Indeed, the value $c'(\omega)x$ of the objective function at any point x is not a number but a random value and one has to recall that the set of random values does not admit of an ordering relation. Consequently, for an $\omega_1 \in \Omega$ we can have $c'(\omega_1)x_0 \leqslant$ $\leqslant c'(\omega_1)x$, whereas for another $\omega_2 \in \Omega$, $c'(\omega_2)x_0 \geqslant c'(\omega_2)x$.

Usually, the stochastic problem is replaced by a suitable deterministic problem (called the Deterministic Equivalent) whose optimal solutions are extended to represent solutions to the stochastic problem. Here, we shall consider other problems, e.g. to find the distribution function of the optimum values of Z.

The following sections may be distinguished in the field of stochastic programming.

I. Distribution problems

II. Two-stage programming under uncertainty

III. Chance-constrained programming

Some aspects of these problems will be tackled in the present chapter.

A more detailed classification of stochastic programming problems can be found in ref. [144].

2.1. DISTRIBUTION PROBLEMS

Let $A(\omega)$, $b(\omega)$, $c(\omega)$ be random matrices whose elements are random variables defined on the probability space $\{\Omega, K, P\}$, where Ω is a Borel set, K is the σ-algebra of all the Borel subsets of Ω and P is a probability.

Denote

$$(2.4) \quad D(\omega) = \{x \mid A(\omega)x \leqslant b(\omega),\ x \geqslant 0\},\ \omega \in \Omega,$$

where $x = (x_1, \ldots, x_n)'$ is an unknown column vector. When A and b do not contain random elements we shall also write $D(A,b)$.

$$(2.5) \quad Z(A(\omega), b(\omega), c(\omega)) = \max_{x \in D(\omega)} (\min)\, c'(\omega)x.$$

For convenience instead of $Z(A(\omega), b(\omega), c(\omega))$ we shall simply write $Z(\omega)$. Depending on circumstances, we shall specify whether the problem is of maximum or of minimum.

If relation (2.5) defines a random variable, then the linear stochastic program (2.4)–(2.5) is said to have the optimum value. The value of $Z(\omega)$ cannot be known beforehand. However, before having the exact values of the coefficients, we wish to know the distribution function and/or certain typical values (the mean and the variance) of $Z(\omega)$ or/and the probability of the optimum value lying in a given interval subject to some *a priori* distributions of the triplet $(A(\omega), b(\omega), c(\omega))$ defining the natural state. Such problems are called *distribution problems* and play a central role in stochastic programming.

The determination of the distribution function of the optimum value was first (1955) formulated by Tintner [155] as a problem of "passive stochastic programming". Later on, Madansky [102] discussed the same problem in terms of "wait and see". In 1955, Mihoc [110] suggested a method for solving the problem of the distribution of the optimum. Bereanu [14], [18] determined exactly the distribution function of the optimum value $Z(\omega)$ using parametric programming and the notion of "decision region" from linear stochastic programming [179]. In fact, Bereanu considers that the random elements of the problem are linear functions of certain random variables which are treated as parameters. In another work [20], Bereanu uses the Laplace transform and determines the distribution function of the optimum value. He also derived approximate formulae for the distribution function of the optimum value using the Cartesian Integration Method. Such methods are quite general but the numerical methods employed restrict the number of random elements. Recently [24], Bereanu approached the generalized distribution problem and introduced, besides ω, the parameter t. This parameter allowed the unified treatment of stochastic programming. A complete presentation of the distribution problem of stochastic linear programming is given in ref. [180], together with numerical methods and recent computational experience in solving the distribution problem.

Before approaching the probability law of the optimum value induced by the probability laws of the coefficients, two lemmas will be proved.

LEMMA 2.1. *$D(A^+, b^-) \subseteq D(A, b) \subseteq D(A^-, b^+)$; where $A^- \leqslant A \leqslant A^+$, $b^- \leqslant b \leqslant b^+$.*

Indeed, let $x \in D(A, b)$, i.e.

$$\sum_{j=1}^{n} a_{ij}x_j \leqslant b_i, \qquad (i = 1, \ldots, m).$$

Minorizing the left-hand side and majorizing the right-hand side yields

$$\sum_{j=1}^{n} a_{ij}^- x_j \leqslant \sum_{j=1}^{n} a_{ij}\, x_j \leqslant b_i \leqslant b_i^+, \qquad (i = 1, \ldots, m).$$

The first and last terms of this chain of inequalities imply that $x \in D(A^-, b^+)$.

The first inclusion may be proved in a similar manner.

LEMMA 2.2. $Z(A^-, b^+, c^-) \leqslant Z(A, b, c) \leqslant Z(A^+, b^-, c^+)$; $c^- \leqslant c \leqslant c^+$, *where* $Z(A^-, b^+, c^-) = \min_{x \in D(A^-, b^+)} c^{-\prime}x$; $Z(A^+, b^-, c^+) = \min_{x \in D(A^+, b^-)} c^{+\prime}x$.

The proof derives from the obvious chain of inequalities

$$\min_{x \in D(A^-, b^+)} c^{-\prime}x \leqslant \min_{x \in D(A,b)} c^{-\prime}x \leqslant \min_{x \in D(A,b)} c'x \leqslant \min_{x \in D(A, b)} c^{+\prime}x \leqslant$$

$$\leqslant \min_{x \in A(A^+, b^-)} c^{+\prime}x.$$

From Lemma 2.2 it results that the support of the probability law for the random function $Z(A, b, c)$ is the interval $[Z(A^-, b^+, c^-)$; $Z(A^+, b^-, c^+)]$ under the assumption that $A^-, b^-, c^-(A^+, b^+, c^+)$ is the lower (upper) bound of matrices A, b, c, respectively. Similar expressions may also be obtained for maximization of function Z.

If only the coefficients of the objective function are random variables and $D(A, b) \neq \emptyset$ and bounded, then the stochastic linear program (2.4)–(2.5) always has an optimum value [22]. But note that generally problem (2.4)–(2.5) does not have an optimum value since, for a positive probability, $D(\omega)$ may be empty and/or $Z(\omega)$ may be unbounded. Bereanu [21] showed that there exists a class of stochastic linear programs (*positive stochastic linear programs*) which always yield an optimum value.

If problem (2.4)–(2.5) has an optimum value, $Z(\omega)$ is a random variable defined on the probability space $\{\Omega_0, K_0, P_0\}$ [115], where

$$\Omega_0 = \{\omega \,|\, D(\omega) \neq \emptyset\} \subseteq \Omega,$$

$$K_0 = K \cap \Omega_0,$$

P_0 is the probability induced by P on K_0, i.e. $P_0(A) = \dfrac{P(A)}{P(\Omega_0)}, A \in K_0$.

The following theorem gives the necessary and sufficient condition for the existence of the optimum value of a linear stochastic program.

THEOREM 2.1 (Bereanu [22]). *The linear stochastic program* (2.4)—(2.5) *has an optimum value if and only if the following implications are almost certain to occur on* Ω (*i.e. with probability 1*) :

(2.6) $A(\omega)x \leqslant 0, \ x \geqslant 0 \Rightarrow c(\omega)x \leqslant 0,$

(2.7) $yA(\omega) \geqslant 0, \ y \geqslant 0 \Rightarrow yb(\omega) \geqslant 0.$

Proof. The linear stochastic program (2.4)—(2.5) has an optimum value if and only if $Z(\omega)$ is almost surely (a.s.) a measurable bounded function on Ω. This occurs if and only if, almost surely, the set $D(\omega) \neq \emptyset$ and the optimum is bounded. From the duality theorems of linear programming we derive that the stochastic linear program has an optimum value if and only if the sets $D(\omega)$ and $E(\omega) = \{y \mid yA \geqslant c(\omega), y \geqslant 0\}$ are almost surely nonempty. By virtue of Gale's Theorem [71], there occurs either of the following alternatives : either the matrix inequality $xA \leqslant b$ has a non-negative solution or the system $Ay \geqslant 0$, $by < 0$ has non-negative solutions. This implies that the conditions $D(\omega) \neq \emptyset$, $E(\omega) \neq \emptyset$ are equivalent with (2.6)—(2.7) and, hence, the proof is complete.

In what follows we shall show that

$$Z(\omega) = \min_{x \in D(\omega)} c'(\omega)x$$

is a random variable.

THEOREM 2.2. *The optimum value* $Z(\omega)$ *is a random variable*

$$Z : \Omega \to R \cup \{-\infty, \infty\}$$

if one sets :

$$Z(\omega_0) = +\infty \text{ when } D(\omega_0) \text{ is empty;}$$

and

$$Z(\omega_0) = -\infty \text{ when } Z(\omega_0) \text{ is unbound.}$$

Proof. Let $B_l (1 \leqslant l \leqslant C_n^m)$ be all the square matrices of order m (the bases) that can be extracted from the matrix A. Denote by a_j the column vectors of the matrix B_l and let

$$I = \{j : a_j \in B_l\}, \ J = \{j : a_j \notin B_l\}.$$

Each base B_l will induce a partition of the matrices A, c and x:

$$A = (A_I, A_J);\ A_I = B_l;\ A_J = \{a_j : j \in J\},$$

$$c = (c_I, c_J);\ c_I = \{c_i : i \in I\};\ c_J = \{c_j : j \in J\},$$

$$x = (x_I, x_J);\ x_I = \{x_j : j \in I\};\ x_J = \{x_j : j \in J\}.$$

Define the events

$$U_l = \{\omega|\ |B_l(\omega)| \neq \emptyset,\ B_l^{-1}(\omega)\, b(\omega) \geqslant 0;$$

$$c_I(\omega)\, B_l^{-1}(\omega)\, A_J(\omega) \leqslant c_J(\omega)\}.$$

and

$$U_l(a) = U_l \cap \{\omega : c_I(\omega) B_l^{-1}(\omega)\, b(\omega) < a\},$$

where U_l is the following event: the base B_l is optimum and $|B_l(\omega)|$ is the determinant of matrix B_l.

Since for $\omega \in U_l$ the optimal solution is

$$x_I(\omega) = B_l^{-1}(\omega)\, b(\omega),$$

it follows that

$$\{\omega : Z(\omega) < a\} \cap U_l = \{\omega : c_I(\omega)\, B_l^{-1}(\omega)\, b(\omega) < a\} \cap U_l =$$

$$= U_l(a).$$

But

$$\{\omega : -\infty < Z(\omega) < a\} = \bigcup_{l \in I} U_l(a).$$

Since all the functions participating in the definitions of events U_l and $U_l(a)$ are random variables, it follows that $U_l(a) \in K$, and hence

$$\text{(2.7.a)} \quad \{\omega : -\infty < Z(\omega) < a\} \in K.$$

Also

$$\text{(2.7.b)} \quad \{\omega : Z(\omega) = +\infty\} = \bigcup_l \{\omega|\ |B_l(\omega)| \neq \emptyset,$$

$$B_l^{-1}(\omega)\, b(\omega) \geqslant 0\},$$

i.e.

$$\{\omega : Z(\omega) = +\infty\} \in K.$$

As

$$\{\omega : Z(\omega) = -\infty\} = [\{\omega : Z(\omega) = +\infty\} \cup \bigcup_{n \in N^*} \{\omega : -\infty < Z(\omega) < n\}],$$

then

$$\text{(2.7.c)} \quad \{\omega : Z(\omega) = -\infty\} \in K.$$

The proof of the theorem derives from (2.7.a)—(2.7.c).

Note that in the preceding proof one had to use the notion of base solution. Kall [89] gives a different proof of the same theorem without using base solutions.

Let us now assume that only the vector c is random and it has the mixt probability density $f(c)$. Let C be the probability space defined by n-vectors $c = (c_1, \ldots, c_n)$. The space C can be partitioned by the sets

$$S_{1i} = \{c \mid c_{Bi} B_i^{-1} a_j - c_j \leqslant 0,\ j \in J\}.$$

If $f(c)$ is a continuous function, then the probability of event

$$\{c \mid c_{Bi} B_i^{-1} a_j - c_j = 0, \quad j \in J\}$$

is zero.

Since

$$P\{S_{1i} \cap S_{1j}\} = 0, \quad i \neq j$$

and

$$\cup S_{1i} = C,$$

one can use the total probability formula and obtain

$$P\{Z(\omega) < a\} = \sum_i [P\{Z(\omega) < a\} \cap S_{1i}] = \sum_i P[Z(\omega) < a \mid S_{1i}]\, P(S_{1i}),$$

where $P[Z(\omega) < a \mid S_{1i}]$ denotes the conditional probability of the event $\{Z(\omega) < a\}$, given the event S_{1i}.

By definition,

$$P(S_{1i}) = \int_{U_i} \cdots \int f(c) \prod_{i=1}^{n} dc_i,$$

where $U_i = \{c | c \in S_{1i}\}$.

Hence,

(2.7.d) $$P\{Z(\omega) < a\} = \sum_i P[\{Z(\omega) < a\} \cap S_{1i}] =$$

$$= \sum_i \int_{[c_{Bi}\, x_B^i < a] \cap U_i} \cdots \int f(c) \prod_{i=1}^{n} dc_i \cdot$$

Hence, if only the vector c is random, we can find the distribution function of the optimum by evaluating the above integral. (But note that in general is rather difficult to give an explicit formula for the distribution function of the optimum value). The evaluation of the above integrals is discussed by Ewbank, Foote and Kumin [61]. The computation of the integral (2.7.d) may be, sometimes, difficult since the regions of integration are involved and sometimes not even simply connected. Reference [61] gives several transformations which may simplify these regions of integration in special cases.

Bereanu considers *stochastic linear programs with simple randomization* as stochastic linear programs with random coefficients as a function of only one random variable. Assuming that the random vector c is of the form $\bar{c} + t(\omega)\bar{d}$, where $\bar{c}$ and $\bar{d}$ are constant vectors, A and b non-random, and $t(\omega)$ a random variable with whose values range over the finite interval $[\delta_1, \delta_2]$ and with a distribution function $T(z)$ continuous and strictly increasing, Bereanu [14] derives the following.

THEOREM 2.3 (Bereanu [14]). *The distribution function $F(z)$ of the random variable*

(2.7.e) $$f(\omega) = \min \{(\bar{c}' + t(\omega)\, \bar{d}')\, x \mid Ax = b,\ x \geqslant 0\}$$

is

$$F(z) = \sum_{s=1}^{p} H_s(z),$$

where

$$u_s(z) = \frac{z - \bar{c}' x^s}{\bar{d}' x^s},$$

and

$$\underset{1\leqslant s\leqslant p}{H_s(z)} = \begin{cases} T(\lambda_s) - T(\lambda_{s-1}) \text{ if } \begin{cases} \bar{d}'x^s > 0 \text{ and } u_s(z) \geqslant \lambda_s, \\ \bar{d}'x^s < 0 \text{ and } u_s(z) \leqslant \lambda_{s-1}, \\ \bar{d}'x^s = 0 \text{ and } \bar{c}'x^s < z \end{cases} \\ \left.\begin{matrix} T(u_s(z)) - T(\lambda_{s-1}) \\ T(\lambda_s) - T(u_s(z)) \end{matrix}\right\} \text{if } \lambda_{s-1} < u_s(z) < \lambda_s \text{ and } \begin{cases} \bar{d}'x^s > 0, \\ \bar{d}'x^s < 0, \end{cases} \\ 0 \qquad \text{if } \begin{cases} \bar{d}'x^s > 0 \text{ and } u_s(z) \leqslant \lambda_{s-1}, \\ \bar{d}'x^s < 0 \text{ and } u_s(z) > \lambda_s, \\ \bar{d}'x^s = 0 \text{ and } \bar{c}'x^s \geqslant z. \end{cases} \end{cases}$$

Proof. Consider the events

$$A = \{\omega \mid f(\omega) < z\}$$

and

$$B_j = \{\omega \mid \lambda_{j-1} \leqslant t(\omega) < \lambda_j\}, \ (j = 1, \ldots, p),$$

where $\lambda_j (1 \leqslant j \leqslant p - 1)$ are the critical values of the linear parametric program when $t(\omega)$ is replaced by the parameter $\lambda \in [\delta_1, \delta_2]$. Denote by x^j, $1 \leqslant j \leqslant p$ the characteristic solutions and put $\lambda_0 = \delta_1$, $\lambda_p = \delta_2$. The segments $[\lambda_j, \lambda_{j+1}]$ $(j = 1, \ldots, p - 1)$ are the decision regions of the stochastic linear programming problem (2.7.e). Indeed these segments are non-overlapping and cover $[\delta_1, \delta_2]$. Moreover, under the assumption that the random variable $t(\omega)$ has a continuous distribution function, the intersection of two such decision regions has zero probability.

Hence,

$$\bigcup_{i=1}^{p} B_i = \Omega, \quad B_i \cap B_j = \emptyset, \quad i \neq j$$

$$P(B_j) > 0, \ (1 \leqslant j \leqslant p),$$

and one can use the total probability formula

$$F(z) = P(A) = \sum_{j=1}^{p} P(A \mid B_j)\, P(B_j) = \sum_{j=1}^{p} P(A \cap B_j).$$

If the event B_j takes place, then

$$f(\omega) = (\bar{c}' + t(\omega)\,\bar{d}')\,x^j,$$

and hence

$$F(z) = \sum_{j=1}^{p} P\{[\omega \mid (\bar{c}' + t(\omega)\,\bar{d}')x^j < z] \cap$$

$$\cap\, [\omega \mid (\lambda_{j-1} \leqslant t(\omega) < \lambda_j]\} = \sum_{j=1}^{p} H_j(z).$$

This is tantamount to solving the system of inequalities

$$(\bar{c}' + t\bar{d}')\,x^j < z,$$

$$\lambda_{j-1} \leqslant t \leqslant \lambda_j.$$

The following situations may occur:

$$\lambda_{j-1} \leqslant t < \min\left(\lambda_j, \frac{z - \bar{c}'x^j}{\bar{d}'x^j}\right), \text{ if } \bar{d}'x^j > 0,$$

$$\max\left(\lambda_{j-1}, \frac{z - \bar{c}'x^j}{\bar{d}'x^j}\right) \leqslant t < \lambda_j, \text{ if } \bar{d}'x^j < 0,$$

$$t \in [\lambda_{j-1}, \lambda_j), \text{ if } \bar{d}'x^j = 0, \text{ and } \bar{c}'x^j < z,$$

$$\{t\} = \emptyset, \text{ if } \bar{d}'x^j = 0, \text{ and } \bar{c}'x^j \geqslant z.$$

Whence the expression for $H_j(z)$ given in the theorem.

The results obtained for the random vector of the form $\bar{c} + t(\omega)\bar{d}$ can be also extended [21] to the vector

$$c(t) = c^0 + t_1 c^1 + \ldots + t_r c^r,$$

where c^i, $(i = 0, \ldots, r)$ are constant vectors and $t = (t_1, \ldots, t_r)$ is a random vector with finite components of known probability distribution.

The case of vector b having random components is treated either by passing to the dual problem, i.e. c random (see the previous problem) or directly. Both results are similar.

Bereanu [180] reports computing experience in distribution problems with the computer program STOPRO which gives the probability distribution function of the optimal value as well as its expectation and variance and the probability that the optimal solution for a stochastic linear programming with stochastic right-hand side lies, component-wise, between given bounds. The distribution of the random variables involved can be arbitrarily normal, exponential, uniform or given as a histogram with up to 100 points. A summary of the examples treated is given in Table 2.1.

Table 2.1

Problem	Dimensions	Number of coefficients	Density	Number of rand. coeff.	Running time (sec)		Central Processing Unit (CPU) time (sec.)	Conventional Unit
					1 distr.	6 distr.		
1	8×13	62	59.61	5	0.51	0.69	0.5	1.50
2	102×181	1010	5.47	35	0.53	0.67	1.8	1.03
3	118×343	1300	3.21	4	0.61	0.71	4.2	1.09
4	226×508	3266	2.84	185	0.61	0.89	6.0	1.02

These examples show that the running time of STOPRO does not depend upon the dimensions of the problem or the number of random coefficients. This is due to the fact that the number of decision regions is independent of the problem dimensions. But the parametric procedures of linear programming are based on iterations of the simplex algorithm.

In order to make the computational results independent of the computer, Bereanu [180] takes as unit of measure the number of iterations required to obtain the optimal solution for the zero value of the parameter. The running time necessary for the solution of a stochastic programming problem is then determined as a function of this conventional unit.

An exact method for determining the distribution function of the optimum value is

The Complete Description Method. Although the use of this method requires lengthy calculations it is the most general method for finding the distribution function of the optimum value.

Assume that in problem (2.1)—(2.3), expressions (2.2) are equality constraints. Then $F(a) = P\{\omega \mid Z(\omega) < a\}$, $a \in R$.

In compliance with Kronecker-Capelli's theorem, the set (2.2)—(2.3) has a solution if and only if rank $A(\omega) =$ $=$ rank $(A(\omega), b(\omega))$. Consider the event

$$E = \{\omega \mid \text{rank } A(\omega) = \text{rank } (A(\omega), b(\omega))\}.$$

In this case

$$(2.8) \quad F(a) = P_E\{\omega \mid Z(\omega) < a\}.\ P(E).$$

Consider a partition of the certain event

$$\Omega = \bigcup_{r=0}^{m} A_r,\ A_i \cap A_j = \emptyset, \text{ where } A_r = \{\omega \mid \text{rank } A(\omega) = r\}.$$

Using the total probability formula,

$$(2.9) \quad P_E\{\omega \mid Z(\omega) < a\} = \sum_{r=0}^{m} P_{E \cap A_r}\{\omega \mid Z(\omega) < a\} \cdot P(A_r).$$

If the rank of matrix A is r there exist C_m^r possibilities of choosing the r linearly independent rows, which lead to C_m^r systems of linearly independent rows. Consider the k-th system $(k = 1, \ldots, C_m^r)$ and define the partition $E \cap A_r = \bigcup_{k=1}^{C_m^r} L_k$ $(L_i \cap \cap L_j = \emptyset)$, where $L_k = \{\omega \in A_r \cap E \mid$ the rows of $A(\omega)$ belonging to the k-th system are linearly independent whereas those belonging to systems $1, \ldots, k-1$ are not$\}$.

Applying again the total probability formula,

$$(2.10) \quad P_{E \cap A_r}\{\omega \mid Z(\omega) < a\} = \sum_{k=1}^{C_m^r} P_{E \cap A_r \cap L_k}\{\omega) \mid Z(\omega) < a\} \cdot P(L_k).$$

Then (2.8), (2.9) and (2.10) yield

$$(2.11) \quad F(a) = P\{\omega \mid Z(\omega) < a\} =$$

$$= P(E) \sum_{r=0}^{m} P(A_r) \sum_{k=1}^{C_m^r} P(L_k)\ P_{E \cap A_r \cap L_k}\{\omega \mid Z(\omega) < a\},$$

i.e. the distribution of the optimum.

Let us assume now that the triplet (A, b, c) takes discrete values (A_i, b_i, c_i), $i \in I$, with probabilities $p_i \geqslant 0$; $\sum_{i \in I} p_i = 1$. The Complete Description Method can be again used, but the discrete distribution of $Z(A, b, c)$ can be obtained in a direct manner.

Let

$$Z_i(A_i, b_i, c_i) = \min\{c_i'x \mid A_i(\omega)x = b_i(\omega),\ x \geqslant 0\},$$

$$I_{+\infty} = \{i \in I \mid Z_i(A_i, b_i, c_i) \text{ has no solution}\}$$

$$I_{-\infty} = \{i \in I \mid Z_i(A_i, b_i, c_i) \text{ is not finite}\}.$$

In this case the following formulae are true :

$$P(Z(A, b, c) = -\infty) = \sum_{i \in I_{-\infty}} p_i,$$

$$P(Z(A, b, c) = +\infty) = \sum_{i \in I_{+\infty}} p_i,$$

$$F(a) = P(Z_i(A_i, b_i, c_i) < a) = \sum_{i \mid Z_i(A_i, b_i, c_i) < a} p_i.$$

Hence,

$$\mathbf{E}Z = \sum_i Z_i p_i,$$

$$\mathbf{E}Z^2 = \sum_i Z_i^2 p_i,$$

where $\mathbf{E}(\xi)$ is the mean value of the random variable ξ. This last case can be illustrated by the following numerical

Example. Find the distribution function for the optimum of the following stochastic programming problem :

$$Z(A, b, c) = \max(c_1x_1 + c_2x_2 + c_3x_3)$$

subject to

$$a_{11}x_1 + a_{12}x_2 + a_{13}x_3 \leqslant b_1.$$

$$a_{21}x_1 + a_{22}x_2 + a_{23}x_3 \geqslant b_2,$$

$$a_{31}x_1 + a_{32}x_2 + a_{33}x_3 \leqslant b_3,$$

$$x_1, x_2, x_3 \geqslant 0,$$

where A, b, c are discrete independent random variables with the following distributions :

$$A = \begin{pmatrix} a_{11} & a_{12} & a_{13} \\ a_{21} & a_{22} & a_{23} \\ a_{31} & a_{32} & a_{33} \end{pmatrix} = \begin{pmatrix} A_1 = \begin{pmatrix} 4 & -7 & 3 \\ 1 & 2 & -5 \\ 0 & 3 & 1 \end{pmatrix} & A_2 = \begin{pmatrix} 3 & -2 & -1 \\ 1 & 1 & 5 \\ 4 & 1 & 3 \end{pmatrix} \\ 0.2 & 0.8 \end{pmatrix}$$

$$b = (b_1, b_2, b_3) = \begin{pmatrix} b_1 = (1, -5, 3) & b_2 = (7, 6, 4) \\ 0.4 & 0.6 \end{pmatrix},$$

$$c = (c_1, c_2, c_3) = \begin{pmatrix} c_1 = (1, 3, 4) & c_2 = (2, -3, 1) & c_3 = (1, 3, -4) \\ 0.2 & 0.5 & 0.3 \end{pmatrix}.$$

The triplet (A, b, c) can take 12 values. The solutions associated to deterministic programming problems are given in Table 2.2. The probability that the optimum value takes a given value, say 7.16, is

$$P(Z(A, b, c) = 7.16) = P(A = A_1, \ b = b_1, \ c = c_1) =$$

$$= P(A = A_1)\, P(b = b_1)\, P(c = c_1) = 0.2 \cdot 0.4 \cdot$$

$$\cdot 0.2 = 0.016.$$

Table 2.2

No	(A, b, c)	Optimal solution	$Z(A, b, c)$	Iteration	Probability of getting $Z(A, b, c)$
1	(A_1, b_1, c_1)	$x_1 = 0.29$; $x_2 = 0.57$; $x_3 = 1.29$	7.16	4	0.016
2	(A_1, b_1, c_2)	$x_1 = 2$; $x_2 = 1$; $x_3 = 0$	1	3	0.040
3	(A_1, b_1, c_3)	$x_1 = 2$; $x_2 = 1$; $x_3 = 0$	5	3	0.024
4	(A_2, b_1, c_1)	$x_1 = 0$; $x_2 = 3$; $x_3 = 0$	9	3	0.064
5	(A_2, b_1, c_2)	$x_1 = 0.46$; $x_2 = 0$; $x_3 = 0.38$	1.3	3	0.160
6	(A_2, b_1, c_3)	$x_1 = 0$; $x_2 = 3$; $x_3 = 0$	9	2	0.096
7	(A_2, b_2, c_1)	$x_1 = 0$; $x_2 = 1$; $x_3 = 1$	7	3	0.096
8	(A_2, b_2, c_2)	$x_1 = 0.12$; $x_2 = 0$; $x_3 = 0.18$	1.42	3	0.24
9	(A_2, b_2, c_3)	$x_1 = 0$; $x_2 = 1$; $x_3 = 1$	−1	3	0.144
10	(A_1, b_2, c_1)	$x_1 = 3.94$; $x_2 = 1.3$; $x_3 = 0.11$	8.28	4	0.024
11	(A_1, b_2, c_2)	$x_1 = 4.08$; $x_2 = 1.33$; $x_3 = 0$	4.17	5	0.060
12	(A_1, b_2, c_3)	$x_1 = 4.08$; $x_2 = 1.33$; $x_3 = 0$	8.07	4	0.036

Hence the distribution of $Z(A, b, c)$ is

$$Z(A, b, c) =$$

$$= \begin{pmatrix} -1 & 1 & 1.3 & 1.42 & 4.17 & 5 & 7 & 7.16 & 8.07 & 8.28 & 9 \\ 0.144 & 0.04 & 0.16 & 0.24 & 0.06 & 0.024 & 0.096 & 0.016 & 0.036 & 0.024 & 0.16 \end{pmatrix}.$$

The distribution function of the optimum value is

$$F(a) = \begin{cases} 0 & \text{if } a \leqslant -1 \\ 0.144 & \text{if } -1 < a \leqslant 1 \\ 0.184 & \text{if } 1 < a \leqslant 1.3 \\ 0.344 & \text{if } 1.3 < a \leqslant 1.42 \\ 0.584 & \text{if } 1.42 < a \leqslant 4.17 \\ 0.644 & \text{if } 4.17 < a \leqslant 5 \\ 0.668 & \text{if } 5 < a \leqslant 7 \\ 0.764 & \text{if } 7 < a \leqslant 7.16 \\ 0.780 & \text{if } 7.16 < a \leqslant 8.07 \\ 0.816 & \text{if } 8.07 < a \leqslant 8.28 \\ 0.840 & \text{if } 8.28 < a \leqslant 9 \\ 1 & \text{if } a > 9. \end{cases}$$

Note that in this case

$$P(Z(A, b, c) = -\infty) = P(Z(A, b, c) = +\infty) = 0.$$

Because of the lengthy calculations involved, the Complete Description Method is often avoided in finding the distribution of the optimum.

There exist other less accurate methods for determining the distribution function :

— the Simulation Method

— the Discretization Method

and

— the Incomplete Description Method.

1) *The Simulation Method.* This method, involving a large amount of computer work, consists in constructing a sequence of optimum values $Z(A_n, b_n, c_n)$ which constitutes an approximation for the distribution function of $Z(A, b, c)$.

The method is based on a theorem of probability convergence.

2) *The Discretization Method.* This method consists in approximating the continuous random variables by discrete variables for which the distribution function of the optimum can be readily calculated with the aid of the method already presented. But the discrete distribution function thus obtained will only be an approximation of the continuous distribution function of the initial problem.

The Discretization Method can be applied to problems defined by such random variables as the vectors b and c because the optimum of a problem is not a continuous function of the matrix A [188], [218].

3) *The Incomplete Description Method.* Consider that only the vector c is random and it has a density $f(c)$. The set of optimal solutions is made up of the corners of the convex polyhedron of the feasible solutions x^r $(r = 1, \ldots, k)$. A corner x^r is optimal for $c \in \gamma_r$, which is a polyhedral convex cone with its peak at the origin. Hence for the optimal solution x^r,

$$P(x^* = x^r) = P(c \in \gamma_r) = \int_{\gamma_r} f(c)\mathrm{d}c.$$

The mainstay of the Incomplete Description Method is Fortet's observation [70] that if c is a continuous variable whose values are concentrated around a magnitude c^*, it appears justified that the distribution induced on the set of optimal solutions $x^*(c)$ is sufficiently concentrated around the value $x^*(c^*)$.

Hence, the base B_l associated to c^* will be optimal with a high probability. The bases neighbouring B_l will also have a high probability of being optimal, whereas those far away from B_l will have a much lower probability. That is why we must examine only the base B_l and those neighbouring it, i.e. an incomplete set.

Several attempts have been made to estimate the distribution of the optimum value in an asymptotical manner. (Prékopa [118] and Dragomirescu [55]).

Consider the linear program

(2.12) $\min \bar{Z} = \bar{c}'x,$

subject to

$$\bar{A}x = \bar{b},\ x \geqslant 0,$$

where $\bar{A}$, $\bar{b}$, $\bar{c}$ are the mean values of A, b, c.
A solution to problem (2.12) is

$$\bar{x} = \bar{B}^{-1}\,\bar{b},$$

and a solution of the dual problem is

$$\bar{y} = \bar{B}^{-1}\,\bar{c}^{B}.$$

If $\bar{B}$ is a matrix obtained from $\bar{A}$ we shall denote by B the matrix obtained from A and having the same column as $\bar{B}$.

THEOREM 2.4 (Dragomirescu [55]). *If the linear program* (2.12) *has an optimum unique and non-degenerated solution* $\bar{x}$, *then there exists a neighbourhood* $V \subset R^{mn+m+n}$ *of* $\bar{f} = (\bar{A}, \bar{b}, \bar{c})$ *such that for each* $f = (A, b, c) \in V$ *the optimum value of the linear program :*

$$\text{(2.12')} \quad \min Z = c'x$$

subject to

$$Ax = b, x \geqslant 0,$$

can be written as

$$\begin{aligned}
\text{(2.13)} \quad Z = \bar{Z} &+ [((c-\bar{c})^{B})'\bar{x} + \bar{y}'(b-\bar{b}) - \bar{y}'(B-\bar{B})\bar{x}] + \\
&+ [((c-\bar{c})^{B})'\bar{B}^{-1}(b-\bar{b}) - ((c-\bar{c})^{B})'\,\bar{B}^{-1}(B-\bar{B})\bar{x} - \\
&- \bar{y}'(B-\bar{B})\,\bar{B}^{-1}(b-\bar{b}) + \bar{y}'\bar{B}^{-1}(B-\bar{B})\bar{B}^{-1}(B - \\
&- \bar{B})\,\bar{x}] + [-((c-\bar{c})^{B})'\,B^{-1}(B-\bar{B})\bar{B}^{-1}(b-\bar{b}) + \\
&+ ((c-\bar{c})^{B})'\,\bar{B}^{-1}(B-\bar{B})\,\bar{B}^{-1}(B-\bar{B})\,\bar{x} + \\
&+ \bar{y}'(B-\bar{B})\,\bar{B}^{-1}(B-\bar{B})\,\bar{B}^{-1}(b-\bar{b}) - \\
&- \bar{y}'(B-\bar{B})\,\bar{B}^{-1}(B-\bar{B})\,\bar{B}^{-1}(B-\bar{B})\,\bar{x}] + \dots
\end{aligned}$$

Proof. If in the algebra of $m \times m$ matrices, for a given norm,

(*) $\|B\bar{B}^{-1} - I\| \leqslant 1,$

then

(2.14) $B^{-1} = \bar{B}^{-1}[I+(B-\bar{B})\bar{B}^{-1}]^{-1} = \bar{B}^{-1}\{I-(B-\bar{B})\bar{B}^{-1}+$

$+[(B-\bar{B})\bar{B}^{-1}]^2 - [(B-\bar{B})\bar{B}^{-1}]^3+[(B-\bar{B})\bar{B}^{-1}]^4 \ldots$

or

(2.15) $B^{-1} = \bar{B}^{-1} - \bar{B}^{-1}(B-\bar{B})\bar{B}^{-1}+\bar{B}^{-1}(B-\bar{B})\bar{B}^{-1}(B-$

$-\bar{B})\bar{B}^{-1} - \bar{B}^{-1}(B-\bar{B})\bar{B}^{-1}(B-\bar{B})\bar{B}^{-1}(B-$

$-\bar{B})\,\bar{B}^{-1} + \ldots$

But the optimal base $\bar{B}$ is non-degenerated and $\bar{x}$ is unique, i.e. the following conditions are satisfied :

$$\bar{B}^{-1}\,\bar{b} > 0$$

$$(c^B)'\,\bar{B}^{-1}\bar{a}_j - \bar{c}_j < 0, \; j \in \{j : \bar{a}_j \notin \bar{B}\}.$$

Thus, there exists a neighbourhood V of $\bar{f} = (\bar{A}, \bar{b}, \bar{c})$ such that for each $f = (A, b, c) \in V$ the condition (*) is satisfied and additionally

(2.16) $B^{-1}\,b > 0$

and

(2.17) $(c^B)'B^{-1}\,a_j - c_j < 0,$

since all the functions appearing in the expressions (2.14), (2.16) and (2.17) are continuous functions in (A, b, c).

Expressions (2.16) and (2.17) imply that for each $f = (A, b, c) \in V$, B is an optimal base for problem (2.12′).

But $Z = (c^B)'\,B^{-1}b$. This means that by multiplying on the left relation (2.15) by $(c^B)'$, and on the right by b yields :

$$Z = \bar{Z} + ((c-\bar{c})^B)'\,\bar{B}^{-1}(b-\bar{b}) + ((c-\bar{c})^B)'\bar{B}^{-1}\,\bar{b} +$$

$$+ (\bar{c}^B)'\,\bar{B}^{-1}(b-\bar{b}) - ((c-\bar{c})^B)'\,\bar{B}^{-1}.$$

$$(B - \bar{B})\bar{B}^{-1}(b - \bar{b}) - (\bar{c}^B)' \bar{B}^{-1}(B - \bar{B}) \bar{B}^{-1} \bar{b} -$$

$$- ((c - \bar{c})^B)' B^{-1}(B - \bar{B})\bar{B}^{-1} \bar{b} - (\bar{c}^B)' B^{-1}(B -$$

$$- \bar{B}) \bar{B}^{-1}(b - \bar{b}) + ((c - \bar{c})^B)' \bar{B}^{-1}(B - \bar{B})\bar{B}^{-1}(B-$$

$$- \bar{B}) \bar{B}^{-1}(b - \bar{b}) + \dots$$

i.e. which is an alternative version of (2.13). The proof is complete.

Let

$$B_0 = \bar{B} + \delta(B - \bar{B}),$$

$$b_0 = \bar{b} + \delta(b - \bar{b}),$$

$$c_0 = \bar{c} + \delta(c - \bar{c}), \qquad \delta \in [0, 1],$$

$$\rho = (c_0^B)' \bar{B}^{-1} b_0 - (c_0^B)'\bar{B}^{-1}B_0\bar{x} - \bar{y}'B_0\bar{B}^{-1} b_0 +$$

$$+ \bar{y}'\bar{B}^{-1}B_0\bar{B}^{-1}B_0\bar{x},$$

$$\bar{\sigma}^2 = \mathbf{D}^2[(c^B)'\bar{x} + \bar{y}'b - \bar{y}'B\bar{x}],$$

$$\sigma^2 = \max_{\substack{1 \leqslant i \leqslant m \\ 1 \leqslant j \leqslant n}} \{\mathbf{D}^2 a_{ij}, \mathbf{D}^2 b_i, \mathbf{D}^2 c_j\},$$

where $\mathbf{D}^2(\xi)$ denotes the variance of the random variable ξ.

THEOREM 2.5 (Prékopa [118]). *Let* $(f_k = (A_k, b_k, c_k))_{k \in N^*}$ *be a sequence of stochastic linear programs such that* f_k, $k \in N^*$ *are normal equidistributed in* R^{mn+m+n} *with parameters*

$(\mathbf{E}A_k = \bar{A},\ \mathbf{E}b_k = \bar{b},\ \mathbf{E}c_k = \bar{c},\ k \in N^*)$.

If:

a) $\bar{f} = (\bar{A}, \bar{b}, \bar{c})$ *satisfies the conditions of Theorem 2.4.*

b) $\lim_{k \to \infty} \sigma_k = 0$, *where* σ_k *is obtained from* σ *by substituting* (A, b, c) *by* (A_k, b_k, c_k), $k \in N^*$.

c) $\frac{\rho_k}{\bar{\sigma}}$ *converges in probability towards zero*; ρ_k *is obtained from* ρ *by substituting* (A, b, c) *by* (A_k, b_k, c_k), $k \in N^*$.

Then the random variables $\frac{Z_n - \bar{Z}}{\bar{\sigma}}$ *is asymptotically normal.*

Proof. From a) and Theorem 2.4, there exists a neighbourhood V of $\bar{f} = (\bar{A}, \bar{b}, \bar{c})$ such that for any $f = (A, b, c) \in V$ one can write the series expansion (2.13). Consider only the first order terms:

$$(2.18) \quad Z = \bar{Z} + [((c - \bar{c})^B)'\bar{x} + \bar{y}'(b - \bar{b}) - \bar{y}'(B - \bar{B})\bar{x}] + \rho.$$

Denoting by H_k the event $\{(A_k, b_k, c_k) \in V\}_{n \in N^*}$, then (2.18) implies that for $n \in N^*$ and $\omega \in H_k$ one can write

$$\frac{Z_k(\omega) - \bar{Z}}{\bar{\sigma}} - \frac{\rho_k(\omega)}{\bar{\sigma}} = \frac{1}{\bar{\sigma}} [((c_k - \bar{c})^B)'\bar{x} + \bar{y}'(b_k -$$

$$- \bar{b}) - \bar{y}'(B_k - \bar{B})\bar{x}].$$

Hence this variable is normal, as it represents a linear combination of normal variables. From the expression of $\bar{\sigma}^2$, it derives that the variable is of type $N(0,1)$. Then b) implies that $P(H_k) \xrightarrow[k \to \infty]{} 1$ and c) implies that $\frac{Z_k - \bar{Z}}{\bar{\sigma}}$ is asymptotically normal.

Finally, we shall discuss the estimation of the probability that the optimum value takes values in a given interval.

By virtue of Lemmas 2.1 and 2.2., one can estimate the probability that the optimum value lies in a given interval.

Let A, b, c be random matrices and

$$a_{ij}(t) = \mathbf{E}a_{ij} + t\mathbf{D}a_{ij}$$

$$b_i(t) = \mathbf{E}b_i + t\mathbf{D}b_i$$

$$c_j(t) = \mathbf{E}c_j + t\mathbf{D}c_j$$

$$Z(t) = \min \{c'(t)x \mid A(t)x \geqslant b(t),\ x \geqslant 0\}.$$

For two given numbers $t_1 < t_2$, Lemma 2.2 shows that the event

$$E(t_1, t_2) = \{\omega \mid A(t_2) \leqslant A(\omega) \leqslant A(t_1),\ b(t_1) \leqslant b(\omega) \leqslant$$

$$\leqslant b(t_2),\ c(t_1) \leqslant c(\omega) \leqslant c(t_2)\}$$

implies the event

$$\{\omega \mid Z(t_1) \leqslant Z(\omega) \leqslant Z(t_2)\}.$$

Hence

$$(2.19) \quad P\{Z(t_1) \leqslant Z \leqslant Z(t_2)\} \geqslant P\{E(t_1, t_2)\}.$$

Thus, one obtains a lower approximation of the probability that Z takes values between two given limits $Z(t_1)$ and $Z(t_2)$.

If all the random elements are normal and independent [82], [83], then

$$P\{E(t_1, t_2)\} = [\Phi(t_2) - \Phi(t_1)]^k,$$

where Φ is the Laplace function* and k is the number of random components in (A, b, c).

If $t_2 = -t_1 = t \geqslant 0$ and the variables are normal and dependent, $P\{E(-t, t)\}$ can be expressed by a distribution χ_k^2 with k degrees of freedom

$$P\{E(-t, t)\} \geqslant P\{\chi_k^2 \leqslant t^2\}.$$

For independent random variables that are not necessarily normally distributed, Chebyshev's inequality yields

$$P\{\mathbf{E}a_{ij} - t\mathbf{D}a_{ij} \leqslant a_{ij} \leqslant \mathbf{E}a_{ij} + t\mathbf{D}a_{ij}\} \geqslant \left(1 - \frac{1}{t^2}\right),$$

i.e.

$$P\{E(-t, t)\} \geqslant \left(1 - \frac{1}{t^2}\right)^k.$$

---

* i.e. the probability distribution function of the standard normal distribution $N(0, 1)$.

2.2. MINIMUM-RISK PROBLEMS IN STOCHASTIC PROGRAMMING

In mathematical programming problems, where the vector c is random, the following problems are of interest:

1) What is the probability that the economic function $c'x$ exceeds (or is smaller than) a given value?

2) What is the largest or smallest value that the objective function of a given probability may assume? This means that for a given value α we must determine the greatest λ such that $P(c'x > \lambda) = \alpha$ or $P(c'x < \lambda) = \alpha$.

3) The value of parameter λ which is certain to be a lower or an upper bound of the optimum values, i.e. that $\lambda = \lambda^*$ such that $P[\min c'x \leqslant \lambda^*] = 1$ or $P[\max c'x \geqslant \lambda^*] = 1$.

Bereanu [16], [17], [18] deals with the determination of the solution of a programming problem with stochastic economic function for which assuming a given value of the parameter λ, $\max P(c'x \leqslant \lambda)$, where the random vector c is of the form

$$c(\omega) = c^0 + \sum_{i=1}^{r} c^i \tau_i(\omega),$$

and the components of c are independent, normal random variables.

The solution x_λ, which satisfies the constraints of the problem and maximizes $P(c'(\omega)x \leqslant \lambda)$ is called by Bereanu the Minimum-Risk Solution. The same notion is introduced by Charnes and Cooper and the corresponding problem is called the P-model. Simon [127] refers to satisfying rather than optimising a given objective.

Akin to the minimum risk problem is the Problem of Kataoka [91], [92], which minimizes k such that $P(c'x \leqslant k) \geqslant \alpha$.

Consider the function

$$W_\alpha(x) = \min \{k \in R \mid P[f(x, \xi) \leqslant k] \geqslant \alpha\},$$

where $f(x, \xi)$ is a real function of the vector x and the stochastic vector ξ. The function $W_\alpha(x)$ is the quantile of order α for $f(x, \xi)$. The properties of $W_\alpha(x)$ are given in the following

LEMMA 2.3. *The function $W_\alpha(x)$ is non-decreasing with respect to α.*

Proof. Let us show that $\alpha_1 \geqslant \alpha_2$ implies that $W_{\alpha_1}(x) \geqslant W_{\alpha_2}(x)$. Indeed, if $\alpha_1 \geqslant \alpha_2$, then

$$\{y : P[f(x, \xi) \leqslant y] \geqslant \alpha_1\} \subset \{y : P[f(x, \xi) \leqslant y] \geqslant \alpha_2\},$$

i.e. $W_{\alpha_1}(x) \geqslant W_{\alpha_2}(x)$, since $W_{\alpha_1}(x)$ is a minimum over a set more restricted than the one over which $W_{\alpha_2}(x)$ is taken. The proof is complete.

LEMMA 2.4. *The function $W_\alpha(x)$ is left continuous with respect to α.*

Proof. It is enough to show that for any increasing sequence $\{\beta_n\}$ converging towards $\overline{\alpha}$,

$$\lim_{n\to\infty} W_{\beta_n}(x) = W_{\overline{\alpha}}(x). \tag{2.20}$$

From the choice of $\{\beta_n\}$ and since $W_\alpha(x)$ is non-decreasing in α (lemma 2.3.), it follows that

$$W_{\beta_n}(x) \leqslant W_{\beta_m}(x) \leqslant W_{\overline{\alpha}}(x), \; n \leqslant m,$$

i.e. $\{W_{\beta_n}(x)\}$ is convergent since it is increasing and bounded from above by $W_{\overline{\alpha}}(x)$.

Assume that (2.20) is not satisfied, i.e.

$$\lim_{n\to\infty} W_{\beta_n}(x) = \gamma < W_{\overline{\alpha}}(x).$$

Then there exists an $\varepsilon > 0$ such that

$$\lim_{n\to\infty} W_{\beta_n}(x) = \gamma < W_{\overline{\alpha}}(x) - \varepsilon. \tag{2.21}$$

From the definition of $W_{\overline{\alpha}}(x)$,

$$P[f(x, \xi) \leqslant W_{\overline{\alpha}}(x) - \varepsilon] = a < \overline{\alpha}$$

and since $\{\beta_n\}$ is increasing and convergent towards $\bar{\alpha}$, in any neighbourhood of $\bar{\alpha}$, in particular $(a, \bar{\alpha}]$, there exists an infinity of terms in the sequence $\{\beta_n\}$, i.e. there exists an N such that, for any $n \geqslant N$, $a < \beta_n < \bar{\alpha}$.

This means that

$$W_{\bar{\alpha}}(x) - \varepsilon \leqslant W_{\beta_n}(x) \leqslant \gamma, \quad n \geqslant N,$$

which contradicts hypothesis (2.21), i.e. (2.20) is true. The proof is complete.

When the random vector c has normal distribution, Kataoka's problem becomes a deterministic problem with objective function $k = \sum_j m_j x_j + \tau(x' Vx)^{1/2}$ (τ is such chosen that

$$\alpha = \frac{1}{\sqrt{2\pi}} \int_{-\infty}^{\tau} \exp\left(-\frac{1}{2} y^2\right) \mathrm{d}y, \quad m = (m_1, \ldots, m_n),\ m_k = \mathbf{E} c_k,\ k =$$

$= 1, \ldots, n$, $V = (V_{ij})$ is the covariance matrix, $V_{ij} = \mathbf{E}[(c_i - \bar{c}_i)(c_j - \bar{c}_j)]$) which is convex as the sum of a linear (convex) function and a convex function, as will be shown below.

THEOREM 2.6. *The function* $\varphi(x) = \sqrt{x'Vx}$ *is convex.*

Proof. Let us show that for any x, y and $\lambda \in [0, 1]$,

$$\varphi[\lambda x + (1 - \lambda)y] \leqslant \lambda\varphi(x) + (1 - \lambda)\,\varphi(y).$$

Indeed, let $\psi(x, y, \lambda) = \varphi[\lambda x + (1 - \lambda)\, y] - \lambda\varphi(x) - (1 - \lambda)\,\varphi(y)$
Then:

$$\begin{aligned} \operatorname{Sign} \psi(x, y, \lambda) &= \operatorname{Sign}\{\varphi[\lambda x + (1 - \lambda)y] - \\ &\quad - \lambda\varphi(x) - (1 - \lambda)\varphi(y)\} = \\ &\operatorname{Sign}\{\varphi^2[\lambda x + (1 - \lambda)y] - (\lambda\varphi(x) + (1 - \lambda)\varphi(y))^2\} = \\ &= \operatorname{Sign}\{\lambda^2 x'Vx + 2\lambda(1 - \lambda)x'Vy + (1 - \lambda)^2 y'Vy - \end{aligned}$$

$$- \lambda^2 x'Vx - 2\lambda(1-\lambda)\sqrt{x'Vx}\cdot\sqrt{y'Vy} - (1-$$

$$-\lambda)^2 y'Vy\} = \text{Sign}\{2\lambda(1-\lambda)(x'Vy - \sqrt{x'Vx}\sqrt{y'Vy})\}.$$

The matrix V is positive semi-definite, i.e. for any arbitrary t

$$(tx + y)' V(tx + y) = t^2 x'Vx + 2tx'Vy + y'Vy \geqslant 0,$$

i.e. $\Delta \leqslant 0$ which means that $(x'Vy)^2 \leqslant (x'Vx)\,(y'Vy)$ (Δ is the discriminator of the trinomial). Then

$$x'Vy \leqslant \sqrt{x'Vx}\sqrt{y'Vy},$$

since $x'Vx$ and $y'Vy$ are non-negative. The sign of function ψ is thus non-positive, i.e. $\psi(x, y, \lambda) \leqslant 0$ which means that $\varphi(x)$ is a convex function. The proof is complete.

Using Geoffrion's method [73], one can reduce both the Minimum-Risk Problem and Kataoka's Problem to a problem of quadratic parametric programming. Bergthaller [25] uses the transformation method for reducing the Minimum-Risk Problem to an easier one. (Details about this will be given in Chapter 4). Charnes et al. [37] deal with the determination of that vector x for which the probability that all the restrictions are satisfied with at least p is as large as possible. More specifically, they use the following model.

$$\max\ p$$

subject to

$$P\left(\sum_{j=1}^{n} a_{ij}\, x_j \geqslant b_i\right) \geqslant p, \qquad (i = 1, \ldots, n).$$

An approximated determination of x is given in [124].

If the distribution of the random vector c is not entirely known, there are several ways of proceeding. First, using Chebyshev's inequality, one can cast the stochastic objective function into a non-stochastic one. Second, the efficiency function Z can be replaced by an utility function $U(Z)$. If the utility function $U(Z)$ is analytical, then it can be expanded in a Taylor series about some value, say its mean $\bar{Z}$, and hence the mean of the utility function will be written as

$$\mathbf{E}U(Z) = U(\bar{Z}) + \frac{1}{2} r \sigma_Z^2 + O(\cdot),$$

where r is a constant defined by $\left.\frac{\partial^2 U}{\partial Z^2}\right|_{Z=\bar{Z}}$ and $O(\cdot)$ refers to the remainder of terms.

If the utility function is quadratic, the terms $O(\cdot)$ vanish and if $r < 0$ (i.e. $U(Z)$ is concave) a typical quadratic programming problem is obtained.

Likewise, if the utility function is linear and Z is normally distributed (the vector c is obviously normally distributed), one can formulate an equivalent problem of quadratic programming (Prékopa [118]).

The utility function $U(Z)$ does not need to be quadratic in order to lead to a quadratic program if the higher order terms $\left.\frac{\partial^i U}{\partial Z^i}\right|_{Z=\bar{Z}}$ can be neglected for $i \geqslant 3$. For instance, Freund [68] uses the following form of utility function

$$U(Z) = 1 - \exp(-aZ),$$

where a is a positive constant measuring the risk-aversion.

In this case one has to minimize

$$\mathbf{E}(U(Z)) = \frac{1}{D\sqrt{2\pi}} \int_{-\infty}^{+\infty} (1 - \exp(-aZ) \cdot \exp\left[\frac{-(Z-M)^2}{2D^2}\right] \mathrm{d}Z =$$

$$= 1 - \frac{1}{D\sqrt{2\pi}} \int_{-\infty}^{+\infty} \exp\left[-aZ - \frac{(Z - M^2)}{2D^2}\right] \mathrm{d}Z =$$

$$= 1 - \exp\left(-aM + \frac{1}{2} a^2 D^2\right] \times$$

$$\times \frac{1}{D\sqrt{2\pi}} \int_{-\infty}^{+\infty} \exp\left[\frac{-(Z - M + aD^2)^2}{2D^2}\right] dZ =$$

$$= 1 - \exp\left(- aM + \frac{1}{2} a^2D^2\right),$$

where $M = \mathbf{E}(Z) = m'x$, $D^2 = \mathbf{D}^2(Z) = x^2Vx$.

The problem reduces to a deterministic program

$$\min\left(aM - \frac{1}{2} a^2D^2\right),$$

subject to

$$Ax \leqslant b,$$

$$x \geqslant 0,$$

which is quadratic in x.

Note that the maximization of $\mathbf{E}U(Z) = \mathbf{E}(1 - \exp(-\alpha Z))$ (i.e. minimization of $\mathbf{E}\exp(-\alpha Z)$) always leads to the moment-generating function of the random variable Z. This means that even if Z is not normal, the moment-generating function exists, and then a mathematical programming problem can always be defined. For instance, if Z has the Gamma-density with parameters β and θ such that $\mathbf{E}Z = \mu = \theta/\beta$ and $\mathbf{D}^2Z = \sigma_Z^2 = \theta/\beta^2$, then the maximization of the mean utility leads to the following nonlinear program

$$\max f(x) = (\mu/\sigma_Z)^2 \log(1 + \alpha(\sigma_2/\mu)) = \theta \log\left(1 + \alpha \frac{1}{\sqrt{\theta}}\right),$$

subject to

$$Ax \leqslant b,$$

$$x \geqslant 0.$$

The objective function is now a little more complicated than a quadratic one.

2.3. TWO-STAGE PROGRAMMING UNDER UNCERTAINTY

Let us now consider a problem in which a decision is first taken before the value of the random variables are known and then, after the random events have happened and their value is known, a second decision is used in order to minimize the "penalties" that may appear due to any infeasibility.

Thus, consider

$$\min c'x$$

subject to

$$Ax = b, \quad x \geqslant 0,$$

where c and x are n-vectors, b is an m-vector and A an $m \times n$ matrix.

Assume first that b is not precisely known and only its distribution, with finite mean $\mathbf{E}b$, is given. Then it is impossible to determine $x \geqslant 0$ such that $Ax = b$. The difference between Ax and b is a random variable. One can assume that there exists a penalty for any difference between Ax and b. The problem is to choose x so as to minimize the sum of $c'x$ and the mean of the penalties. Several assumptions may be adopted regarding these penalties. Thus, Beale [5] and Dantzig [49] consider the fairly general case

$$\min c'x + \mathbf{E} \min_y d'y$$

subject to

$$Ax + By = b,$$

$$x, y \geqslant 0.$$

Here d and y are n_1-vectors and B is an $m \times n_1$ matrix. This model means that one has to determine a vector $x \geqslant 0$ "here and now" before finding b and once the value of b is known, one determines a "recourse y" from the following second-stage program:

$$\min_y d'y$$

subject to

$$By = b - Ax, \quad y \geqslant 0.$$

The variable y is the variable used to write $By = b - Ax$ for an x determined in the first stage, and b known. The initial problem is transformed into a deterministic program [247]

$$\min \; c'x + Q(x), \quad x \geqslant 0,$$

where $Q(x) = \mathbf{E}Q(x, b)$, and $Q(x, b)$ is the minimum value obtained in the second stage of the program. (If the constraints are inconsistent we shall take $Q(x, b) = \infty$ and if they are unbounded from below $Q(x, b) = -\infty$).

The case $B = [I, -I]$ and $d' = [f', g']$, where I is the identity matrix of order m and f and g are m-vectors representing the scale factors of the penalty, was studied by Beale [5] and Wets [159], [160], [161]. The latter calls it the Complete Problem. Walkup and Wets [158] call it Simple Recourse Program since by Complete Problem they mean a more general case (see Section 2.4. below).

If one assumes the penalties proportional to the difference between Ax and b and introduces the variables

$$y_i^+ = \begin{cases} b_i - A_i x & \text{if} \quad b_i - A_i x \geqslant 0 \\ 0 & \text{otherwise} \end{cases}$$

$$y_i^- = \begin{cases} A_i x - b_i & \text{if} \quad A_i x - b_i \geqslant 0, \\ 0 & \text{otherwise} \end{cases}$$

then the two-stage programming problem becomes

$$\min [c'x + \mathbf{E}(f'y^+ + g'y^-)],$$

subject to

$$y^+ - y^- = b - Ax$$

$$x, y^+, y^- \geqslant 0,$$

where A_i is the i-th row of A and $f' = (f_1, \ldots, f_m)$, $g' = (g_1, \ldots, g_m)$ is the penalty for a unit undersupply (oversupply) of the i-th item to which the i-th constraint refers.

Charnes, Kirby and Raike [40] obtain the efficiency function

$$\left(c' - \frac{1}{2}(f' - g')\,A\right) x + \frac{1}{2}\mathbf{E}\,(f + g)'\,|b - Ax|,$$

and call it of the constrained median type.

Consider now the second stage of the program

$$\min\,(f'y^+ + g'y^-),$$

subject to

$$y^+ - y^- = b - Ax,$$

$$y^+,\ y^- \geqslant 0,$$

where b and x are known. Its constraints are never contradictory.

The dual problem reads as

$$\max \sum_{i=1}^{m} z_i(b_i - A_i\, x)$$

subject to

$$-\,g_i \leqslant z_i \leqslant f_i, \quad (i = 1, \ldots, m)$$

and its solution is

$$z_i = \begin{cases} -\,g_i & \text{if } \ b_i - A_i x < 0 \\ \quad f_i & \text{if } \ b_i - A_i x > 0 \\ \text{any value} & \text{if } \ b_i - A_i x = 0. \end{cases}$$

The constraints of the dual problem are compatible if and only if $f + g > 0$.

The mean value of the optimal z_i is

$$\mathbf{E}(z_i) = - g_i P(b_i < A_i x) + f_i P(b_i > A_i x) + \xi P(b_i - A_i x = 0) =$$

$$= - g_i P(b_i < A_i x) + f_i P(b_i > A_i x) = - g_i F(A_i x) +$$

$$+ f_i P(- b_i < - A_i x) = - g_i \int_{-\infty}^{A_i x} \mathrm{d}F_i(b_i) - f_i \int_{-\infty}^{-A_i x} \mathrm{d}F_i(b_i) =$$

$$= - g_i \int_{-\infty}^{A_i x} \mathrm{d}F_i(b_i) + f_i \int_{A_i x}^{\infty} \mathrm{d}F_i(b_i) = f_i - (f_i + g_i) \int_{-\infty}^{A_i x} \mathrm{d}F_i(b_i),$$

where $F_i(b_i)$ is the marginal distribution function of b_i.

Wets [160] proved that this function is convex in x.

The minimum of the second-stage program equals the maximum of its dual which is a sum of $z_i^0(b_i - A_i x)$, where z_i^0 is the optimal value of the dual variable.

Similarly one computes

$$Q_i(x) = f_i \mathbf{E}\, b_i - f_i y_i + (f_i + g_i) \int_{-\infty}^{A_i x} F_i(b_i)\, \mathrm{d}b_i.$$

Therefore, the objective function of the two-stage problem is

$$c'x + \Sigma \left\{ f_i \mathbf{E} b_i - f_i y_i + (f_i + g_i) \int_{-\infty}^{A_i x} F_i(b_i)\, \mathrm{d}b_i \right\}.$$

Since $F_i(b_i)$ is a monotonic, non-decreasing function, the integral is convex in y_i. This means that the problem is one of convex programming and the Kuhn-Tucker saddlepoint theorem applies.

Williams [163] gives a characterization of the optimal solution and finds an upper bound of the error induced by assigning to b a given value b^* and then solving the problem as a deterministic one.

Assume that b_i has an uniform distribution in the interval (b_i^0, b_i^1). In this case, the minimization of $\Sigma_i Q_i(x)$ under linear constraints is a qadratic programming problem.

For other approximate expressions of $F_i(b_i)$, see Beale [6] and Wets [161].

We revert now to the general form of the second-stage program as considered by Beale [5] and Dantzig [49]. In the particular case studied above (i.e. $B = [I, -I]$), the constraints of the second-stage program always had a solution. But in general, the choice of x can sometimes lead to a second-stage program with contradictory constraints.

The vector x is feasible if it satisfies all the constraints that are independent of y (if any) and if the second-stage program has a solution $y \geqslant 0$ for any value of b.

Let K be the region of feasible vectors x.
Then, we have

THEOREM 2.7. *If* $\min \{u'(b_{\min} - Ax) \mid u'B \geqslant 0,\ u \geqslant 0\} > 0$, *then* $x \in K$, *where* u *is the unknown vector of the dual problem and* $b_{\min}$ *is a bound to* b.

The reciprocal theorem holds only if $b_{\min}$ is feasible.

The optimality of a vector $\bar{x}$ is defined in the following

THEOREM 2.8 (Dantzig and Madansky [50]). *A necessary requirement for the vector* $\bar{x}$ *to be optimal in the two-stage program is*

$$[c' - \mathbf{E}\bar{z}(b, x)' A]\,\bar{x} \leqslant [c' - \mathbf{E}\bar{z}(b, x)' A]x,$$

where $\bar{z}(b, x)$ *is the optimal vector of the dual problem of the second-stage program.*

The sufficient condition for optimality is given by

THEOREM 2.9 (Dantzig and Madansky [50]). *If* $\bar{x}$ *is feasible and if*

$$[c' - \mathbf{E}\bar{z}(b, x)' A]\bar{x} \leqslant [c' - \mathbf{E}\bar{z}(b, \bar{x})' A]x$$

for all feasible x, *then* $\bar{x}$ *is optimal.*

A necessary and sufficient requirement for optimality is given by Elmaghraby [60].

The case when A and b are random is called by Walkup and Wets [158] the Fixed Recourse Case. An example to this is Tintner's Active Approach [156].

If all the elements (A, b, c) are random, the definition of a certain feasible $x \geqslant 0$ is subject to the following assumptions concerning the second-stage program.

a) x is always feasible (strong feasibility);

b) x is feasible with probability 1 (weak feasibility).

Walkup and Wets [158] refer to Case b) as the Complete Recourse Problem. They also established [158] the necessary and sufficient conditions for the strong and weak feasibilities to coincide.

Another solution to the two-stage programming problem under uncertainty was suggested by Dempster [51]. He considers the minimization of $c'x + d(b - Ax) + |D(b - Ax)|$, where D is a matrix, and finds the sufficient condition for a finite optimal solution.

Likewise, he considers the minimization of

$$c'x + d(b - Ax) + \mathbf{E}(b - Ax)' H(b - Ax)$$

where H is a symmetric, positive-definite matrix, and finds that, if the constraints are linear, the problem reduces to a quadratic programming one.

If the distribution of the random elements is unknown and it is known that it belongs to a given family, one can use the minimax approach and choose as optimal that x which minimizes the largest mean of the sum of the objective function and the penalties. This situation, which allows one to choose a distribution for x rather than a fixed value, is treated together with penalty functions in general by Iosifescu and Theodorescu [85], [86], and Záckovà [166]. These approaches employ notions and results from the theory of games.

Thus, Iosifescu and Theodorescu [85], [86] use a zero-sum two-person game in which the player is the programmer and the adversary is fictitious. The strategy of the programmers are the vectors $x \geqslant 0$ and those of the adversary the triplets $(A(\omega), b(\omega), c(\omega))$. After a pair of strategies x and $(A(\omega), b(\omega), c(\omega))$ has been chosen the win of the programmer is

$$g(x, A(\omega), b(\omega), c(\omega)) = - c'(\omega)x - \sum_{i=1}^{m} \varphi_i(b_i(\omega) - \\ - \sum_{j=1}^{n} a_{ij}(\omega)\, x_j).$$

Thus, the problem reduces to solving a zero-sum two-person game.

2.4. THE COMPLETE PROBLEM

In what follows we shall consider the standard form of two-stage linear programming under uncertainty:

$$\min c'x + \mathbf{E}_\xi\, d'y$$

subject to

$$Ax = b$$

$$Tx + My = \xi, \quad \xi \in \{\Omega, \mathrm{K}, \mathrm{P}\},$$

$$x \geqslant 0, \; y \geqslant 0$$

where A, T, M are matrices of orders $m \times n$, $\bar{m} \times n$ and $\bar{m} \times \bar{n}$, respectively, c and d are vectors of dimensions n and $\bar{n}$, respectively, and $\xi = (\xi_1, \ldots, \xi_{\bar{m}})$ is a random vector numerically defined on the probability space $\{\Omega, K, P\}$ $(\Omega \subset R^{\bar{m}})$ and having $\bar{m}$ components, whereas $\mathbf{E}_\xi$ denotes the mean with respect to ξ. The probability space of ξ_i is $\{\Omega_i, K_i, P_i\}$.

Wets [159] approaches the case when matrix M can be partitioned (after some rearrangement of rows and columns) as $M = (I, -I)$, where I is the unit matrix.

Similarly, one devides the vectors d and y in (d^+, d^-) and (y^+, y^-), respectively. Hence, we obtain

$$\min c'x + \mathbf{E}_\xi(d^+y^+ + d^-y^-) \tag{2.22}$$

subject to

$$Ax = b$$

$$Tx + Iy^+ - Iy^- = \xi, \quad \xi \in \{\Omega, K, P\}$$

$$x \geqslant 0, \; y^+ \geqslant 0, \; y^- \geqslant 0.$$

This is called the *complete problem.*

Now we shall try to give a deterministic equivalent to problem (2.22) and an algorithm for its solution.

THEOREM 2.10. *Problem* (2.20) *is equivalent to*

$$(2.23) \quad \min Z(x) = c'x + Q(x)$$

subject to

$$Ax = b$$

$$x \geqslant 0,$$

where

$$(2.24) \quad Q(x) = \mathbf{E}_\xi\{Q(x, \xi)\}.$$

and

$$(2.25) \quad Q(x, \xi) = \min\{(d^+y^+ + d^-y^-) \mid y^+ - y^- =$$

$$= \xi - Tx, \ y^+ \geqslant 0, \ y^- \geqslant 0\}.$$

Proof. The theorem requires to show that Problems (2.22) and (2.23) have the same domain of feasible solutions and that their objective functions coincide.

Indeed, the decision variable of problem (2.22) is x. For a given x, the variable y is determined from the second-stage program

$$(2.26) \quad \min(d^+y^+ + d^-y)$$

subject to

$$y^+ - y^- = \xi - Tx$$

$$y^+, y^- \geqslant 0.$$

This y is always feasible since for whatever x, i.e. for whatever $\xi - Tx$, there exists always an y^+ and y^- such that

$$y^+ - y^- = \xi - Tx.$$

Thus, problems (2.22) and (2.23) have the same domain of feasible solutions.

The coincidence of the objective functions is given by expressions (2.24) and (2.25) and the proof is complete.

THEOREM 2.11. *Problem* (2.23) *is a convex programming problem.*

Indeed, $Q(x, \xi)$ is a convex function of x. The operator $\mathbf{E}_\xi$ applied on $Q(x, \xi)$ leads to a positive linear combination of positive functions convex in x; hence $Q(x)$ is convex too. The objective function of problem (2.23) is the sum of a linear (hence convex) function and a convex function, thus a convex function. Additionally, the domain of feasible solutions is a convex set. The proof is complete.

Let $\chi = Tx$, i.e. of components $\chi_i = T_i x$, $(i = 1, \ldots, \overline{m})$, where T_i is the i-th row of matrix T. For $\chi = Tx$, $Q(\chi) = Q(x)$. The function $Q(\chi)$ is separable [159], i.e.

$$Q(\chi) = \sum_{i=1}^{m} Q_i(\chi_i), \quad \chi = (\chi_1, \chi_2, \ldots, \chi_{\overline{m}}).$$

For a given x (hence, a given χ) and a given ξ, problem (2.26) becomes

$$(2.27) \quad Q(\chi, \xi) = \min_{y_i^+,\, y_i^-} \sum_{i=1}^{\overline{m}} (d_i^+ y_i^+ + d_i^- y_i^-)$$

subject to

$$y_i^+ - y_i^- = \xi_i - \chi_i, \ (i = 1, \ldots, \overline{m}),$$

$$y_i^+, y_i^- \geqslant 0.$$

The dual to problem (2.27) is

$$(2.28) \quad Q(\chi, \xi) = \max \sum_{i=1}^{\overline{m}} \pi_i(\xi_i, \chi_i)\,(\xi_i - \chi_i)$$

subject to

$$-d_i^- \leqslant \pi_i(\xi_i, \chi_i) \leqslant d_i^+, \ (i = 1, \ldots, \overline{m}).$$

The dual problem will have a solution if for any $i, [-d_i^-, d_i^+] \neq 0$, i.e. $\tilde{d}_i = d_i^+ + d_i^- \geqslant 0$. This expression represents the condition for problem (2.27) to have a finite optimal solution, and will be considered true throughout the remainder of the book.

The optimal solution to problem (2.28) is easily written as

a) $\pi_i^*(\xi_i, \chi_i) = d_i^+$ if $\xi_i - \chi_i > 0$;

b) $\pi_i^*(\xi_i, \chi_i) = -d_i^-$ if $\xi_i - \chi_i < 0$;

c) $\pi_i^*(\xi_i, \chi_i)$ can assume any value within the interval $[-d_i^-, d_i^+]$ if $\xi_i - \chi_i = 0$.

The value $\pi_i^*(\xi_i, \chi_i) = -q_i^-$ will be used in the following.

Let us now examine several properties of the functions $Q_i(\chi_i)$, which may be used in deriving concrete solutions to problem (2.23).

Let

$$\pi_i(\chi_i) = \mathbf{E}_{\xi_i}[\pi_i^*(\chi_i, \xi_i)] = d_i^+ P(\xi_i > \chi_i) - d_i^- P(\xi_i < \chi_i) = d_i^+(1 - F_{\xi_i}(\chi_i)) - d_i^- F_{\xi_i}(\chi_i) = d_i^+ - (d_i^+ + d_i^-) F_{\xi_i}(\chi_i) = d_i^+ - \tilde{d}_i F_{\xi_i}(\chi_i) = d_i^+ - \tilde{d}_i \int_{\xi_i \leqslant \chi_i} \mathrm{d}F_i(\xi_i),$$

where $F_i(\xi_i)$ is the distribution of ξ_i. Then

$$Q_i(\chi_i) = -d_i^- \int_{\xi_i \leqslant \chi_i} (\xi_i - \chi_i)\, \mathrm{d}F_i(\xi_i) + d_i^+ \int_{\xi_i > \chi_i} (\xi_i - \chi_i)\, \mathrm{d}F_i(\xi_i) = d_i^+ \int_{\xi_i \in \Omega_i} (\xi_i - \chi_i)\, \mathrm{d}F_i(\xi_i) - \tilde{q}_i \int_{\xi_i \in \chi_i} (\xi_i - \chi_i) \mathrm{d}F_i(\xi_i) = d_i^+ \overline{\xi}_i - \psi_i(\chi_i) - \pi_i(\chi_i)\, \chi_i,$$

where $\bar{\xi}_i = \mathbf{E}_{\xi_i}\{\xi_i\}$ and $\psi_i(\chi_i) = \tilde{d}_i \int_{\xi_i \leqslant \chi_i} \xi_i \, dF_i(\xi_i)$.

Hence,

$$(2.29)\quad Q(\chi) = \sum_{i=1}^{\bar{m}} Q_i(\chi_i) = \sum_{i=1}^{\bar{m}} d_i^+ \bar{\xi}_i - \sum_{i=1}^{\bar{m}} [\psi_i(\chi_i) + \pi_i(\chi_i)\,\chi_i].$$

In order to study the differentiability of function $Q_i(\chi_i)$, we shall assume that $[\alpha_i, \beta_i]$ is the support of ξ_i. If ξ_i has no lower bound, it is obvious that $\alpha_i = -\infty$, whereas if it has no upper bound, $\beta_i = +\infty$. The following situations may arise:
1° $\chi_i < \alpha_i$. In this case, $\{\xi_i | \xi_i \leqslant \chi_i\} = \emptyset$ and hence

$$\pi_i(\chi_i) = d_i^+,$$

$$\psi_i(\chi_i) = 0,$$

$$Q_i(\chi_i) = d_i^+\bar{\xi}_i - d_i^+\chi_i.$$

Then

$$\frac{dQ_i(\chi_i)}{d\chi_i} = -\,d_i^+ = -\,\pi_i(\chi_i),$$

i.e. the function $Q_i(\chi_i)$ is linear on $(-\infty, \alpha_i)$.
2° $\alpha_i \leqslant \chi_i \leqslant \beta_i$. Now, $\{\xi_i | \xi_i \leqslant \chi_i\} = \{\xi_i | \alpha_i \leqslant \xi_i \leqslant \chi_i\}$, and hence

$$\pi_i(\chi_i) = d_i^+ - \tilde{d}_i \int_{\alpha_i}^{\chi_i} dF_i(\xi_i)$$

$$\psi_i(\chi_i) = \tilde{q}_i \int_{\alpha_i}^{\chi_i} \xi_i \, dF_i(\xi_i)$$

$$Q_i(\chi_i) = d_i^+\bar{\xi} - d_i^+\chi_i - \tilde{d}_i \int_{\alpha_i}^{\chi_i} (\xi_i - \chi_i)\, dF_i(\xi_i).$$

Note that here $Q_i(\chi_i)$ depends on $F_i(\xi_i)$. Hence,

$$\frac{\mathrm{d}Q_i(\chi_i)}{\mathrm{d}\chi_i} = -d_i^+ + \tilde{d}_i \int_{\alpha_i}^{\chi_i} \mathrm{d}F_i(\xi_i) = -\pi_i(\chi_i).$$

3° $\chi_i > \beta_i$. Now $\{\xi_i \mid \xi_i \leqslant \chi_i\} = \Omega_i$ and

$$\pi_i(\chi_i) = -d_i^-,$$

$$\psi_i(\chi_i) = \tilde{d}_i \bar{\xi}_i,$$

$$Q_i(\chi_i) = d_i^+ \bar{\xi}_i - \tilde{d}_i \bar{\xi}_i + d_i^- \chi_i = d_i^- \chi_i - d_i^- \bar{\xi}_i.$$

Then

$$\frac{\mathrm{d}Q_i(\chi_i)}{\mathrm{d}\chi_i} = d_i^- = -\pi_i(\chi_i)$$

and hence $Q_i(\chi_i)$ is linear on $(\beta_i, +\infty)$.

From what was stated above it follows that if $F_i(\xi_i)$ is a continuous distribution, then $Q_i(\chi_i)$ is differentiable and

$$\text{(2.30)} \quad \frac{\mathrm{d}Q_i(\chi_i)}{\mathrm{d}\chi_i} = -\pi_i(\chi_i) \quad \text{on } R.$$

Additionally, $Q_i(\chi_i)$ are also continuous.

2.5. WETS' ALGORITHM

Let us now consider an algorithm given by Wets [159] for solving problem (2.22) under the assumption that the distributions $F_i(\xi_i)$ are continuous and $\tilde{d}_i \geqslant 0$ (which means that (2.22) has a solution).

We have already seen that the deterministic equivalent to problem (2.22) is problem (2.23). Based on expressions (2.29), one can write

$$Q(x) = \sum_{i=1}^{\bar{m}} d_i^+ \bar{\xi}_i - \sum_{i=1}^{\bar{m}} [\psi_i(T_i x) + \pi_i(T_i x)\, T_i x].$$

Since $\sum_{i=1}^{\overline{m}} d_i^+ \xi_i$ is a constant it can be omitted in the objective function and hence problem (2.23) becomes

$$(2.31)\quad \min \hat{z}(x) = \sum_{j=1}^{n} c_j x_j - \sum_{i=1}^{\overline{m}} [\psi_i(T_i x) + \pi_i(T_i x) T_i x]$$

subject to

$$\sum_{j=1}^{n} a_{ij} x_j = b_i \qquad (i = 1, \ldots, m)$$

$$x_j \geqslant 0, \qquad (j = 1, \ldots, n).$$

In order to solve problem (2.31), we shall consider first some propositions.

Proposition 2.1. $\dfrac{\mathrm{d}}{\mathrm{d}x}\hat{z}(x) = c - \pi(\chi)T.$

This is straightforward from expression (2.30) under the assumption that $\chi = Tx$.

Proposition 2.2. *If* $\bar{x} \in K = \{x \mid Ax = b, x \geqslant 0\}$ *and* $[c - \pi(\chi)T](x - \bar{x}) \geqslant 0$ *for any* $x \in K$, *then* $\hat{z}(x)$ *reaches its minimum in* $\bar{x}$.

The proof springs from the fact that function $\hat{z}(x)$ is convex and hence

$$\hat{z}(x) - \hat{z}(\bar{x}) \geqslant [c - \pi(\chi)T](x - \bar{x}) \geqslant 0,$$

i.e.

$$\hat{z}(x) \geqslant \hat{z}(\bar{x})$$

for any $x \in K$.

Proposition 2.3. *If*

$$[c - \pi(\chi)T]\, x > [c - \pi(\chi)T]\bar{x}, \qquad x, \bar{x} \in K,$$

then there exists an $x^0 \in (x, \bar{x}]$ *such that* $\hat{z}(x^0) < \hat{z}(x)$.

Proof. By virtue of the hypothesis,

$$[c - \pi(\chi)T]x > [c - \pi(\chi)T]\,(\lambda x + (1 - \lambda)\bar{x}),$$

$$\lambda \in [0, 1].$$

Assume by *reductio ad absurdum* that

$$\hat{z}(x) \leqslant \hat{z}(\lambda x + (1 - \lambda)\bar{x}), \quad \lambda \in [0, 1]. \tag{2.32}$$

and let

$$\delta(\lambda) = \hat{z}(\lambda x + (1 - \lambda)\bar{x}).$$

Then

$$\frac{\mathrm{d}}{\mathrm{d}\lambda}\,\delta(\lambda)_{\lambda=1} = [c - \pi(x)T](x - \bar{x}) > 0,$$

and hence there exists a $\lambda^0 \in [0,1]$ such that

$$\delta(\lambda^0) < \delta\,(1) = \hat{z}(x).$$

Choose

$$x^0 = \lambda^0 x + (1 - \lambda^0)\,\bar{x},$$

and hence

$$\hat{z}(x^0) < \hat{z}(x)$$

which contradicts hypothesis (2.32).

PROPOSITION 2.4. *If for* $x \in K$

$$\hat{z}(x) > \hat{z}(x^0) = \min_{x \in K} \hat{z}(x),$$

then there exists an $\bar{x}$ *such that*

$$[c - \pi(\chi)T]x > [c - \pi(\chi)T]\bar{x}.$$

The proof derives from the hypothesis and from the fact that $\hat{z}(x)$ is a convex function.

Hence

$$0 > \hat{z}(x^0) - \hat{z}(x) \geqslant [c - \pi(\chi)T](x^0 - x).$$

The proof is complete.

The last two propositions indicate the manner in which problem 2.22 must be solved.

The optimality test for this algorithm is given by

PROPOSITION 2.5. $\hat{z}(x^0) = \min_{x \in K} \hat{z}(x)$, *if and only if*

$$[c - \pi(\chi)T]\, x^0 = \min_{x \in K} [c - \pi(\chi^0)T]\, x,$$

where $\chi^0 = Tx^0$.

Proof. If $[c - \pi(\chi^0)T]x^0 \leqslant [c - \pi(\chi^0)T]x$ for any $x \in K$ then, according to Proposition 2.2, x^0 is optimal for the function $\hat{z}(x)$. Conversely, let us assume that $\hat{z}(x^0) \leqslant \hat{z}(x)$ for any $x \in K$ and there exists an $x^0 \in K$ such that

$$(2.33) \quad [c - \pi(\chi^0)T]x^0 > [c - \pi(\chi^0)T]x.$$

According to Proposition 2.3, there exists an $\bar{x} \in (x^0, x]$ such that $\hat{z}(\bar{x}) < \hat{z}(x^0)$, which contradicts the hypothesis that there exists an $x \in K$ such that (2.33) is true.

We can now sketch the solution to problem (2.31).

1) Choose an $x_1 \in K$, compute $\chi_1 = Tx_1$ and solve the linear programming problem

$$(2.34) \quad \min_x \{[c - \pi(\chi_1)T]x \mid Ax = b,\ x \geqslant 0\}.$$

Let $\hat{x}_1$ be the optimal solution.

a) If $\hat{x}_1 = x_1$, then by virtue of Proposition 2.5, $\hat{x}_1$ is the optimal solution for problem (2.31).

b) If $\hat{x}_1 \neq x_1$, then according to Proposition 2.4,

$$[c - \pi(\chi_1)T](x_1 - \hat{x}_1) > 0$$

and by virtue of Proposition 2.3, there exists an $x_2 \in (x_1, \hat{x}_1]$ such that $\hat{z}(x_2) < \hat{z}(\hat{x}_1)$.

The vector x_2, which improves the value of $\hat{z}$, is obtained as is shown below.

Consider the problem

$$\min_{\lambda \in [0,1]} \delta(\lambda) = \hat{z}(\lambda x_1 + (1-\lambda)\hat{x}_1)$$

and let λ_1 be its minimum value. Take $x_2 = \lambda_1 x_1 + (1 - \lambda_1)\hat{x}_1$.

2) For x_2 one resumes the procedure from stage 1), i.e. one calculates $\chi_2 = Tx_2$ and solves (2.34) using $\pi(\chi_2)$ in the expression of the objective function.

Repeating this process, one either obtains the solution x^k which is the solution for problem (2.31) or a sequence $\{x^k\}$ which approaches the solution x_0 of problem (2.31).

2.6. CHANCE-CONSTRAINED PROGRAMMING

The chance-constrained programming problem was first formulated by Charnes, Cooper and Symonds [33] and by Charnes and Cooper [34] and subsequently developed by Ben-Israel, Kataoka [91], Naslund [112], Van de Panne and Popp [157], and Miller and Wagner [111].

This problem is formulated as

$$\text{optimum } f(c, x) = c'x$$

subject to

$$P(Ax \leqslant b) \geqslant \alpha,$$

$$x \geqslant 0,$$

where α is a column vector having m components ranging between 0 and 1.

A non-negative vector x is feasible for this problem if and only if

$$P\left\{\sum_{j=1}^{n} a_{ij}\, x_j \leqslant b_i\right\} \geqslant \alpha_i, \qquad (i = 1, \ldots, m)$$

such that the complementary probability $1 - \alpha_i$ represents the admissible risk that the random variables assume values such that

$$\sum_{j=1}^{n} a_{ij}\, x_j > b_i.$$

If $a_{i1}, \ldots, a_{in}, b_i$ are constants for a certain value of i, then α_i is nonsignificant and the i-th constraint can take the form

$$\sum_{j=1}^{n} a_{ij}\, x_j \leqslant b_i.$$

The chance constraints can also appear as

$$P\left(\sum_{j=1}^{n} a_{ij}\, x_j \leqslant b_i,\;\; i = 1, \ldots, m\right) \geqslant \alpha,$$

i.e. as a threshold imposed on the aggregated probability that the constraints are satisfied. Such approaches may be found in the papers of Balintfy [4], Jagannathan [88] and Wagner [111].

The objective function is usually assumed to be the mean of $c'x$:

$$\mathbf{E}(c'x) = \sum_{j=1}^{n} \mathbf{E}(c_j)\, x_j.$$

Due to its probabilistic aspect, the problem can be solved in two different manners.

a) We calculate x before the value of the random variables are known (such rules are called [38] zero-order rules). In this case we find deterministic constraints which are equivalent to the chance-constraints.

b) We wait for the random variables to be known but decide beforehand how to use these values (nonzero-order rules). In this case the chance -constraints are replaced by other equivalent constraints which hold with probability 1. (These constraints are called certainty equivalents [38].)

In case b) if certain random variables can be observed before some of the elements of x are specified, we can formulate the problem by choosing a decision rule $x = \Phi(A, b, c)$ instead of specifying all the elements of x directly. The function Φ is to be chosen from a prescribed class of functions.

The function Φ is generally assumed to be linear. In practice, one can assume that A (and c) is not a random variable, and hence the decisions are of the form

$$x = Db \left(\text{i.e. } x_j = \sum_{k=1}^{m} d_{jk} b_k \right),$$

where D must be optimally determined.

In dynamic (sequential decision) problems, the matrix D is of triangular shape (first mentioned by Charnes and Kirby [39]). Such a situation arises when b_j represents the demand during the period j. Then, in period j (i.e. when x_j is established) the previous decisions $(x_1, x_2, \ldots, x_{j-1})$ and the values taken by the random variables $b_1, b_2, \ldots, b_{j-1}$ up to this period are known, whereas for the subsequent periods only the probability thresholds are known. The objective is to maximize the mean profit on the set of n periods.

Hillier [78] approached the case when the decision variables are either bounded continuous or 0 and 1 and when some or all parameters are eventually dependent random variables. The objective function is taken $\mathbf{E}(cx)$. Thus the problem under consideration is

$$\max \mathbf{E}(cx) = \sum_{j=1}^{n} \mathbf{E}(c_j) x_j$$

subject to

$$P(Ax \leqslant b) \geqslant \alpha$$

$$0 \leqslant x_j \leqslant 1, \; j \in C$$

$$x_j = 0 \text{ or } 1, \; j \in D,$$

where $C \cap D = \emptyset$ or $C \cup D = \{1, 2, \ldots, n\}$.

Both exact and approximate solutions are presented in ref. [78], many of them based on certain linear inequalities which allow the approximation of this problem by several ordinary linear problems.

First, the chance constraints are transformed into deterministic constraints. For instance, if for the constraint

$$P\left(\sum_{j=1}^{n} a_{ij} x_j - b_i \leqslant 0 \right) \geqslant \alpha_i$$

one knows $\mathbf{E}(a_{i1}), \ldots, \mathbf{E}(a_{in}), \mathbf{E}(b_i), V_i$ the mean values and the correlation matrix of $a_{i1}, a_{i2}, \ldots, a_{in}, b_i$, the distribution of $\sum_{j=1}^{n} a_{ij} x_j - b_i$ and if $F(.)$ is the cummulated distribution of

$$\left[\sum_{j=1}^{n} a_{ij} x_j - b_i - \mathbf{E}\left(\sum_{j=1}^{n} a_{ij} x_j - b_i\right)\right] / \mathbf{D}\left(\sum_{j=1}^{n} a_{ij} x_j - b_i\right),$$

then the deterministic constraint becomes

$$\mathbf{E}\left(\sum_{j=1}^{n} a_{ij} x_j - b_i\right) + K_{\alpha_i} \mathbf{D}\left(\sum_{j=1}^{n} a_{ij} x_j - b_i\right) \leqslant 0$$

which may reduce to

$$\sum_{j=1}^{n} \mathbf{E}(a_{ij}) x_j + K_{\alpha_i}\left([x', -1]\, V_i \begin{bmatrix} x \\ -1 \end{bmatrix}\right)^{1/2} \leqslant \mathbf{E}(b_i),$$

where K_β is defined by $F(K_\beta) = \beta \in [0, 1]$. If $D = \emptyset$, then all the decision variables are continuous.

Kataoka showed that $\left([x', -1]\, V_i \begin{bmatrix} x \\ -1 \end{bmatrix}\right)^{1/2}$ is a convex function. Hence, if $K_{\alpha_i} \geqslant 0$ for $i = 1, \ldots, n$, the problem reduces to an ordinary convex programming problem for the solution of which one can use the algorithms given by Rosen [119] Zoutendijk [170], [171], Kelley [94], Fiacco and McCormick [64].

In stochastic programming with chance constraints, one must often find deterministic constraints equivalent to the given chance constraints and then solve the deterministic problem thus obtained [33]. The form of the deterministic constraints depends on the distribution pattern assumed by the random elements.

Let us first consider the case of only one random variable.

1) If the vector b in the constraint $P(ax \geqslant b) \geqslant \alpha$ is random and has the distribution $F_b(z) = P(b \leqslant z)$, and if we denote by B_α the least value such that $F_b(B_\alpha) = \alpha$ (i.e. $B_\alpha = F_b^{-1}(\alpha)$), then we have the following

THEOREM 2.12. *The constraint* $P(ax \geqslant b) \geqslant \alpha$ *is equivalent to the deterministic constraint*

$$ax \geqslant B_\alpha.$$

Example. Consider that in the constraint $P(ax \geqslant b) \geqslant \alpha$ the vector b follows a Gamma-distribution with parameters d and p :

$$P(X \leqslant x) = \int_{-a}^{x} f(t)\,\mathrm{d}t = \frac{1}{\Gamma(p+1)} \int_{0}^{x^*} \mathrm{e}^{-t} t^p \,\mathrm{d}t = I(u, p),$$

where

$$x^* = \frac{p(d+x)}{d}, \; u = \frac{x^*}{(p+1)^{1/2}}$$

and $I(u, p)$ is the incomplete Gamma function [the values of this function may be found in the Tables of the Incomplete Γ-Function, compiled by Pearson (ed.), Cambridge University Press, London (1946)].

Hence, the constraint $P(ax \geqslant b) \geqslant \alpha$ becomes

$$I[p(d+ax)/d(p+1)^{1/2},\, p] \geqslant \alpha$$

and, since the distribution is monotonously increasing, the following deterministic equivalent is obtained :

$$p(d+ax)/d(p+1)^{1/2} \geqslant I^{-1}(\alpha).$$

The values of $I^{-1}(\alpha)$ are obtained using that value of u from the tables of incomplete Gamma function such that $I(u, p) = \alpha$.

2) If in the constraint $P(ax \geqslant b) \geqslant \alpha$, a is random and follows the distribution $F_a(z) = P(a \leqslant z)$ and if a_0 is the greatest value such that $F_a(a_0) = 1 - \alpha$ (i.e. $a_0 = F_a^{-1}(1-\alpha)$), then we have

THEOREM 2.13. *The constraint* $P(ax \geqslant b) \geqslant \alpha$ *is equivalent to the deterministic constraint*

$$a_0 x \geqslant b.$$

For constraints like $P(\sum_j a_{ij}x_j \geqslant b_i) \geqslant \alpha_i$, where b_i is random, we have

THEOREM 2.14. *The constraint* $P(\sum_j a_{ij}x_j \geqslant b_i) \geqslant \alpha_i$ *is equivalent to the constraint* $\sum_j a_{ij}x_j \geqslant F_{b_i}^{-1}(\alpha_i)$.

If a_{ij} is random the substitution of the random constraints by deterministic equivalents is only possible if the distribution of a sum of variables having the same distribution belongs to the same family. In Lukacs' [101] formulation these distributions are *stable*. Three stable distributions are so far known : normal, Cauchy and that studied by Lévy.

If the distribution of a_{ij} is not stable, other methods must be found to obtain the deterministic equivalents. For example, Charnes, Kirby and Raike [40] consider for all j, those γ_j for which $\sum_{j=1}^{n} \gamma_j x_j \geqslant b$ for given x_j. Then, for one constraint alone, we have

THEOREM 2.15. *$P(\sum_j a_j x_j \geqslant b) \geqslant \alpha$ is equivalent to the requirement that the total probability of $a_j \geqslant \gamma_j$ for all such γ_j be at least α, i.e. $P(a_j \geqslant \gamma_j, \forall_j) \geqslant \alpha$.*

This theorem cannot be used for calculation purposes. We shall now give a sufficient condition for the chance constraints to occur.

THEOREM 2.16. *If for a given x there exist γ_j such that*

$$\sum_j \gamma_j x_j \geqslant b$$

and

$$\prod_j P(a_j \geqslant \gamma_j) \geqslant \alpha,$$

then

$$P(\sum_j a_j x_j \geqslant b) \geqslant \alpha.$$

For the necessary condition, one can state

THEOREM 2.17. *If $P(\sum_j a_j x_j \geqslant b) \geqslant \alpha > 0$ when $x_1 + x_2 + \ldots + x_n = 1$, then there exists δ_j $(j = 1, \ldots, n)$ such that $\sum_j \delta_j x_i \geqslant b$ and $1 - \prod_j P(a_j < \delta_j) \geqslant \alpha$.*

So far, we have examined only a few aspects of the stochastic programming with only one objective function. Further exploration into the methods for solving this problem are still required.

Chapter 3

Approaches to stochastic programming with multiple objective functions

We shall outline several approaches to the stochastic programming problems with multiple objective functions. It was shown in Chapter 1 that the solution of the mathematical programming problems with several deterministic objective functions raises several difficulties which stem from the fact that the solution vector is not uniquely defined. Further difficulties appear in solving the stochastic programming problems with one objective function. This explains why in solving the stochastic programming problems with multiple objective functions one is confronted with almost intractable difficulties and ambiguities.

Consider the following stochastic programming problem with multiple objectives functions :

$$(3.1)\quad \gamma_k(\omega) = \min_{x \in D(\omega)} (\max) \{Z_k(\omega, x) = c^{k\prime}(\omega)x\}, \; (k = 1, \ldots, r)$$

subject to

$$(3.2)\quad D(\omega) = \{x \mid A(\omega)x \leqslant b(\omega),\; x \geqslant 0\},$$

where

$$A(\omega) = (a_{ij}(\omega)),\;\; 1 \leqslant i \leqslant m,\;\; 1 \leqslant j \leqslant n,$$

$$b(\omega) = (b_i(\omega)),\;\; 1 \leqslant i \leqslant m$$

and $c^k(\omega) = (c_l^k(\omega))$, $1 \leqslant k \leqslant r, 1 \leqslant l \leqslant n$ are matrices of real random variables defined on the probability space $\{\Omega, K, P\} : \Omega$

is a Borel set in $R^s(s = mn + m + rn)$, K is the σ-algebra of all the Borel subsets of Ω and P is a probability; x is the n-dimensional unknown vector. In this case there does not exist only one system of objective functions Z_k $(k = 1,\ldots, r)$ but to each $\omega \in \Omega$ there corresponds a different system $Z_i(\cdot, \omega)$ of such functions.

If $P\{\omega | -\infty < \gamma_k(\omega) < \infty\} = 1$, then $\gamma_k(\omega)$ $(k = 1,\ldots r)$ is a random variable.

Since the optimal values of $\gamma_k(\omega)$ are random variables they cannot be known beforehand.

However, we wish to determine the probability distribution and/or certain moments (e.g. the mean value and the variance) of the random vector

$$\gamma(\omega) = (\gamma_1(\omega), \gamma_2(\omega),\ldots, \gamma_r(\omega))$$

subjected to a certain *a priori* probability distribution of random elements contained in the triplet $\{A(\omega), b(\omega), (c^1(\omega), \ldots \ldots, c^r(\omega))\}$.

Knowledge of the distribution F_γ of vector $\gamma(\omega)$ allows us to determine the probability of those $\omega \in \Omega$ such that the values of the objective functions lie between some given limits, i.e.

$$P(\{\omega : \underline{Z}_1 \leqslant Z_1(\omega) < \bar{Z}_1,\ldots, \underline{Z}_r \leqslant Z_r(\omega) < \bar{Z}_r\}).$$

LEMMA 3.1. $P(\{\omega : \underline{Z}_1 \leqslant Z_1(\omega) < \bar{Z}_1,\ldots, \underline{Z}_r \leqslant Z_r(\omega) < \bar{Z}_r\}) =$

$$= F_\gamma(\bar{Z}_1,\ldots, \bar{Z}_r) - \sum_{i=1}^{r} F_\gamma(\bar{Z}_1, \bar{Z}_2,\ldots, \underline{Z}_i,\ldots, \bar{Z}_r) +$$

$$+ \sum_{i<j} F_\gamma(\bar{Z}_1, \bar{Z}_2,\ldots, \underline{Z}_i,\ldots, \underline{Z}_j,\ldots, \bar{Z}_r) + \ldots +$$

$$+ (-1)^r F_\gamma(\underline{Z}_1, \underline{Z}_2, \ldots, \underline{Z}_r).$$

Proof. The above formula is a direct consequence of the total probability formula:

$$P\left(\bigcup_{i=1}^{r} A_i\right) = \sum_i P(A_i) - \sum_{i \neq j} P(A_i \cap A_j) +$$

$$+ \sum_{i \neq j \neq k} P(A_i \cap A_j \cap A_k) + \ldots +$$

$$+ (-1)^r P(A_1 \cap A_2 \cap \ldots \cap A_r).$$

The solution of the above and similar problems can be obtained in the following manner:

1) either seek to replace the stochastic programming problem with multiple objective functions by an equivalent deterministic problem with multiple objective functions (by finding deterministic equivalences to each objective function). Next, one uses the methods already presented in Chapter 1.

2) or seek to reduce the problem to a stochastic problem with only one objective function. The solution of this problem is obtained with the aid of the methods presented in Chapter 2.

3.1. CHEBYSHEV'S STOCHASTIC PROBLEM. THE DISTRIBUTION PROBLEM [148]

In what follows we shall adopt the second approach.

For a given elementary event $\omega \in \Omega$, problem (3.1)–(3.2) becomes a deterministic mathematical programming problem with multiple objective functions, i.e.

$$(3.1') \quad \min \{Z_k(x) = c'^k x\}, \qquad (k = 1, \ldots, r),$$

subject to

$$(3.2') \quad Ax \leqslant b; \; x \geqslant 0.$$

Problem (3.1')–(3.2') can reduce to one of a single objective function, using a synthesis function of type 1^0-a (Chapter 1, Section 1.1.), i.e.

$$Z(x) = \max_{1 \leqslant i \leqslant r} \{Z_i(x)\}.$$

Then one considers the problem

$$(3.1'') \quad \min_x \max_{1 \leqslant i \leqslant r} \{Z_i(x)\}$$

subject to

$$(3.2'') \quad Ax \leqslant b; \; x \geqslant 0.$$

This problem is conventionally equivalent to (3.1′)−(3.2′) and its solution serves also as a solution to (3.1′)−(3.2′).

Problem (3.1″)−(3.2″) is a convex programming problem since the domain of feasible solutions $D = \{x \mid Ax \leqslant b,\ x \geqslant 0\}$ is a convex set and the function $Z(x)$ is also convex [57].

Consider the following subdomains of D:

$$D_k = \{x \in D : Z(x) = Z_k(x)\}, \quad k = 1, \ldots, r.$$

LEMMA 3.2 (Bereanu [15]). *The subset D_k of D defined above is a convex set.*

Proof. Let $y, z \in D_k$, i.e.

$$Z(y) = Z_k(y)$$

and

$$Z(z) = Z_k(z).$$

Let us prove that

$$\alpha y + \beta z \in D_k \quad (\alpha, \beta \geqslant 0,\ \alpha + \beta = 1),$$

i.e.

$$Z(\alpha y + \beta z) = Z_k(\alpha y + \beta z).$$

Indeed,

$$Z_k(\alpha y + \beta z) \leqslant \max_{1 \leqslant i \leqslant r} [Z_i(\alpha y + \beta z)] = Z(\alpha y + \beta z). \tag{3.3}$$

Since the function Z is convex and the functions Z_k are linear it follows that

$$Z_k(\alpha y + \beta z) = \alpha Z_k(y) + \beta Z_k(z) = \alpha Z(y) + \beta Z(z) \geqslant$$
$$\geqslant Z(\alpha y + \beta z). \tag{3.4}$$

Equations (3.3) and (3.4) yield

$$Z(\alpha y + \beta z) = Z_k(\alpha y + \beta z).$$

Hence $\alpha y + \beta z \in D_k$.

The domain D was devided up into at most r subdomains such that the function Z coincides with function Z_k on D_k. Conversely, if Z coincides with Z_k on the domain D_k for each $k = 1, \ldots, r$ then

$$Z(x) = \max_{1 \leqslant i \leqslant r} \{Z_i(x)\}.$$

This is straightforward from the following

THEOREM 3.1 (Bereanu [15]). *Let $Z(x)$ be a piecewise linear convex function on the convex set $D \subset R^n$. If there exists a partition $\{D_k\}$ of D and a system of linear functions Z_k such that*

$$Z(x) = Z_k(x), \qquad (x \in D_k,\ 1 \leqslant k \leqslant r),$$

then

1) $Z(x) = \max\limits_{1 \leqslant i \leqslant r} [Z_i(x)], \qquad x \in D;$

2) *the sets D_k are convex.*

Proof. Define the set $\chi \subset R^{n+1}$,

$$\chi = \{(x_1, \ldots, x_n, x_{n+1});\ (x_1, \ldots, x_n) \in D,\ x_{n+1} \geqslant \geqslant Z(x_1, \ldots, x_n)\},$$

It can be easily shown that the set χ is convex. The hyperplanes

$$P_k \equiv \{(x, x_{n+1}) \mid Z_k(x) - x_{n+1} = 0\}, \qquad (k = 1, \ldots, r) \tag{3.5}$$

are the boundary hyperplanes of the set χ.

To show this, let $(x_1^0, \ldots, x_n^0)$ be an interior point of the set D_k and let $x_{n+1}^0 = Z_k(x_1^0, x_2^0, \ldots, x_n^0)$. The point $\bar{x}^0 = (x_1^0, \ldots, x_n^0, x_{n+1}^0)$ belongs to the boundary of χ, i.e. there exists at least a boundary hyperplane to χ through x^0.

Consider a hypersphere S_ε centered in $(x_1^0, \ldots, x_{n+1}^0)$ and of radius ε. Let H_1 be the open half-space $x_{n+1} > Z_k(x)$. Choose ε such that $S_\varepsilon \cap H_1 \subset \operatorname{int} \chi$. Denote by π a hyperplane different from P_k passing through $(x_1^0, \ldots, x_{n+1}^0)$, i.e. intersecting P_k. Hence, there exist points of the hyperplane π in each of the two half-spaces generated by P_k in R^{n+1}. In particular, there exist also points in H_1 and hence in $S_\varepsilon \cap H_1$. Hence π intersects the interior of χ and thus it cannot be a boundary hyperplane to it. However, there exists at least one boundary hyperplane of χ in $(x_1^0, \ldots, x_{n+1}^0)$ [63], i.e. this boundary hyperplane is P_k itself.

Let us assume that the equality $Z(x) = \max_{1 \leqslant k \leqslant r} [Z_k(x)]$ is not true. Then there would exist a set D_k and a point $\bar{x} = (\bar{x}_1, \ldots, \bar{x}_n) \in D_k$ such that

$$(3.6) \qquad x'_{n+1} = Z_k(\bar{x}) < Z_{k'}(\bar{x}) = x''_{n+1} \; (k' \neq k;\ 1 \leqslant k' \leqslant r).$$

Let x_{n+1}^*, x_{n+1}^{**} be two real numbers such that

$$(3.7) \qquad x'_{n+1} < x_{n+1}^* < x''_{n+1} < x_{n+1}^{**}.$$

From (3.6) and (3.7) and from the definition of χ we derive that the points $(\bar{x}_1, \ldots, \bar{x}_n, x_{n+1}^*)$ and $(\bar{x}_1, \ldots, \bar{x}_n, x_{n+1}^{**})$ belong to the interior of χ.

Since $Z_{k'}(\bar{x}) - x_{n+1}^{**} < 0$ and $Z_{k'}(\bar{x}) - x_{n+1}^* > 0$, the hyperplane $Z_{k'}(x) - x_{n+1}$ crosses the set χ, which contradicts the hypothesis that this hyperplane is a boundary hyperplane to this set. Therefore, $Z(x) = \max_{1 \leqslant k \leqslant r} [Z_k(x)]$.

The second point in the theorem is a consequence of Lemma 3.2. and of the first point in the present theorem. The proof is complete.

Considering problem (3.1′)—(3.2′) one can easily verify that

$$D_i = \{x \mid x \in D,\ c'^k x \leqslant c'^i x,\ \ \forall k \in \{1, 2, \ldots, r\} - \{i\} = J - \{i\}\},\ i = 1, \ldots, r.$$

It is also possible that for certain values of i, $D_i = \emptyset$. In order to avoid such situations define the set of indices

$$T = \{i \mid i \in J \text{ and } D_i \neq \emptyset\}.$$

The next Lemma allows us to eliminate the redundant constraints from the definition of D_i.

LEMMA 3.3. *Let $r \in J$, $r \notin T$, $k \in T$ and $x \in D_k$. Then $c'^r x < c'^k x$.*

Proof. Assume the inequality is not strict, hence for $k \in T$ and $x \in D_k$ we have $c'^k x = c'^r x$. Then

$$c'^j x \leqslant c'^r x, \ j \in J - \{r\},$$

which contradicts the hypothesis that $r \notin T$.

Before approaching the probability distribution of the optimum of problem (3.1)—(3.2), we shall deduce a formula similar to that of total probability.

If $A_1, A_2, \ldots, A_n$ are n events that are pairwise incompatible $(A_i \cap A_j = \emptyset, i, j = 1, \ldots, n, i \neq j)$ and if their reunion is equal to $\Omega \left(\bigcup_{i=1}^{n} A_i = \Omega, \ P(A_i) > 0, \ i = 1, \ldots, n \right)$, then irrespective of the event X that may occur depending on one of the events A_i the total probability is given by

$$P(X) = \sum_{i=1}^{n} P(A_i)\, P_{A_i}(X) = \sum_{i=1}^{n} P(A_i \cap X).$$

Let $Y_i = A_i \cap X$ and let $B_1, \ldots, B_m$ be another partition of the certain event $\left(\bigcup_{i=1}^{m} B_i = \Omega, \ P(B_i) > 0, \ B_i \cap B_j = \emptyset, \ i \neq j \right)$. The events B_i condition the occurrence of the event Y_i. According to the total probability formula,

$$P(Y_i) = \sum_{j=1}^{m} P(B_j)\, P_{B_j}(Y_i) = \sum_{j=1}^{m} P(A_i \cap B_j \cap X),$$

hence

$$P(X) = \sum_{i=1}^{n} \sum_{j=1}^{m} P(A_i \cap B_j \cap X).$$

Thus, we have proved the following

LEMMA 3.4. *Let $\{A_i\}_{i=1,\ldots,n}$; $\{B_j\}_{j=1,\ldots,m}$ be two partitions of the certain event. Then, whatever the event X whose occurrence is conditioned by the events A_i or B_j, one has*

$$P(X) = \sum_{i=1}^{n} \sum_{j=1}^{m} P(A_i \cap B_j \cap X). \tag{3.8}$$

Let us now revert to problem (3.1)–(3.2).

Having given the random variables $Z_k(\omega, x)$ one constructs a new random variable

$$\eta(\omega, x) = \max_{1 \leqslant i \leqslant r} Z_k(\omega, x).$$

For convenience, we assume that problem (3.1)–(3.2) is equivalent to the following one-objective mathematical programming problem

$$\xi(\omega) = \min_{x \in D} \eta(\omega, x) = \min_{x \in D} \max_{1 \leqslant i \leqslant r} Z_i(\omega, x), \tag{3.9}$$

subject to

$$Ax \leqslant b; \quad x \geqslant 0. \tag{3.10}$$

In terms of probability distributions, the problem of finding the distribution of the vector $\gamma(\omega)$ was reduced to that of finding the distribution of only one variable $\xi(\omega)$ as defined by the program (3.9)–(3.10). In order to find this distribution function, we shall follow a route similar to that used by Bereanu [16] for ordinary linear problems.

For the sake of simplicity, we assume that the functions $Z_i(\omega, x)$ depend linearly on the same random variable $t(\omega)$ with continuous and strictly increasing distribution $T(z)$:

$$Z_i(x, \omega) = (c'^i + t(\omega)\,\mathrm{d}'^i)\,x.$$

Let $\lambda = t(\omega)$ be a parameter whose values range within a finite interval $[\delta_1, \delta_2]$. For a given function $Z_i(x, \omega)$ let λ_j^i be the characteristic values:

$$\delta_1 \leqslant \lambda_1^i < \lambda_2^i < \ldots < \lambda_{p_i-1}^i \leqslant \delta_2$$

and let $x^{ji}(1 \leqslant j \leqslant p_i)$ be the optimal solutions corresponding to the intervals $(\lambda_{s-1}^i, \lambda_s^i)$, i.e. the optimal solutions of the parametric programs obtained by replacing $t(\omega)$ by the parameter $\lambda \in [\delta_1, \delta_2]$ in problem (3.9)–(3.10).

If $\lambda \in [\delta_1, \delta_2]$ denote $\delta_1 = \lambda_0^i$; $\delta_2 = \lambda_{p_i}^i$.

THEOREM 3.2. *Let $F(z)$ be the distribution function of the random variable $\xi(\omega)$,*

$$u_{ij}(z) = \frac{z - c'^i x^{ji}}{d'^i x^{ji}} \quad \text{and} \quad v_{ijk} = \frac{(c'^i - c'^k)\,x^{ji}}{(d'^k - d'^i)\,x^{ji}}\,.$$

Then

$$F(z) = \sum_{i=1}^{r} \sum_{j=1}^{p_i} H_{ij}(z),$$

where

$$\underset{\substack{i=1,\ldots,r\\ j=1,\ldots,p_i}}{H_{ij}(z)} = \begin{cases} T[\min(u_{ij}(z), \lambda_j^i, \min\limits_{k\in I_1} v_{ijk})] - T[\max(\lambda_{j-1}^i, \max\limits_{k\in I_2} v_{ijk})] \\ \qquad\qquad \text{if } d'^i x^{ji} > 0,\ \mathscr{A} \text{ and } C_1^* \\ T[\min(\lambda_j^i, \min\limits_{k\in I_1} v_{ijk})] - T[\max(u_{ij}(z), \lambda_{j-1}^i, \max\limits_{k\in I_2} v_{ijk})] \\ \qquad\qquad \text{if } d'^i x^{ji} < 0,\ \mathscr{A} \text{ and } C_2 \\ T[\min(\lambda_j^i, \min\limits_{k\in I_1} v_{ijk})] - T[\max(\lambda_{j-1}^i, \max\limits_{k\in I_2} v_{ijk})] \\ \qquad\qquad \text{if } d'^i x^{ji} = 0,\ c'^i x^{ji} < z,\ \mathscr{A} \text{ and } C_3 \\ 0 \text{ if } \begin{cases} d'^i x^{ji} > 0,\ \mathscr{B} \text{ and/or } (\mathscr{A} \text{ and non-}C_1) \\ d'^i x^{ji} < 0,\ \mathscr{B} \text{ and/or } (\mathscr{A} \text{ and non-}C_2) \\ d'^i x^{ji} = 0,\ c'^i x^{ji} < z,\ \mathscr{B} \text{ and/or } (\mathscr{A} \text{ and non-}C_3) \\ d'^i x^{ji} = 0,\ c'^i x^{ji} \geqslant z \text{ and/or } \mathscr{B}. \end{cases} \end{cases}$$

* In applying the above formula we shall bear in mind that, for example, $\min(u_{ij}(z), \lambda_j^i, \min\limits_{k\in I_1} v_{ijk}) = \min\,(u_{ij}(z), \lambda_j^i)$ if $I_1 = \emptyset$.

The assertions C_1, C_2, C_3, $\mathscr{A}$, $\mathscr{B}$, $\mathscr{C}$ will be defined below.

Proof. Consider the events

$$A = \{\omega \mid \xi(\omega) < z\},$$

$$B_j^i = \{\omega \mid \lambda_{j-1}^i \leqslant t(\omega) < \lambda_j^i\}, \ (i = 1, \ldots, r, \ j = 1, \ldots, p_i),$$

$$C_i = \{\omega \mid (c'^k + t(\omega)\, d'^k)\, x \leqslant (c'^i + t(\omega)\, d'^i)\, x, \quad \forall k \in \{1, \ldots, r\} - \{i\}\}.$$

Since

$$P\left(\bigcup_{j=1}^{p_i} B_j^i\right) = 1, \ P(B_j^i) > 0 \ (i = 1, \ldots, r, \ j = 1, \ldots, p_i)$$

and

$$P\left(\bigcup_{i=1}^{r} C_i\right) = 1, \ P(C_i) > 0, \quad (i = 1, \ldots, r),$$

one can apply (3.8) and get

$$F(z) = P(\{\omega \mid \xi(\omega) < z\}) = P(A) = \sum_{i=1}^{r} \sum_{j=1}^{p_i} P(A \cap B_j^i \cap C_i).$$

If the event C_i occurs, then $\max\limits_{1 \leqslant k \leqslant r} Z_k(\omega, x) = Z_i(\omega, x)$ and, hence,

$$\xi(\omega) = \min_{x \in D} \{Z_i(\omega, x) = (c'^i + t(\omega)\, d'^i)\, x\}.$$

If the event B_j^i also occurs, then

$$\xi(\omega) = (c'^i + t(\omega)\, d'^i)\, x^{ji}.$$

Hence

$$F(z) = P(A) = \sum_{i=1}^{r} \sum_{j=1}^{p_i} P\{[\omega \mid (c'^i + t(\omega)\, d'^i)\, x^{ji} < z] \cap \cap [\omega \mid \lambda_{j-1} \leqslant t(\omega) < \lambda_j] \cap [\omega \mid (c'^k + t(\omega)\, d'^k)\, x^{ji} \leqslant \leqslant (c'^i + t(\omega)\, d'^i)\, x^{ji}, \ \forall k \in \{1, \ldots, r\} - \{i\}]\}.$$

If we denote

$$H_{ij}(z) = P\{[\omega \,|\, (c'^i + t(\omega)d'^i)x^{ji} < z] \cap [\omega \,|\, \lambda_{j-1} \leqslant t(\omega) <$$

$$< \lambda_j^i] \cap [\omega \,|\, (c'^k + t(\omega)\, d'^k)\, x^{ji} \leqslant (c'^i + t(\omega)d'^i)x^{ji}$$

$$\forall k \in \{1, \ldots, r\} - \{i\}]\},$$

then

$$F(z) = \sum_{i=1}^{r} \sum_{j=1}^{p_i} H_{ij}(z).$$

In order to find the formula for $H_{ij}(z)$ denote by D^* the set of solutions to the system of inequalities

$$(c'^i + td'^i)\, x^{ji} < z,$$

$$(c'^k + td'^k)\, x^{ji} \leqslant (c'^i + td'^i)\, x^{ji}, \; (k \in \{1, \ldots, r\} - \{i\})$$

$$\lambda_{j-1}^i \leqslant t < \lambda_j^i.$$

Let

$$I_1 = \{k \,|\, k \in \{1, 2, \ldots, r\} - \{i\}, (d'^k - d'^i)\, x^{ji} > 0, \; j = 1, \ldots, p_i\},$$

$$I_2 = \{k \,|\, k \in \{1, 2, \ldots, r\} - \{i\}, (d'^k - d'^i)x^{ji} < 0, j = 1, \ldots, p_i\},$$

$$I_3 = \{k \,|\, k \in \{1, 2, \ldots, r\} - \{i\}, (d'^k - d'^i)x^{ji} = 0, j = 1, \ldots, p_i\}.$$

In order to determine the set D^* we shall consider the following equivalences.

I. The inequality $(c'^i + td'^i)x^{ji} < z$ is equivalent to

a) $t \lesseqgtr \dfrac{z - c'^i x^{ji}}{d'^i x^{ji}}$ if $d'^i x^{ji} \gtrless 0$.

b) $t \in R$ if $d'^i x^{ji} = 0$ and $c'^i x^{ji} < z$.

c) impossibility if $d'^i x^{ji} = 0$ and $c'^i x^{ji} \geqslant z$.

II. Inequalities $(c'^k + td'^k)x^{ji} \leqslant (c'^i + td'^i)x^{ji}$, $k \in T - \{i\}$ are equivalent to

$$\text{a) } t \leqslant \min_{k \in I_1} v_{ijk} \text{ if } \begin{cases} I_1 \neq \emptyset \text{ and } I_2 \cup I_3 = \emptyset \\ I_1 \neq \emptyset, I_2 \neq \emptyset, I_3 \neq \emptyset \text{ and } c'^k x^{ji} \leqslant c'^i x^{ji}, k \in I_3, \\ \text{(the inequality is strict for } k \notin T \text{ (see} \\ \text{Lemma 3.3))} \end{cases}$$

$$\text{b) } t \geqslant \max_{k \in I_2} v_{ijk} \text{ if } \begin{cases} I_2 \neq \emptyset \text{ and } I_1 \cup I_3 = \emptyset, \\ I_1 = \emptyset, I_2 \neq \emptyset, I_3 \neq \emptyset \text{ and } c'^k x^{ji} \leqslant c'^i x^{ji}, \\ k \in I_3, \text{(the inequality is strict for } k \notin T) \end{cases}$$

$$\text{c) } \max_{k^* \in I_2} v_{ijk^*} \leqslant \min_{k \in I_1} v_{ijk} \text{ if } \begin{cases} I_1 \neq \emptyset, I_2 \neq \emptyset, I_3 = \emptyset, (k \in I_1, k^* \in I_2), \\ I_1 \neq \emptyset, I_2 \neq \emptyset, I_3 \neq \emptyset \text{ and } c'^k x^{ki} \leqslant c'^i x^{ji}, \\ k \in I_3, \text{ (the inequalities are strict} \\ \text{for } k \notin \text{T)} \end{cases}$$

d) $t \in R$ if $I_1 \cup I_2 = \emptyset$, $I_3 \neq \emptyset$ and $c'^k x^{ji} \leqslant c'^i x^{ji}$, $k \in I_3$,

(the inequalities are strict for $i \in T$ and $k \in T$)

e) impossibility if

$$\left.\begin{array}{lll} I_1 \cup I_2 = \emptyset, & & I_3 \neq \emptyset \\ I_1 \neq \emptyset, & I_2 = \emptyset, & I_3 \neq \emptyset \\ I_1 = \emptyset, & I_2 \neq \emptyset, & I_3 \neq \emptyset \\ I_1 \neq \emptyset, & I_2 \neq \emptyset, & I_3 \neq \emptyset \end{array}\right\} \text{ and } \begin{cases} c'^k x^{ji} \geqslant c'^i x^{ji} \text{ for } i \in T,\ k \in I_3 \\ \text{and for at least one } k \notin T \\ c'^k x^{ji} > c'^i x^{ji} \text{ for } i \in T,\ k \in I_3 \\ \text{and for at least one } k \in T. \end{cases}$$

In order to simplify notation, we shall denote by $\mathscr{A}$ the statement $\{I_3 = \emptyset$ or $[I_3 \neq \emptyset$ and $c'^k x^{ji} \leqslant c'^i x^{ji}$, $k \in I_3$, the inequalities being strict for $i \in T$ and $k \notin T]\}$
and by $\mathscr{B}$ the statement

$$I_3 \neq \emptyset \text{ and } \left\{ \begin{array}{l} c'^k x^{ji} \geqslant c'^i x^{ji} \text{ for } i \in T,\ k \in I_3 \text{ and for at least} \\ \text{one } k \notin T. \\ c'^k x^{ji} > c'^i x^{ji} \text{ for } i \in T,\ k \in I_3 \text{ and for at least} \\ \text{one } k \in T. \end{array} \right\}$$

We shall also consider the statements C_1, C_2 and C_3:

$$C_1 : \min\,(u_{ij}(z),\ \lambda_j^i,\ \min_{k \in I_1} v_{ijk}) > \max\,(\lambda_{j-1}^i,\ \max_{k \in I_2} v_{ijk}),$$

$$C_2 : \min\,(\lambda_j^i, \min_{k\in I_1} v_{ijk}) > \max\,(u_{ij}(z), \lambda_{j-1}^i, \max_{k\in I_2} v_{ijk}),$$

$$C_3 : \min\,(\lambda_j^i, \min_{k\in I_1} v_{ijk}) > \max\,(\lambda_{j-1}^i, \max_{k\in I_2} v_{ijk}),$$

together with their negations non-C_1, non-C_2 and non-C_3.

Combining the above cases leads to the following possibilities :

I. $d'^i x^{ji} > 0$, $\mathscr{A}$ and C_1

$$H_{ij}(z) = T[\min(u_{ij}(z), \lambda_j^i, \min_{k\in I_1} v_{ijk}) - T[\max\,(\lambda_{j-1}^i, \max_{k\in I_2} v_{ijk})]$$

II. $d'^i x^{ji} > 0$, $\mathscr{B}$ and/or ($\mathscr{A}$ and non-C_1)

$$H_{ij}(z) = 0$$

III. $d'^i x^{ji} < 0$, $\mathscr{A}$ and C_2.

$$H_{ij}(z) = T[(\min(\lambda_j^i, \min_{k\in I_1} v_{ijk})] - T[\max\,(u_{ij}(z), \lambda_{j-1}^i, \max_{k\in I_2} v_{ijk})]$$

IV. $d'^i x^{ji} < 0$; $\mathscr{B}$ and/or ($\mathscr{A}$ and non-C_2)

$$H_{ij}(z) = 0.$$

V. $d'^i x^{ji} = 0$, $c'^i x^{ji} < z$, $\mathscr{A}$ and C_3.

$$H_{ij}(z) = T[\min\,(\lambda_j^i, \min_{k\in I_1} v_{ijk})] - T[\max(\lambda_{j-1}^i, \max_{k\in I_2} v_{ijk})].$$

VI. $d'^i x^{ji} = 0$, $c'^i x^{ji} < z$, $\mathscr{B}$ and/or ($\mathscr{A}$ and non-C_3)

$$H_{ij}(z) = 0.$$

VII. $d'^i x^{ji} = 0$, $c'^i x^{ji} \geqslant z$ and/or $\mathscr{B}$

$$H^{ij}(z) = 0.$$

Summarizing, we get the proof of the theorem.

Example. Consider the following Chebyshev problem

$$\min_x \max \begin{Bmatrix} Z_1(x) = (2-t)x_2 - 3x_3 + 2tx_5 \\ Z_2(x) = (1-t)x_1 - x_2 - tx_3 + (1+t)x_4 \end{Bmatrix}$$

on the domain D defined by the equations

$$x_1 + 3x_2 - x_3 + 2x_5 = 7$$

$$-2x_2 + 4x_3 + x_4 = 12$$

$$-4x_2 + 3x_3 + 8x_5 + x_6 = 10$$

$$x_i \geqslant 0 \ (i = 1, 2, \ldots, 6).$$

Assume that $t(\omega)$ has the normal distribution N (3, 1). With a probability of 0.9987, $t(\omega) \in [0 ; 6]$, i.e. one can neglect the values of $t(\omega)$ outside this interval.

Denote by $\xi(\omega)$ the random variable defined as

$$\xi(\omega) = \min_{x \in D} \max \begin{Bmatrix} Z_1(x, \omega) = (2 - t(\omega))x_2 - 3x_3 + 2t(\omega)x_5 \\ Z_2(x, \omega) = (1 - t(\omega))x_1 - x_2 - t(\omega)x_3 + (1 + t(\omega))x_4 \end{Bmatrix}$$

and let $F(z)$ be its distribution function.

Consider the first parametric programming problem

$$\min_{x \in D} Z_1(x, t) = (2 - t)\, x_2 - 3x_3 + 2tx_5,$$

Its characteristic value in the interval $[0, 6]$ is $t = \dfrac{1}{2}$ and the characteristic solutions are

$$x_1^{11} = 10,\ x_2^{11} = 0,\ x_3^{11} = 3,\ x_4^{11} = 0,\ x_5^{11} = 0,\ x_6^{11} = 1 \text{ for } t \in \left[0, \frac{1}{2}\right]$$

and

$$x_1^{21} = 0,\ x_2^{21} = 4,\ x_3^{21} = 5,\ x_4^{21} = 0,\ x_5^{21} = 0,\ x_6^{21} = 11, \text{ for } t \in \left[\frac{1}{2}, 6\right].$$

Consider the second parametric programming problem

$$\min_{x \in D} Z_2(x, t) = (1 - t)\, x_1 - x_2 - tx_3 + (1 + t)x_4.$$

Its characteristic value in the interval $[0, 6]$ is $t = \frac{7}{4}$ and the characteristic solutions are

$$x_1^{12} = 0,\ x_2^{12} = 4,\ x_3^{12} = 5,\ x_4^{12} = 0,\ x_5^{12} = 0,\ x_6^{12} = 11$$

$$\text{for } t \in [0, 7/4]$$

and

$$x_1^{22} = 10,\ x_2^{22} = 0,\ x_3^{22} = 3,\ x_4^{22} = 0,\ x_5^{22} = 0,\ x_6^{22} = 1$$

$$\text{for } t \in [7/4, 6].$$

The values of the efficiency functions for these solutions are

$$Z_1(x^{11}) = Z_1(x^{22}) = -9,$$

$$Z_1(x^{21}) = Z_1(x^{12}) = -7 - 4t,$$

$$Z_2(x^{12}) = Z_2(x^{21}) = -4 - 5t,$$

and

$$Z_2(x^{22}) = Z_2(x^{11}) = -13t + 10,$$

respectively.

Since there exist only two functions, the problem is meaningful for $T = J = \{1,2\}$. Otherwise one will find that on the whole domain of feasible solutions one function will be greater than the other.

According to the formulae in Theorem 3.2, the distribution function of $\xi(\omega)$ is

$$F(z) = \sum_{i=1}^{2} \sum_{j=1}^{2} H_{ij}(z) = H_{11}(z) + H_{12}(z) + H_{21}(z) + H_{22}(z),$$

where

$$H_{11}(z) = P\left\{[\omega | -9 < z] \cap \left[\omega | 0 \leqslant t(\omega) < \frac{1}{2}\right] \cap \right.$$

$$\left. \cap\ [\omega | -13t(\omega) + 10 \leqslant -9]\right\} = 0$$

since the inequation $-13t + 10 \leqslant -9$ has the solution $t \in \left[+\frac{19}{13}, \infty\right)$, which is disjoint from the interval $\left[0, \frac{1}{2}\right]$,

$$H_{12}(z) = P\left\{[\omega \mid -7 - 4t(\omega) < z] \cap \left[\omega \mid \frac{1}{2} \leqslant t(\omega) < 6\right] \cap \right.$$

$$\left. \cap \ [\omega \mid -4 - 5t(\omega) \leqslant -7 - 4t(\omega)]\right\} =$$

$$= \begin{cases} T(6) - T(3) & \text{if } \frac{-7-z}{4} \leqslant 3 \text{ (i.e. } z \geqslant -19) \\ T(6) - T\left(\frac{-7-z}{4}\right) & \text{if } 6 > \frac{-7-z}{4} > 3 \text{(i.e.} -31 < z < -19) \\ 0 & \text{if } \frac{-7-z}{4} \geqslant 6 \text{ (i.e. } z \leqslant -31), \end{cases}$$

$$H_{21}(z) = \left\{P[\omega \mid -4 - 5t(\omega) < z] \cap \left[\omega \mid 0 \leqslant t(\omega) < \frac{7}{4}\right] \cap \right.$$

$$\left. \cap \ [\omega \mid -7 - 4t(\omega) \leqslant -4 - 5t(\omega)]\right\}$$

$$= \begin{cases} T(7/4) - T(0) & \text{if } \frac{-4-z}{5} \leqslant 0 \text{ (i.e. } z \geqslant -4) \\ T(7/4) - T\left(\frac{-4-z}{5}\right) & \text{if } 0 < \frac{-4-z}{5} < \frac{7}{4} \\ \quad \left(\text{i.e.} -\frac{51}{4} < z < -4\right) & \\ 0 & \text{if } \frac{-4-z}{5} \geqslant 7/4 \left(\text{i.e. } z \leqslant -\frac{51}{4}\right) \end{cases}$$

$$H_{22}(z) = P\{[\omega | -13t(\omega) + 10 < z] \cap [\omega | 7/4 \leqslant t(\omega) < 6] \cap$$
$$\cap [\omega | -9 \leqslant -13t(\omega) + 10]\} = 0$$

since the inequation $-9 \leqslant -13t + 10$ has the solution $t \in \left(-\infty, \frac{19}{13}\right]$, i.e. an interval disjoint from $\left[\frac{7}{4}, 6\right]$.

Hence, for all the values of z, the distribution function of $\xi(\omega)$ will be

$$F(z) = \begin{cases} 0 & \text{if } z \leqslant -31 \\ T(6) - T\left(-\dfrac{7-z}{4}\right) & \text{if } -31 \leqslant z \leqslant -19 \\ T(6) - T(3) & \text{if } -19 \leqslant z \leqslant -\dfrac{51}{4} \\ T(6) + T(7/4) - T(3) - T\left(\dfrac{-4-z}{5}\right) & \text{if } -\dfrac{51}{4} \leqslant z \leqslant -4 \\ T(6) + T(7/4) - T(0) - T(3) & \text{if } z \geqslant -4. \end{cases}$$

For instance, for $z = -21$, $F(-21) = 0.0655$, for $z = -5$, $F(z) = 0.4718$ and for $z = -15$, $F(z) = 0.1138$.

3.2. STOCHASTIC FRACTIONAL PROGRAMMING. THE DISTRIBUTION PROBLEM

Another approach to the stochastic programming consists in replacing two objective functions by one of the form $h(Z_1, Z_2) = \dfrac{Z_1(x, \omega)}{Z_2(x, \omega)}$ (see Chapter 1, Section 1.1, functions of type 1°b). Thus, the stochastic programming problem with two objective functions is replaced by one with only one fractional objective function:

$$\gamma(\omega) = \underset{x}{\text{optimum}} \frac{Z_1(x, '\omega)}{Z_2(x, \omega)}$$

subject to

$$Ax = b,$$

$$x \geqslant 0.$$

In order to find the distribution of $\gamma(\omega)$, assume that Z_1 and Z_2 are linear functions of the same random variable $t(\omega)$, i.e.

$$Z_1(\omega) = (c' + (t(\omega)c_1')x,$$

$$Z_2(\omega) = (d' + t(\omega)d_1')x,$$

where A is an $m \times n$-matrix $(m < n)$, c, c_1, d, d_1, x and 0 are n-vectors and b is an m-vector. $t(\omega)$ is a random variable having a continuous distribution function $T(z)$.

For the stochastic fractional programming problem

$$\gamma(\omega) = \underset{x}{\text{optimum}} \frac{(c' + t(\omega)c_1')x}{(d' + t(\omega)d_1')x}$$

subject to

$$Ax = b$$

$$x \geqslant 0$$

assume that

a) The set $D = \{x \mid Ax = b, x \geqslant 0\}$ is non-empty and bounded.

b) The denominator of the objective function keeps the same sign (say positive) on D, hence $P\{\omega \mid (d' + t(\omega)d_1')x > 0\} = 1$.

c) Each feasible base solution is non-degenerated.

Indeed, with the exception of certain trivial cases, the denominator does not change in sign on D[169].

The distribution function of the random variable $\gamma(\omega)$ will be determined in a manner similar with that used by Bereanu [14].

Assume that $t(\omega) = \lambda$ is a parameter ranging in a finite interval $[\delta_1, \delta_2]$.

It is already known [169], [174] that for the parametric fractional problem, the interval $[\delta_1, \delta_2]$ may be devided into a finite number of so-called "critical regions" which are defined by the various combinations of variables forming optimal bases.

Let λ_j $(1 \leqslant j \leqslant p-1)$ be the characteristic values

$$\delta_1 \leqslant \lambda_1 < \lambda_2 < \ldots < \lambda_{p-1} \leqslant \delta_2.$$

and let $x^j (1 \leqslant j \leqslant p)$ be the optimal solutions corresponding to the intervals $(\lambda_{s-1}, \lambda_s)$, $s = 1, \ldots, p$. Denote $\delta_1 = \lambda_0$ and $\delta_2 = \lambda_p$. Since the random variable $t(\omega)$ is supposed to have a continuous distribution function, the intersection of two such critical regions has zero probability.

THEOREM 3.3. *Let $F(z)$ be the distribution function of the random variable $\gamma(\omega)$ and let* $u_j(z) = \dfrac{(zd' - c')x^j}{(c_1' - zd_1')x^j}$. *Then*

$$F(z) = \sum_{j=1}^{p} H_j(z),$$

where

$$\underset{(1 \leqslant j \leqslant p)}{H_j(z)} = \begin{cases} T(\lambda_j) - T(\lambda_{j-1}) \text{ if } \begin{cases} (c_1' - zd_1')x^j > 0 \text{ and } u_j(z) \geqslant \lambda_j \\ (c_1' - zd_1')x^j < 0 \text{ and } u_j(z) \leqslant \lambda_{j-1} \\ (c_1' - zd_1')x^j = 0 \text{ and } c'x^j < zd'x^j \end{cases} \\ 0 \qquad \begin{cases} (c_1' - zd_1')x^j > 0 \text{ and } u_j(z) \leqslant \lambda_{j-1} \\ (c_1' - zd_1')x^j < 0 \text{ and } u_j(z) \geqslant \lambda_j \\ (c_1' - zd_1')x^j = 0 \text{ and } c'x^j \geqslant zd'x^j \end{cases} \\ \begin{matrix} T(u_j(z)) - T(\lambda_{j-1}) \\ T(\lambda_j) - T(u_j(z)) \end{matrix} \text{ if } \lambda_{j-1} < u_j(z) < \lambda_j, \text{ and} \begin{cases} (c_1' - zd_1')x^j > 0 \\ (c_1' - zd_1')x^j < 0. \end{cases} \end{cases}$$

Proof. Consider the events $A = \{\omega | \gamma(\omega) < z\}$ and $B = \{\omega | \lambda_{j-1} \leqslant t(\omega) < \lambda_j\}$. Since $\bigcup_{i=1}^{p} B_j = \Omega$ and $P(B_j) > 0$ $(1 \leqslant j \leqslant p)(B_i \cap \cap B_j = \emptyset,\ i \neq j)$, one can use the total probability formula

$$P(A) = \sum_{j=1}^{p} P(A | B_j) \cdot P(B_j),$$

i.e.

$$F(z) = P(A) = \sum_{j=1}^{p} P(A \cap B_j) = \sum_{j=1}^{p} H_j(z),$$

where

$$H_j(z) = P\left\{\left[\omega | \frac{(c' + t(\omega)c_1')x^j}{(d' + t(\omega)d_1')x^j} < z\right] \cap \right.$$

$$\left. \cap\ [\omega |\ \lambda_{j-1} \leqslant t\,(\omega) < \lambda_j]\right\}.$$

a) If $(c_1' - zd_1')x^j > 0$, then $\dfrac{(c' + t(\omega)c_1')x^j}{(d' + t(\omega)d_1')x^j} < z$ when

$$t < \frac{(d'z - c')x^j}{(c_1' - zd_1')x^j},$$

such that

$$H_k(z) = 0, \text{ if } u_j(z) \leqslant \lambda_{j-1}$$

$$H_k(z) = T(\lambda_j) - T(\lambda_{j-1}) \text{ if } u_j(z) \geqslant \lambda_j$$

and

$$H_k(z) = T(u_j(z)) - T(\lambda_{j-1}), \text{ if } \lambda_{j-1} < u_j(z) < \lambda_j;$$

b) If $(c_1' - zd_1')\,x^j < 0$, then

$$\frac{(c' + t(\omega)c_1')x^j}{(d' + t(\omega)d_1')x^j} < z \text{ when } t > \frac{(d'z - c')x^j}{(c_1' - zd_1')x^j},$$

such that

$$H_k(z) = 0 \text{ if } u_j(z) \geqslant \lambda_j,$$

$$H_k(z) = T(\lambda_j) - T(\lambda_{j-1}), \text{ if } u_j(z) \leqslant \lambda_{j-1}$$

and

$$H_k(z) = T(\lambda_j) - T(u_j(z)), \text{ if } \lambda_{j-1} < u_j(z) < \lambda_j;$$

c) If $(c_1' - zd_1')x^j = 0$, then $H_j(z) = 0$, if $c'x^j \geqslant zd'x^j$ and $H_j(z) = T(\lambda_j) - T(\lambda_{j-1})$ if $c'x^j < zd'x^j$.

The mean value of the random variable $\gamma(\omega)$ can be calculated using the following formula:

$$\mathbf{E}\gamma(\omega) = \sum_{j=1}^{p} \int_{\lambda_{j-1}}^{\lambda_j} \frac{(c' + tc_1')x^j}{(d' + td_1')x^j} \, \mathrm{d}T(\lambda).$$

Example. Consider the problem

$$\max \frac{tx_1 - x_2}{tx_1 + x_2 + 1}$$

subject to

$$2x_1 + x_2 + x_3 = 4,$$

$$x_1 - x_2 + x_4 = 1,$$

$$x_1, x_2, x_3, x_4 \geqslant 0.$$

Assume that $t(\omega)$ has a normal distribution $N\left(0, \frac{1}{6}\right)$. Let $T(z)$ be its distribution function. For a probability of 0.9987, $t(\omega) \in \in \left[-\frac{1}{2}, \frac{1}{2}\right]$; hence one can neglect the values of $t(\omega)$ outside this interval.

Denote by $\xi(\omega)$ the random variable defined as

$$\xi(\omega) = \max_x \frac{t(\omega)x_1 - x_2}{t(\omega)x_1 + x_2 + 1}$$

and by $F(z)$ its distribution function.

The characteristic value of the parametric programming problem defined above, in the interval $\left[-\frac{1}{2}, \frac{1}{2}\right]$, is $t = 0$ and the characteristic solutions are $x_1^1 = 0$, $x_2^1 = 0$, $x_3^1 = 4$, $x_4^1 = 1$, for $\lambda \in [-1/2, 0]$ and $x_1^2 = 1$, $x_2^2 = 0$, $x_3^2 = 2$, $x_4^2 = 0$, for $\lambda \in [0, 1/2]$.

Then

$$H_1(z) = P\{[\omega | 0 < z] \cap [\omega | -1/2 \leqslant t(\omega) < 0]\} = \begin{cases} T(0) - T(-1/2) & \text{if } z > 0 \\ 0 & \text{if } z \leqslant 0 \end{cases}$$

$$H_2(z) = P\left\{\left[\omega \,|\, \frac{t(\omega)}{t(\omega)+1} < z\right] \cap [\omega | 0 \leqslant t(\omega) < 1/2]\right\} =$$

$$= P\{[\omega | t(1-z) < z] \cap [\omega | 0 \leqslant t(\omega) < 1/2]\} =$$

$$= \begin{cases} T(1/2) - T(0) & \text{if } z = 1 \\ T(1/2) - T(0) & \text{if } z < 1 \text{ and } \dfrac{z}{1-z} \geqslant 1/2 \\ T\left(\dfrac{z}{1-z}\right) - T(0) & \text{if } z < 1 \text{ and } 0 < \dfrac{z}{1-z} < 1/2 \\ 0 & \text{if } z < 1 \text{ and } \dfrac{z}{1-z} \leqslant 0 \\ T(1/2) - T(0) & \text{if } z > 1 \text{ and } \dfrac{z}{1-z} \leqslant 0 \\ T(1/2) - T\left(\dfrac{z}{1-z}\right) & \text{if } z > 1 \text{ and } 0 < \dfrac{z}{1-z} < 1/2 \\ 0 & \text{if } z > 1 \text{ and } \dfrac{z}{1-z} \geqslant 1/2. \end{cases}$$

Performing a simple calculation yields:

$$H_2(z) = \begin{cases} T(1/2) - T(0) & \text{if } z \geqslant \frac{1}{3} \\ T\left(\frac{z}{1-z}\right) - T(0) & \text{if } 0 < z < \frac{1}{3} \\ 0 & \text{if } z \leqslant 0. \end{cases}$$

Hence the distribution function of $\xi(\omega)$ is

$$F(z) = H_1(z) + H_2(z) =$$

$$= \begin{cases} 0 & \text{if } z \leqslant 0, \\ T\left(\frac{z}{1-z}\right) - T(-1/2) & \text{if } 0 < z < \frac{1}{3}, \\ T(1/2) - T(-1/2) & \text{if } z \geqslant \frac{1}{3}. \end{cases}$$

3.3. GOAL PROGRAMMING. THE STOCHASTIC CASE

3.3.1. Throughout this Sub-section goal programming problems will be discussed with reference to the stochastic case.

Consider the mathematical programming problem with multiple objective functions,

$$(\max)\ F_1 = c_{11}x_1 + c_{12}x_2 + \ldots + c_{1n}x_n$$

$$(\max)\ F_2 = c_{21}x_1 + c_{22}x_2 + \ldots + c_{2n}x_n$$

$$\cdots\cdots\cdots\cdots\cdots\cdots$$

$$(\max)\ F_r = c_{r1}x_1 + c_{r2}x_2 + \ldots + c_{rn}x_n,$$

subject to

$$Ax = b;\ x \geqslant 0.$$

Assume that the relation between $F = (F_1, \ldots, F_r)'$ and $x = (x_1, \ldots, x_n)'$ contains a random factor $u = (u_1, \ldots, u_r)$ [44]

$$F_1 = c_{11}x_1 + c_{12}x_2 + \ldots + c_{1n}x_n + u_1 = c_1'x + u_1$$

$$F_2 = c_{21}x_1 + c_{22}x_2 + \ldots + c_{2n}x_n + u_2 = c_2'x + u_2$$

$$\cdots\cdots\cdots\cdots\cdots\cdots\cdots\cdots\cdots$$

$$F_r = c_{r1}x_1 + c_{r2}x_2 + \ldots + c_{rn}x_n + u_r = c_r'x + u_r.$$

For the sake of simplicity, we assume that the random vector u follows a normal distribution, that it has zero mean-value vector and non-singular covariance matrix V and the known matrix $C = (c_{ij})_{\substack{i=1,\ldots,r \\ j=1,\ldots,n}}$ is at least of rank r.

One can easily prove that $F = (F_1, \ldots, F_r)'$ is a random vector of mean-value vector $(c_{11}x_1 + c_{12}x_2 + \ldots + c_{1n}x_n, \ldots, c_{r1}x_1 + c_{r2}x_2 + \ldots + c_{rn}x_n)'$ and covariance matrix V. Hence its distribution density is

$$f(y_1, \ldots, y_r) = (2\pi)^{-r/2}|V|^{-1/2}e^{-Q/2},$$

where Q is a quadratic form defined as

$$Q = (F - Cx)'V^{-1}(F - Cx).$$

An optimality criterion can be obtained by reconsidering the relation

$$\bar{F}_i = \{F_i = c_i'x\},$$

where $\bar{F}_i$ is the required level for F_i.

Due to the existence of a perturbation factor u the above equality cannot hold. Then we choose Y^* in the space E^r such that $\bar{F} = (\bar{F}_1, \bar{F}_2, \ldots, \bar{F}_r) \in Y^*$ and determine the vector x such that this maximizes the probability that $(F_1(x), F_2(x), \ldots, F_r(x))$ belongs to Y^*. One gets the following model [44]:

$$(3.11) \quad \max_x P(F(x) \in Y^*)$$

subject to

$$Ax = b\,;\ x \geqslant 0.$$

Since u is normal the domain Y^* is usually, an ellipsoid in E^r centered in $(\bar{F}_1, \ldots, \bar{F}_r)$ and expressed as

$$Y^* = \{y = (y_1, \ldots, y_r) \,|\, Q^* = (y - \bar{F})'V^{-1}(y - \bar{F}) \leqslant c^2\},$$

where c is a convenient scalar.

The quadratic form Q has a distribution χ^2 with r degrees of freedom. Hence, Y^* can be interpreted as a confidence region for $F(x)$ corresponding to level α. This means that if one chooses a point x^0 such that $\mathrm{E}F(x^0) = Cx^0 = \bar{F}$, then

$$P(F(x) \in Y^*) = \alpha.$$

The model (3.11) is equivalent to the following model

$$\text{(3.12)} \quad \min_x R(x) = (\bar{F} - Cx)'V^{-1}(\bar{F} - Cx)$$

subject to

$$Ax = b \quad x \geqslant 0.$$

Indeed, this becomes obvious if one treats the expression $R^*(x) = (2\pi)^{-r/2}|V|^{-1/2}\exp\left(-\dfrac{R(x)}{2}\right)$ as the maximum verisimility function of the event

$$\bar{F} = Cx + u$$

to be maximized, i.e. maximize its logarithm

$$\text{(3.12')} \quad \max\,[-R(x)] = -\min R(x)$$

which leads to model (3.12). Re-write model (3.12) as

$$\begin{aligned} R(x) &= (\bar{F} - Cx)'V^{-1}(\bar{F} - Cx) = \\ &= \bar{F}'V^{-1}\bar{F} - \bar{F}'V^{-1}Cx - x'C'V^{-1}\bar{F} + x'C'V^{-1}Cx = \\ &= \bar{F}'V^{-1}\bar{F} + x'C'V^{-1}Cx - 2\bar{F}'V^{-1}Cx. \end{aligned}$$

and introduce the following notation:

$$k = \bar{F}'V^{-1}\bar{F}, \ B = C'V^{-1}C, \ p = -C'V^{-1}\bar{F}.$$

Then

$$R(x) = k + x'Bx + 2p'x.$$

Hence,

THEOREM 3.4. *The optimal solution of problem* (3.11) *is obtained by solving the quadratic programming problem*

$$(3.13) \quad \min_x \, [k + x'Bx + 2p'x],$$

subject to

$$Ax = b, \ x \geqslant 0.$$

Note that the solution to model (3.13) can be obtained by any of the quadratic programming algorithms available in the literature.

The matrix B is positive definite if the number of objective functions (r) is equal to $n (r = n)$ and positive semidefinite if $r < n$, whereas V is positive definite by definition.

Next, we shall assume that u_i are independent random variables. In this case, the covariance matrix V is diagonal. Let

$$V = \begin{pmatrix} v_{11}^2 & 0 & \dots & 0 \\ 0 & v_{22}^2 & \dots & 0 \\ \cdot & \cdot & \cdot & \cdot \\ 0 & 0 & \dots & v_{rr}^2 \end{pmatrix}.$$

Then V^{-1}, also a diagonal matrix, will be

$$V^{-1} = \begin{pmatrix} \frac{1}{v_{11}^2} & 0 & \dots & 0 \\ 0 & \frac{1}{v_{22}^2} & \dots & 0 \\ \cdot & \cdot & \cdot & \cdot \\ 0 & 0 & \dots & \frac{1}{v_{rr}^2} \end{pmatrix}.$$

The model (3.11) becomes

$$\min_x R(x) = \sum_{i=1}^{r} \left[\left(\frac{1}{v_{ii}^2} \right) (\bar{F}_i - C_i x)^2 \right],$$

subject to

$$Ax = b, \qquad x \geqslant 0.$$

One notes that this model coincides with the deterministic model 3″ presented in Chapter 1, Section 1.3, if it is assumed that $\alpha_i = \frac{1}{v_{ii}^2}$.

This means that if u_i are independent random variables, then the solution of the stochastic model (3.11) is similar to that of the deterministic model 3″ (Chapter 1, Section 1.3) provided that the weighting factors are $\alpha_i = \frac{1}{v_{ii}^2}$.

3.3.2. Assume that in $F(x) = Cx$ the elements of C are random variables. Let $\bar{F} = (\bar{F}_1, \bar{F}_2, \ldots, \bar{F}_r)$ be the levels to be reached by the r objective functions. For each x there exist deviations of Cx from $\bar{F}$, denoted by $d^+(x)$ and $d^-(x)$ (see Chapter 1).

Consider that the optimization criterion chosen is that of minimizing the mean value of the norm vector $\bar{F} - Cx$. If the norm $\|\cdot\|_p$ is considered, then we obtain the following model

(3.14) $\min \mathbf{E}(\|\bar{F} - Cx\|_p)$,

subject to

$$Ax = b, \ x \geqslant 0.$$

Let us examine model (3.14) for various values of p. Thus, for $p = 1$, one gets the model

$$\min_x \mathbf{E}\left(\sum_{k=1}^{r} (d_k^+(x) + d_k^-(x)) \right),$$

subject to

$$F_k(x) + d_k^- - d_k^+ = \bar{F}_k, \qquad (k = 1, \ldots, r),$$

$$Ax = b,$$

$$x, d_k^-, d_k^+ \geqslant 0. \qquad (k = 1, \ldots, r)$$

For $p = 2$ one obtains the model

$$(3.15) \quad \min_x \mathbf{E}\left(\sum_{i=1}^{r} |\bar{F}_i - C_i x|^2\right)$$

subject to

$$Ax = b, \quad x \geqslant 0,$$

where C_i is the row i of matrix C.

In order to find a deterministic equivalent of problem (3.15), assume that the elements of matrix C are independent random variables with mean values $\bar{c}_{ij}$ and variances σ_{ij}^2, i.e.

$$\bar{c}_{ij} = \mathbf{E}c_{ij},$$

$$\sigma_{ij}^2 = \mathbf{D}^2(c_{ij}).$$

Hence, the mean value of the variable

$$z_i = \bar{F}_i - C_i x = \bar{F}_i - \sum_{j=1}^{n} c_{ij} x_j$$

is

$$\bar{z}_i = \bar{F}_i - \sum_{j=1}^{n} \bar{c}_{ij} x_j$$

and the variance is

$$\sigma_{z_i}^2 = \sum_{j=1}^{n} \sigma_{ij}^2 x_j^2.$$

One writes

$$\mathbf{E}\left(\sum_{i=1}^{r} |\bar{F}_i - C_i x|^2\right) = \mathbf{E}\left(\sum_{i=1}^{r} z_i^2\right) = \sum_{i=1}^{r} \mathbf{E}(z_i^2) =$$

$$= \sum_{i=1}^{r} (\bar{z}_i^2 + \sigma_{z_i}^2) = \sum_{i=1}^{r}\left(\bar{F}_i - \sum_{j=1}^{n} \bar{c}_{ij} x_j\right)^2 + \sum_{i=1}^{r}\sum_{j=1}^{n} \sigma_{ij}^2 x_j^2 =$$

$$= \sum_{i=1}^{r}\left(\bar{F}_i - \sum_{j=1}^{n} \bar{c}_{ij} x_j\right)^2 + \sum_{j=1}^{n}\left(\sum_{i=1}^{r} \sigma_{ij}\right)^2 x_j^2.$$

Then one can state

THEOREM 3.5. *The optimal solution of model* (3.15) *is obtained by solving the quadratic programming problem*

$$\min\left[\sum_{i=1}^{r}\left(\bar{F}_i - \sum_{j=1}^{n} \bar{c}_{ij} x_j\right)^2 + \sum_{j=1}^{n}\left(\sum_{i=1}^{r} \sigma_{ij}\right)^2 x_j^2\right],$$

subject to

$$Ax = b, \; x \geqslant 0.$$

Additionally, let $p_j = \left(\sum_{i=1}^{r} \sigma_{ij}\right)^2$ and the following two matrices

$$D = \begin{pmatrix} & C & \\ p_1 & & \\ \cdots & \cdots & \cdots \\ & p_j & \\ \cdots & \cdots & \cdots \\ & & p_n \end{pmatrix}, \qquad H = \begin{pmatrix} F \\ \left.\begin{matrix} 0 \\ \vdots \\ 0 \end{matrix}\right\} n \text{ elements} \end{pmatrix}.$$

It can be easily verified that

$$\left[\sum_{i=1}^{r}\left(\bar{F}_i - \sum_{j=1}^{n} \bar{c}_{ij} x_j\right)^2 + \sum_{j=1}^{n}\left(\sum_{i=1}^{r} \sigma_{ij}\right)^2 x_j^2\right] = \|Dx - H\|_2.$$

Thus, model (3.15) reduces to

$$\min_x \|Dx - H\|_2$$

subject to

$$Ax = b, \; x \geqslant 0.$$

As shown in Chapter 1 this model can be solved by use of the method of generalized inverses.

3.3.3. Let us reconsider now the problem in 3.3.1 in a different version.

Assume that there exists a fixed level $\bar{F} = (\bar{F}_1, \ldots, \bar{F}_r)$ to be reached by the objective functions contained in the vector $F(x) = (F_1(x), \ldots, F_r(x))$.

In 3.3.1 we assumed that a domain Y^* is given such that $\bar{F} \in Y^*$ and then tried to determine x such that the probability that $F \in Y^*$ is maximized.

Now we shall adopt a dual approach. Consider that the elements of C are random. Choose a domain Y^* such that

$$Y^* = \{(y_1, \ldots, y_r) \in R^r \mid \varepsilon_{1i} \leqslant y_i \leqslant \varepsilon_{2i} \; i = 1, \ldots, r\}$$

and set two limiting thresholds for the probability that $F(x) \in Y^*$

$$\alpha_1 = (\alpha_{1i}) \qquad (i = 1, \ldots, r)$$

$$\alpha_2 = (\alpha_{2i}) \qquad (i = 1, \ldots, r).$$

We have to determine the extent to which Y^* can be decreased provided that the probability $F(x) \in Y^*$ ranges within the given limits. Thus, the problem is

$$\text{(3.16)} \quad \min \|\varepsilon_2 - \varepsilon_1\|$$

subject to

$$P\{C_i x \geqslant \bar{F}_i - \varepsilon_{1i}\} \geqslant \alpha_{1i}, \qquad (i = 1, \ldots, r)$$

$$P\{C_i x \leqslant \bar{F}_i + \varepsilon_{2i}\} \geqslant \alpha_{2i}, \qquad (i = 1, \ldots, r)$$

$$\varepsilon_{1i}, \; \varepsilon_{2i} \geqslant 0$$

or

(3.16′) $\min \|\varepsilon_2 - \varepsilon_1\|$

subject to

$$P\{\bar{F}_i - C_i x \leqslant \varepsilon_{1i}\} \geqslant \alpha_{1i} \qquad (i = 1, \ldots, r)$$

$$P\{C_i x - \bar{F}_i \leqslant \varepsilon_{2i}\} \geqslant \alpha_{2i},$$

$$\varepsilon_{1i}, \varepsilon_{2i} \geqslant 0 \qquad (i = 1, \ldots, r)$$

Let $z_i = \bar{F}_i - C_i x = \bar{F}_i - \sum_{j=1}^{n} c_{ij} x_j$ and let $\bar{z}_i$, $\sigma_{z_i}^2$ be the mean and variance of z_i. The constraints of model (3.16) or of the equivalent model (3.16′) can be written as

$$\frac{\bar{F}_i - \varepsilon_{1i} - \bar{z}_i}{\sigma_{z_i}} \leqslant \Phi^{-1}(1 - \alpha_{1i}), \quad (i = 1, \ldots, r)$$

$$\frac{\bar{F}_i + \varepsilon_{2i} - \bar{z}_i}{\sigma_{z_i}} \geqslant \Phi^{-1}(\alpha_{2i}), \qquad (i = 1, \ldots, r).$$

The model (3.16) can then be written as

$$\min \|\varepsilon_2 - \varepsilon_1\|$$

subject to

$$\frac{\bar{F}_i - \varepsilon_{1i} - \sum_{j=1}^{n} \bar{c}_{ij} x_j}{\left(\sum_{j=1}^{n} \sigma_{ij}^2 x_j^2\right)^{1/2}} \leqslant \Phi^{-1}(1 - \alpha_{1i}), \ (i = 1, \ldots, r)$$

$$\frac{\bar{F}_i + \varepsilon_{2i} - \sum_{j=1}^{n} \bar{c}_{ij} x_j}{\left(\sum_{j=1}^{n} \sigma_{ij}^2 x_j^2\right)^{1/2}} \geqslant \Phi^{-1}(\alpha_{2i}), \quad (i = 1, \ldots, r).$$

$$\varepsilon_{1i}, \varepsilon_{2i} \geqslant 0$$

If $\alpha_{1i} > \frac{1}{2}$ and $\alpha_{2i} > \frac{1}{2}$, then the constraints of the model become nonlinear and define a convex set.

Thus we have found a deterministic equivalent for this variant of goal programming.

3.3.4. In what follows we shall assume that the functions F_i are well determined, but the thresholds $\bar{F}$ are not specified at the moment of decision. Assume that the elements of $\bar{F}$ are independent random variables with known distributions $F_i(\cdot)$, $(i = 1, \ldots, r)$.

Consider that [13]

(3.17) $\min \|\varepsilon\|$

is subject to

$$P\{C_i x \geqslant \bar{F}_i - \varepsilon_i\} \geqslant \alpha_{1i}, \qquad (i = 1, \ldots, r)$$

$$P\{C_i x < \bar{F}_i + \varepsilon_i\} \geqslant \alpha_{2i}, \qquad (i = 1, \ldots, r)$$

$$\varepsilon \geqslant 0$$

where α_{1i}, α_{2i} are the lower bounds of the probabilities and $\varepsilon = (\varepsilon_1, \varepsilon_2, \ldots, \varepsilon_r)$. Obviously, $0 \leqslant \alpha_{ji} \leqslant 1$, $(j = 1, 2;$ $i = 1, \ldots, r)$. Let

$$F_i^{-1}(\theta) = \inf \{\eta : F_i(\eta) \geqslant \theta\}$$

and

$$\hat{F}_i^{-1}(\theta) = \sup \{\eta : F_i(\eta) \leqslant \theta\}, \qquad 0 \leqslant \theta \leqslant 1.$$

Introduce the vectors

$$F^{-1}(\alpha_1) = (F_i^{-1}(\alpha_{1i})), \qquad (i = 1, \ldots, r),$$

$$\hat{F}^{-1}(e - \alpha_2) = (\hat{F}_i^{-1}(1 - \alpha_{2i})), \qquad (i = 1, \ldots, r)$$

where e is an r-order vector whose components are all equal to 1.

Let us find the deterministic equivalent of (3.17). Write

$$P\{C_i x \geqslant \overline{F}_i - \varepsilon_i\} = P\{\overline{F}_i \leqslant C_i x + \varepsilon_i\} =$$

$$= F_i(C_i x + \varepsilon_i) \geqslant \alpha_{1i}$$

or

$$C_i x + \varepsilon_i \geqslant F_i^{-1}(\alpha_{1i}).$$

Similarly,

$$P\{C_i x < \overline{F}_i + \varepsilon_i\} = P\{\overline{F}_i > C_i x - \varepsilon_i\} = e -$$

$$- F_i(C_i x - \varepsilon_i)$$

or

$$C_i x - \varepsilon_i \leqslant \hat{F}_i^{-1}(e - \alpha_{2i}).$$

Hence the deterministic equivalent to (3.17) is

(3.18.1) $\min \|\varepsilon\|$,

subject to

(3.18.2) $C_i x + \varepsilon_i \geqslant F_i^{-1}(\alpha_{1i})$,

(3.18.3) $C_i x - \varepsilon_i \leqslant \hat{F}_i^{-1}(e - \alpha_{2i})$,

(3.18.4) $\varepsilon_i \geqslant 0$. $\qquad i = 1, \ldots, r$

Subtracting (3.18.3) from (3.18.2), we obtain

(3.18′) $\varepsilon \geqslant \dfrac{1}{2}[F^{-1}(\alpha_1) - \hat{F}^{-1}(e - \alpha_2)] = \varepsilon_0.$

Observe that if $\varepsilon_0 \geqslant 0$, restriction (3.18.4) is redundant. One has $\varepsilon_0 \geqslant 0$ only if $\alpha_1 > e - \alpha_2$, i.e.

$$\alpha_{1i} + \alpha_{2i} > 1, \qquad (i = 1, \ldots, r).$$

This condition is ordinarily satisfied in practice since α_{1i} and α_{2i} are chosen such that their values are close to 1.

Then (3.18′) yields

$$\varepsilon = \varepsilon_0 + \delta, \; \delta \geqslant 0\text{-}$$

Introducing this expression in (3.18.2) and (3.18.3) yields

$$-\delta \leqslant C_i x - \delta_0 \leqslant \delta,$$

where

$$\delta_0 = \frac{1}{2}[F^{-1}(\alpha_1) + \hat{F}^{-1}(e - \alpha_2)] \text{ and } \delta \geqslant 0.$$

Thus we have proved the following

THEOREM 3.6. *If* $\alpha_1 > e - \alpha_2$ *and if* $\bar{F}_i$ *are independent random variables, then problem* (3.17) *is equivalent to the following deterministic problem*

$$\min \|\varepsilon_0 + \delta\|,$$

subject to

$$-\delta \leqslant Cx - \delta_0 \leqslant \delta,$$

$$\delta \geqslant 0.$$

COROLLARY. *If the norm* $\|\cdot\|$ *is monotonous (i.e. if for each* $x, y \in R^n \, |x| \leqslant |y| \Rightarrow \|x\| \leqslant \|y\|$; $|x|$ *is the vector* $(|x_i|)$ $i = 1, \ldots$ $\ldots, n$), *then :*

a) *The value* $\|\varepsilon_0\|$ *is a lower bound to the optimal solution of problem* (3.17);

b) *If the system*

$$Cx = \delta_0$$

is compatible, then its solution is an optimal solution of (3.17) *with optimal error* ε_0 *and optimal value given by* $\|\varepsilon_0\|$.

3.4. GROUP DECISION–MAKING IN STOCHASTIC PROGRAMMING

Consider the r-objective function stochastic programming problem

$$(3.19)\quad \min Z_k(\omega, x) = c_k'(\omega)x, \qquad (k = 1, \ldots, r),$$

subject to

$$Ax \leqslant b;\ x \geqslant 0,$$

where $c_k(\omega) = (c_{k1}(\omega), \ldots, c_{kn}(\omega))$, $(k = 1, \ldots, r)$.

For a given $\omega \in \Omega$, the above problem becomes an r-objective function deterministic problem.

One can construct a synthesis function by the weighted sum of the r functions using the weights $\alpha_k (k = 1, \ldots, r)$:

$$F^* = \sum_{k=1}^{r} \alpha_k Z_k = \sum_{k=1}^{r} \alpha_k \sum_{j=1}^{n} c_{kj} x_j = \sum_{j=1}^{n} \left(\sum_{k=1}^{r} \alpha_k c_{kj} \right) x_j =$$

$$= \sum_{j=1}^{n} c_j(\alpha) x_j = c'(\alpha) x.$$

The weights α_k are usually normalized such that

$$0 \leqslant \alpha_k \leqslant 1,\ \sum_{k=1}^{r} \alpha_k = 1.$$

Hence, the problem becomes

$$(3.19')\quad \min_{x \in D} \sum_{k=1}^{r} \alpha_k c_k'(\omega)\, x = \min_{x \in D} \sum_{k=1}^{r} \alpha_k Z_k(\omega, x).$$

In practice, the above summation is often meaningless, since the functions Z_k represent entirely different magnitudes. To overcome this difficulty one may transform the functions Z_k

into utility functions through the von Neumann-Morgenstern method. Thus, one writes

$$Z'_k(\omega, x) = \alpha_k Z_k(\omega, x) + \beta_k,$$

where α_k and β_k are given as the solutions of the linear set

$$\alpha_k X_k + \beta_k = 1$$

$$\alpha_k Y_k + \beta_k = 0,$$

where

$$X_k = \underset{\omega \in \Omega,\, x \in D}{\text{infinimum}}\ Z_k(\omega, x)$$

$$Y_k = \underset{\omega \in \Omega,\, x \in D}{\text{supremum}}\ Z_k(\omega, x).$$

The r utility functions Z'_k can be algebraically summated and hence one can maximize a synthesis function, i.e.

$$\max_x Z^*(\omega, x) = \sum_{k=1}^{n} \alpha_k Z'_k(\omega, x).$$

Thus the r-objective function stochastic programming problem has been reduced to a mathematical programming problem with one stochastic objective function.

One should also note that in a group decision-making process the weights α_k are random variables since the members of the group may assign various weights to individual objective functions. Without loss of generality, in the following argument we shall assume that the coefficients c_k are deterministic and the weights α_k are random variables. Note that these weights may be the result of consultations with one or more individuals in a decision-making group.

If the group is sufficiently small, then α_k can be the mean values of the weights given by each individual. But if the group is rather large, it becomes impossible to consult with each member in turn. If the decision-making collectivity is large enough one has to deal with the so-called welfare functions. Thus, one can assume that α_k are random variables.

Let $c(\alpha) = \sum_{k=1}^{r} \alpha_k c_k$ and let $F_\alpha(\cdot)$ be the distribution function of the random vector $\alpha = (\alpha_1, \ldots, \alpha_r)$. If the weights α_k are known then, according to Theorem 1.5 (Chapter 1, Section 1.2), the solutions of problem

$$\text{(3.19'')} \qquad \underset{x \in D}{\text{optimum}}\, f(x) = \sum_{k=1}^{r} \alpha_k Z_k$$

are efficient solutions.

Knowledge of the distribution function $F_\alpha(\cdot)$ can be achieved in two ways:

1) by use of the statistic analysis

2) or by assuming that $F_\alpha(\cdot)$ is an *a priori* distribution of the weights α_k.

The weights α_k can be also established through the Simulation Method [42]. One can imagine the weights α_k to be generated by a random vector of Dirichlet distribution having the probability density

$$g(\alpha_1, \ldots, \alpha_r) = \frac{\Gamma(\nu_1 + \ldots + \nu_{r+1})}{\Gamma(\nu_1) \ldots \Gamma(\nu_{r+1})} \times$$

$$\times\, \alpha_1^{\nu_r - 1} \ldots \alpha_r^{\nu_r - 1}(1 - \alpha_1 - \ldots - \alpha_r)^{\nu_{r+1} - 1}$$

if

$$(\alpha_1, \ldots, \alpha_r) \in S_r = \{(\alpha_1, \ldots, \alpha_r) \mid \alpha_i \geqslant 0,$$

$$1 \leqslant i \leqslant r;\ \sum_{i=1}^{r} \alpha_i \leqslant 1\}$$

and

$$g(\alpha_1, \ldots, \alpha_r) = 0 \text{ if } (\alpha_1, \ldots, \alpha_r) \notin S_r.$$

Let us assume that only the most probable values of α_i are known, i.e. $M(\alpha_i)$ $(i = 1, \ldots, r)$. One has to solve an indeterminate linear system in $\nu_1, \ldots, \nu_r, \nu_{r+1}$.

$$M(\alpha_1) = \frac{\nu_1}{\nu_1 + \nu_2 + \ldots + \nu_{r+1}}, \ldots, M(\alpha_r) =$$

$$= \frac{\nu_r}{\nu_1 + \nu_2 + \ldots + \nu_{r+1}}.$$

The values $M(\alpha_i)$ are assumed to be the most probable (most frequent) values and they are to be given by the user. The parameters ν_i ($i = 1, 2, \ldots, r + 1$) are real positive values that can be obtained by solving the above system. Using ν_i one then generates the random variables δ_i having Gamma-distribution and a common parameter $\lambda = 1$.

If $\delta_1, \ldots, \delta_{r+1}$ are independent random variables of Gamma type with $\lambda = 1$ and parameters $\nu_1, \ldots, \nu_{r+1}$, then the vector $\alpha = (\alpha_1, \ldots, \alpha_r)$ where the components are

$$\alpha_1 = \frac{\delta_1}{\delta_1 + \ldots + \delta_{r+1}}, \ldots, \alpha_r = \frac{\delta_r}{\delta_1 + \ldots + \delta_{r+1}}$$

will be a random vector of Dirichlet distribution. This vector may be assumed to be the weight vector required to solve the multiple objective problem (3.19″).

In practice, to each generation of $\alpha = (\alpha_1, \ldots, \alpha_r)$, there corresponds a different solution of (3.19″). If the decision-maker is not satisfied with the solution obtained, one generates another $\alpha = (\alpha_1, \ldots, \alpha_r)$ until the satisfactory solution is obtained. If the solution obtained is not satisfactory, the decision-maker can modify the values of $M(\alpha_i)$ ($i = 1, \ldots, r$) in order to generate further solutions.

The repeated simulation procedure presented above is associated to the game-type simulation methods.

For mathematical programming problems with random vector α one can consider the following problems :

1) Find the minimum-risk solution corresponding to level u, i.e. the solution which maximizes the function

$$(3.20)\quad P\{\alpha \mid c(\alpha)x \geqslant u,\ x \in D\}.$$

2) Determine the distribution function of the random variable

$$(3.21) \quad \xi(\alpha) = \max \{c(\alpha)x \mid x \in D\}$$

given the distribution $F_\alpha(\cdot)$ and the values of the probabilities that various possible bases are optimal.

Bereanu shows [23] that under certain conditions the optimal solution of problem (3.20) is independent of the distribution of α or of the value of u.

DEFINITION 3.1. If the solution of (3.20) does not depend on $F_\alpha(\cdot)$, then this is the *distribution-free minimum-risk solution* (d.f.m.r.s.). If it does not depend on u either, it is called *free minimum-risk solution* (f.m.r.s).

DEFINITION 3.2. A basis of D is said to be *stable optimal basis* if it is optimal with probability 1.

If the base does not depend on $F_\alpha(\cdot)$ either, it is called *distribution-free optimal basis.*

A *distribution-free optimal solution* (d.f.o.s) is a solution corresponding to a distribution-free optimal basis. The following theorem establishes the connection between the existence of an f.m.r.s and a d.f.o.s.

THEOREM 3.7 (Bereanu [23]). *Let $\emptyset \neq C \subset R^n$ be a non-empty convex set, let $\mathscr{P}_C$ be the set of probability measures on R^n such that $P \in \mathscr{P}_C$ implies $P(C) = 1$ and let $\mathscr{C}(C)$ be the set of random vectors. Then, for any $c(\alpha) \in \mathscr{C}(C)$, the following statements are equivalent:*

1) $x^* \in D$ *is an* f.m.r.s. *for* $\max P\{\alpha \mid c(\alpha)x \geqslant 0,\ x \in D\}$;

2) $x^* \in D$ *is a* d.f.o.s. *for* $\xi(\alpha) = \sup \{c(\alpha)x \mid x \in D\}$.

DEFINITION 3.3. A system of weights $\alpha = (\alpha_1, \alpha_2, \ldots, \alpha_r)$ is *k-stable* if there exists $\{i_1, \ldots, i_k\} \subset \{1, 2, \ldots, r\}$ such that

$$\sum_{j=1}^{r} \alpha_j c_{ji} \geqslant 0 \text{ if } \sum_{j=1}^{r} c_{ji} \geqslant 0, \qquad i \in \{i_1, \ldots, i_k\}$$

and

$$\sum_{j=1}^{r} \alpha_j c_{ji} \leqslant 0 \text{ if } \sum_{j=1}^{r} c_{ji} \leqslant 0, \qquad i \notin \{i_1, \ldots, i_k\}.$$

Let O_n denote the row n-vector whose elements are all equal to zero, I_n the unity n-matrix, e_i the i-th column of matrix I_n;

$$E_k^n = (e_1, \ldots, e_k, -e_{k+1}, \ldots, -e_n); \; \bar{A}' = (A', -I_n);$$

$$\bar{b}' = (b', O_n);$$

$$R_k^n = \{x \mid x \in R^n, x_i \geqslant 0 \; i = 1, \ldots, k, x_k \leqslant 0, j = $$

$$= k+1, \ldots, n\}, \; k \leqslant n.$$

THEOREM 3.8 (Bereanu [23]). *Problem* (3.20) *where* α *is* k*-stable admits of a free minimum-risk solution* (*f.m.r.s*) x^* *if and only if there exists an* $n \times (m+n)$ *matrix* $A^* \geqslant 0$ *such that* $A^*\bar{A} = E_k^n$ *and* $x^* = E_k^n A^* \bar{b}$.

Proof. Sufficiency. Let $e_{i,k} = e_i$ if $i \leqslant k$ and $e_{i,k} = -e_i$ if $i > k$. The row i of matrix A^* is an optimal solution of the problem

$$\min \{y\bar{b} \mid y\bar{A} = e_{i,k}, y \geqslant 0\}$$

and x^* is an optimal solution of

$$\max \{e_{i,k}x \mid A\bar{x} \leqslant \bar{b}\}, \qquad (i = 1, \ldots, n).$$

For a given $x \in D$, $x_i^* \geqslant x_i$ if $i \leqslant k$ and $x_i^* \leqslant x_i$ if $i > k$. Let $\alpha' \in V(x) = \{\alpha \mid c(\alpha)x \geqslant u\} \neq \emptyset$. Then $c(\alpha)x^* \geqslant c(\alpha)x$, i.e. $\alpha' \in V(x^*)$. Hence, $V(x) \subset V(x^*)$ and thus $P(V(x^*)) \geqslant P(V(x))$. Since x and α were arbitrarily chosen, the sufficiency is proved.

Necessity. Suppose that there exists a free minimum-risk solution x^*. By virtue of Theorem 3.7, this is a d.f.o.s. with respect to the same family of random weights. The solution x^* maximizes $e_{i,k}x$ on $\bar{D} = \{x \mid \bar{A}x \leqslant \bar{b}\}$ for each i. From the duality theorem it follows that for each i there exists a vector $y^i \geqslant 0$ such that

$$y^i\bar{A} = e_{i,k} \text{ and } y^i\bar{b} = e_{i,k}x^*.$$

The matrix A^* has y^i as the i-th row. The proof is complete.

THEOREM 3.9. (Bereanu [23]). *Problem* (3.20) *where* α *is*

k-stable admits of a free minimum-risk solution $x^* = B^{-1}b_B$, where B is an $n \times n$ nonsingular submatrix of A if $(E_k^n B)^{-1} \geqslant 0$ and B' is an optimal basis of the program:

$$(3.22) \qquad \min \{\bar{b}'y \mid \bar{A}'y = \bar{c}',\ y \geqslant 0\},$$

$$\bar{c} \in \operatorname{int} R_k^n.$$

Proof. According to Theorem 3.8, it is sufficient to show that $x^* = B^{-1}b_B$ is an optimal solution of

$$\max \{cx \mid x \in D\} \qquad \forall c \in R_k^n.$$

Since $B'^{-1}\bar{c} \geqslant 0$ and $\bar{c} \in \operatorname{Int} R_k^n$, it follows that $B'^{-1}c \geqslant 0$ $\forall c \in R^n$ and $y'(c) = (cB^{-1}, 0_m)$ is an optimal solution for (3.22) $\forall c \in R_k^n$. Since $b'y(c) = cB^{-1}b_B = cx^*$ one has to show that $x^* \in D$. But this is obvious since $(\bar{A}x^* - b)' = b_B'B'^{-1}\bar{A}' - \bar{b} \leqslant \leqslant 0$, where B' is an optimal basis to problem (3.22).

Bereanu showed [23] that the stability of the coefficients is an essential condition for the existence of a free minimum-risk solution.

In Chapter 1 we defined the set

$$\Lambda = \left\{(\lambda_1, \ldots, \lambda_r) \mid 0 < \lambda_i < 1, i=1, \ldots, r, \sum_{i=1}^{r} \lambda_i = 1\right\}.$$

As shown in Theorem 1.5, x is an efficient point of the maximization problem with multiple objective functions $\max_{x \in D} Cx$ (C is an $r \times n$ matrix) if and only if there exists a $\lambda \in \Lambda$ such that x is an optimal solution of the associated linear problem, the objective function of which is a weighted combination of the r objective functions $P_\lambda : \max_{x \in D} \lambda' Cx$.

The parameter space Λ can be decomposed (Theorem 1.6) into a finite number of polyhedra $\Lambda(x_i)$ corresponding to the different nondominated extreme points of the feasible set D.

The following problem is closely connected with the structure of the set Λ [125]: given a feasible base solution, what is the probability that it is also optimal? Equivalently, given an optimal base solution, what is the probability that it is associated with a subset of Λ?

Obviously, the notion of probability is here used in the geometrical sense, since it deals with length, volume, etc. Before considering the theorem to answer the above mentioned questions, let us make several remarks: For a given $\Lambda \subset E^r$, one can define a σ-ring F of all Lebesgue measurable subsets of Λ.

Let $m(\cdot)$ be the corresponding Lebesgue measure. Hence, one can deal with the measurable space $\{\Lambda, F, m(\cdot)\}$.

Denote by $l = m(\Lambda)$ the Lebesgue measure of the whole space Λ. Define another measure $\mu(\cdot) = \dfrac{m(\cdot)}{l}$ and let $\{\Lambda, F, \mu(\cdot)\}$ be the corresponding space. Since $\mu(\Lambda) = 1$, it follows that $\{\Lambda, F, \mu(\cdot)\}$ can be used as a probability space.

THEOREM 3.10. *Let* $\Lambda = \{(\lambda_1, \ldots, \lambda_r) : \Sigma\, \lambda_i = 1,\ 0 < \lambda_i < 1\}$, *let* Λ_G *be a Lebesgue measurable subset of* Λ *and let* x_B *be a feasible base solution of*

$$\max \left\{ \sum_{i=1}^{r} \lambda_i F_i(x) \,|\, Ax = b,\ x \geqslant 0,\ \lambda_i > 0,\ \sum \lambda_i = 1 \right\}.$$

Then the following probabilities exist:

1) $P(x_B$ *is optimal*$)$
2) $P(x_B$ *is optimal* $|\lambda \in \Lambda_G)$
3) $P(\lambda \in \Lambda_G |\, x_B$ *is optimal*$)$.

Proof. We have proved that the sets $\Lambda(x_B)$ are convex and separated by boundary hyperplanes (Theorem 1.6, Chapter 1, Section 1.2). Hence, the measures $m(\Lambda(x_B))$ and $m(\Lambda_G \cap \Lambda(x_B))$ exist. Denote by Λ_G^* the projection of Λ_G on E^{r-1}. According to the definition of the Lebesgue measure and to a result of ordinary calculation [57]

$$m(\Lambda_G) = \int_{\Lambda_G^*} [1 + (\partial x/\partial \lambda_1)^2 + \ldots + (\partial x/\partial \lambda_{r-1})^2]^{1/2}\, d\lambda_1 \ldots d\lambda_{r-1}.$$

Denote $x \equiv \lambda_1 + \ldots + \lambda_{r-1} - 1$, and hence

$$m(\Lambda_G) = \sqrt{r} \int_{\Lambda_G^*} d\lambda_1 d\lambda_2 \ldots d\lambda_{r-1}.$$

Similarly,

$$m(\Lambda(x_B)) = \sqrt{r} \int_{\Lambda^*(x_B)} d\lambda_1 d\lambda_2 \ldots d\lambda_{r-1},$$

$$m(\Lambda(x_B) \cap \Lambda_G) = \sqrt{r} \int_{\Lambda^*(x_B) \cap \Lambda_G^*} d\lambda_1 d\lambda_2 \ldots d\lambda_{r-1}.$$

Under the above conditions:

$$P(x_B \text{ is optimal}) = P(\lambda \in \Lambda(x_B)) = \mu(\Lambda(x_B)) = m(\Lambda(x_B))/m(\Lambda)$$

$$P(x_B \text{ is optimal} \mid \lambda \in \Lambda_G) = \mu(\Lambda(x_B) \cap \Lambda_G)/\mu(\Lambda_G) = \\ = \frac{m(\Lambda(x_B) \cap \Lambda_G)}{m(\Lambda_G)},$$

$$P(\lambda \in \Lambda_G \mid x_B \text{ is optimal}) = \mu(\Lambda_G \cap \Lambda(x_B))/\mu(\Lambda(x_B)) = \\ = m(\Lambda_G \cap \Lambda(x_B))/m(\Lambda(x_B)).$$

Example [125]. Consider the maximization of the following three functions

$$F_1 = 3.5x_1 + 3x_2$$
$$F_2 = -3x_1 + 5x_3$$
$$F_3 = 2x_1 + 4x_2 + x_3$$

subject to

$$4x_1 + 4.25x_2 + 4.5x_3 \leqslant 12 \cdot 10^6$$
$$x_3 \leqslant 2 \cdot 10^6$$
$$4x_1 - 4.25x_2 \leqslant 2 \cdot 10^6$$
$$x_1 + x_2 + x_3 \geqslant 2.5 \cdot 10^6.$$

Consider the synthesis function

$$F = \lambda_1 F_1 + \lambda_2 F_2 + (1 - \lambda_1 - \lambda_2) F_3,$$

where $\lambda_1, \lambda_2 > 0$ and $\lambda_1 + \lambda_2 < 1$.

Hence,

$$F = (1.5\lambda_1 - 5\lambda_2 + 2)x_1 + (-\lambda_1 - 4\lambda_2 + 4)x_2 + $$
$$+ (-\lambda_1 + 4\lambda_2 + 1)x_3.$$

An initial base solution can be obtained by use of the Simplex algorithm

$$x_B^1 = \{x_2 = 2.5 \cdot 10^6;\ x_4 = 1.375 \cdot 10^6;\ x_5 = 2 \cdot 10^6;$$
$$x_6 = 12.625 \cdot 10^6\}.$$

But the inequality set

$$z_1 - c_1 \geqslant 0;\ z_3 - c_3 \geqslant 0;\ z_7 - c_7 \geqslant 0$$

is incompatible, i.e. there exists no λ such that x_B^1 is optimal.

Another base solution is

$$x_B^2 = \{x_2 = 2.85 \cdot 10^6;\ x_5 = 2 \cdot 10^6;\ x_6 = 14 \cdot 10^6;$$
$$x_7 = 0.32 \cdot 10^6\}.$$

This solution is optimal for the values of λ satisfying the set of constraints

$$z_1 - c_1 \geqslant 0;\ z_3 - c_3 \geqslant 0;\ z_4 - c_4 \geqslant 0;\ 0 < \lambda_1,\ \lambda_2 < 1,\ \lambda_1 + \lambda_2 < 1.$$

Such values make up the set $\Lambda(x_B^2)$ shown in fig. 3.1 in the plane (λ_1, λ_2) or fig. 3.2 in the space $(\lambda_1, \lambda_2, \lambda_3)$.

The next base solution is

$$x_B^3 = \{x_1 = 1.75 \cdot 10^6;\ x_2 = 1.1735 \cdot 10^6;\ x_5 = 2 \cdot 10^6;$$
$$x_7 = 0.4244 \cdot 10^6\}.$$

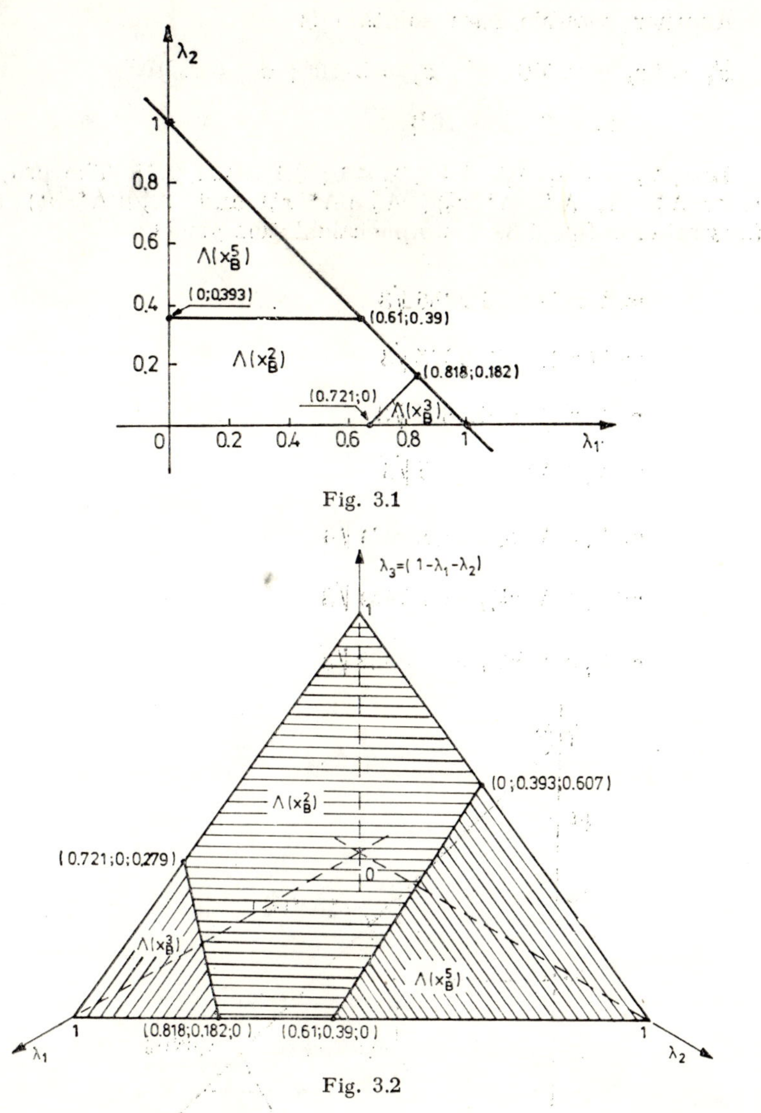

Fig. 3.1

Fig. 3.2

To this solution there corresponds the parameters domain $\Lambda(x_B^3)$. As for x_B^4, there exists no value of λ for which the base solution $x_B^4 = (x_2, x_3, x_5, x_7)$ is optimal.

Another possible base solution is

$$x_B^5 = \{x_2 = 0.706 \cdot 10^6;\ x_3 = 2 \cdot 10^6;\ x_6 = 10 \cdot 10^6;$$
$$x_7 = 0.2058 \cdot 10^6\}.$$

Let $\Lambda_G = \{(\lambda_1, \lambda_2)\ 0.4 \leqslant \lambda_1 \leqslant 1;\ 0.1 \leqslant \lambda_2 \leqslant 0.7\}$. The projections $\Lambda_G^* \cap \Lambda$, $\Lambda_G^* \cap \Lambda^*(x_B^2)$; $\Lambda_G^* \cap \Lambda^*(x_B^3)$ and $\Lambda_G^* \cap \Lambda^*(x_B^5)$ are illustrated in fig. 3.3. A simple calculation yields :

$$m(\Lambda(x_B^2)) = 0.2886\sqrt{3}$$

$$m(\Lambda(x_B^3)) = 0.0275\sqrt{3}$$

$$m(\Lambda(x_B^5)) = 0.1857\sqrt{3}$$

$$m(\Lambda_G \cap \Lambda) = 0.125\sqrt{3}$$

$$m(\Lambda_G \cap \Lambda(x_B^2)) = 0.0974\sqrt{3}$$

$$m(\Lambda_G \cap \Lambda(x_B^3)) = 0.0054\sqrt{3}$$

$$m(\Lambda_G \cap \Lambda(x_B^5)) = 0.0222\sqrt{3}.$$

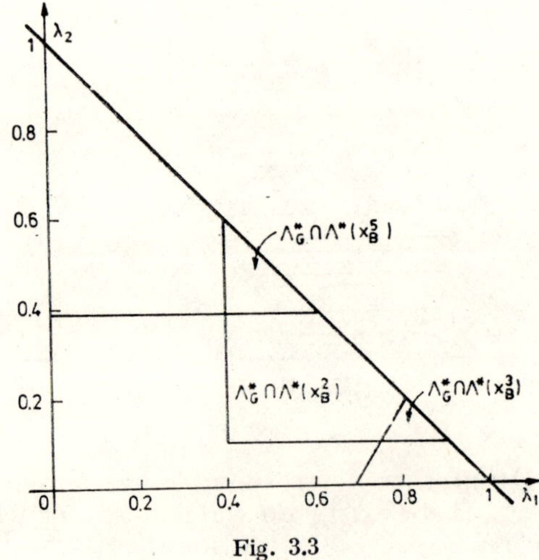

Fig. 3.3

Now one can calculate the probabilities used in Theorem 3.10.

$P(x_B^2 \text{ optimal} \mid \lambda \in \Lambda_G) = \mu(\Lambda_G \cap \Lambda(x_B^2))/\mu(\Lambda_G \cap \Lambda) = 0.776$

$P(\lambda \in \Lambda_G \mid x_B^3 \text{ optimal}) = \mu(\Lambda_G \cap \Lambda(x_B^3))/\mu(\Lambda(x_B^3)) = 0.2$

$P(x_B^3 \text{ optimal}) = m(\Lambda(x_B^3))/m(\Lambda) = 0.027\sqrt{3}/\sqrt{3}/2 = 0.054.$

3.5. EFFICIENT SOLUTIONS IN STOCHASTIC PROGRAMMING

Let us now consider the stochastic programming problem with one objective function

$$(3.23)\quad \gamma(\omega) = \max_{x \in D} c'(\omega)\, x$$

subject to

$$Ax \leqslant b,\; x \geqslant 0,$$

where the constraint coefficients are deterministic and the vector c is random.

As was shown in Chapter 2, the expression $\max c'(\omega)\,x$ becomes meaningless if c becomes random. Thus, we shall introduce the concept of *efficiency* which is the analog of optimality in deterministic mathematical programming with multiple objective functions.

DEFINITION 3.4. A point $x^0 \in D$ is said to be *efficient with probability* 1 if there exists no other vector $x^1 \in D$ almost surely as good as x^0 (with respect to the objective function $c'(\omega)x$) and even better than x^0 with a positive probability.

That is, $x^0 \in D$ is efficient with probability 1 if there exists no $x^1 \in D$ such that

$$P\{\omega \mid c'(\omega)x^0 \leqslant c'(\omega)x^1\} = 1$$

and

$$P\{\omega \mid c'(\omega)x^0 < c'(\omega)x^1\} > 0.$$

We must find the set of points $D_1 \subset D$ which are efficient with probability 1 or at least another suitable subset of D_1.

On the other hand, Tammer [240] assumed that the objective functions are random

$$f(x, y(\omega)) = f_0(x) + \sum_{i=1}^{m} y_i(\omega) f_i(x),$$

where $y(\omega) = (y_1(\omega), \ldots, y_m(\omega))$ is a random vector, and defined the notion of "optimality" as in

DEFINITION 3.5. A point $x^0 \in D$ is said to be ε-efficient (efficient with the probability $1 - \varepsilon$) for problem

$$(3.24)\ \max \{f(x, y(\omega)) | x \in D\},$$

if there exists no $x^1 \in D$ such that

$$P\{\omega | f(x^1, y(\omega)) \geqslant f(x^0, y(\omega))\} \geqslant 1 - \varepsilon$$

and

$$P\{\omega | f(x^1, y(\omega)) > f(x^0, y(\omega))\} > 0.$$

Note that the notion of ε-efficiency is a generalization of the notion of efficiency with probability 1. In order to determine the ε-efficient points one can use a result similar to Theorem 1.5, Chapter 1, Section 1.2. In ref. [240] it is shown that for certain parameters $\lambda \in R^m$, the optimal solutions of the parametric program

$$\max \{f(x, \lambda) | x \in D\}$$

are ε-efficient for the stochastic program (3.24). (For other variants of the notion of efficiency with probability 1, see also refs. [241] and [242]).

Thus, one can calculate all ε-efficient solutions of the stochastic program (3.24) by solving the corresponding parametric programs.

The notion of efficient solution of the stochastic programming problem (3.23) can be introduced in another way, i.e. by associating to it several deterministic problems for which the efficient solutions are those already defined in Chapter 1.

Assume that c has a normal distribution with mean value $\bar{c}$ and covariance matrix Σ. Then, $c'(\omega)x$ is also a normal variable, with mean value $\bar{c}'x$ and variance $x'\Sigma x$.

Thus, the problem can be reformulated in deterministic terms by associating to it one of the following functions :

1) The maximization of the mean value

$$\max_{x\in D} f_1 = \max_{x\in D} \mathbf{E}(c'(\omega)x) = \max_{x\in D} \bar{c}'x\,;$$

2) The minimization of the variance

$$\min_{x\in D} f_2 = \min_{x\in D} \mathbf{D}^2(c'(\omega)x) = \min_{x\in D} x'\Sigma x\,;$$

3) The minimization of the second-order moment

$$\min_{x\in D} f_3 = \min_{x\in D} \mathbf{E}(c'(\omega)x)^2.$$

Other efficiency criteria may be also considered :

4) Maximization of the probability that the value of the function exceeds a given level k (one is satisfied if the objective function takes at least the value k)

$$\max_{x\in D} f_4 = \max_{x\in D} P\{\omega\,|c'(\omega)x \geqslant k\}.$$

This is the minimum-risk model which was independently approached by Charnes and Cooper [38] and by Bereanu [16], [17], [18]. Geoffrion [73] calls it the "aspiration criterion".

5) $\max f_5 = \max f$ such that

$$P\{\omega\,|c'(\omega)x \geqslant f\} \geqslant \alpha.$$

This is the model used by Kataoka [91]. Geoffrion [73] calls it the "fractile criterion".

In what follows we shall show that these last two criteria are a direct consequence of various approaches to the deterministic mathematical programming with multiple objective functions.

In order to solve problem (3.23) we must choose the vector x before knowledge of the values of c is available so that both the mean and the mean square deviation of $c'x$ may be incorporated in the solution, i.e.

$$\max_{x\in D} f_1 = \operatorname{mac}_{x\in D} \bar{c}'x$$

and

$$\min_{x \in D} \tilde{f}_2 = \min_{x \in D} \sqrt{x'\Sigma x}.$$

The function f_1 is equivalent to

$$\min_{x \in D} \tilde{f}_1 = \min_{x \in D} (k - \bar{c}'x),$$

where k is a certain positive constant.

This is the deterministic case with two objective functions, one of which is linear and the other quadratic.

We adopt the definition given in ref. [136]. By the compromise solution x^* between $\tilde{f}_1$, and $\tilde{f}_2$, we understand that solution which minimizes their ratio, i.e.

$$\frac{\tilde{f}_1(x^*)}{\tilde{f}_2(x^*)} = \min_{x \in D} \frac{\tilde{f}_1(x)}{\tilde{f}_2(x)} = \min_{x \in D} \frac{k - \bar{c}'x}{\sqrt{x' \Sigma x}}.$$

This ratio is a nonlinear function. We have thus come to the same result as Bereanu and as Charnes and Cooper.

One can consider as a compromise solution that solution which maximizes a linear combination of f_1 and $\tilde{f}_2$ [136]:

$$\max_{x \in D} [\alpha_1 \bar{c}'x + \alpha_2 (x'\Sigma x)^{1/2}].$$

For $\alpha_1 = 1$ and $\alpha_2 = -\Phi^{-1}(\alpha)$ (Φ is the Laplace function), we again have Kataoka's model since $F_\alpha(c'x) = \bar{c}'x + $ $+ \Phi^{-1}(\alpha)(x'\Sigma x)^{1/2}$, where $F_\alpha(c'x)$ is the α-fractile of the probability distribution of $c'x$.

According to Markowitz [104], if one considers only two functions f_1 and f_3, then there exist parameters a, b and c such that the decisions can be taken on the basis of the function

$$h(f_1, f_3) = a + bf_1 + cf_3.$$

Since $f_2 = f_3 - f_1^2$, it follows from the above relation that

$$h(f_1, f_2) = a + bf_1 + cf_1^2 + cf_2 = h^*(f_1, f_2),$$

which is tantamount to stating that the decisions may be taken depending on the functions f_1 and f_2.

White [162] shows that if one seeks the least variance for a given mean value, then $c < 0$ in the previous expression.

Consider the following problems.

Problem I $\min f_2(x)$ subject to $x \in D$ and $f_1 = \bar{f}_1$,

Problem II $\min f_3(x)$ subject to $x \in D$ and $f_1 = \bar{f}_1$, where $\bar{f}_1$ is a certain mean value.

Since $f_2 = f_3 - f_1^2$, the two problems are equivalent.

We can consider the Lagrangian approach to these problems. For example, the Lagrange function for Problem II is

$$f_4^*(\lambda) = f_3(x) - \lambda f_1(x).$$

One can minimize $f_4^*(\lambda)$ for various values of λ. However, we shall apply the theory of efficient points which is closely connected to the Lagrange-multiplier method.

Let us recall the definition of the efficient points for functions f_1 and f_2.

DEFINITION 3.6. Point x is efficient with respect to functions f_1 and f_3, if there exists no x^* such that

$$f_1(x^*) \geqslant f_1(x)$$

$$f_3(x^*) \leqslant f_3(x)$$

and at least one of the inequalities is strict.

Accordingly, one can write

DEFINITION 3.7. Point x is efficient with respect to functions f_1 and f_2 if there exists no point x^* such that

$$f_1(x^*) \geqslant f_1(x)$$

$$f_2(x^*) \leqslant f_2(x)$$

and at least one of the inequalities is strict.

According to Karlin [90] and Geoffrion [74] let us assume that

$$Q(\alpha) = \max_{x \in D} [\alpha f_1(x) - (1 - \alpha) f_3(x)],$$

where $0 \leqslant \alpha \leqslant 1$, f_1 is concave and f_3 is convex. Let $x(\alpha)$ be its optimal solution.

If $0 < \alpha < 1$ then, according to Theorems 1.5, 1.9 and 1.10, $x(\alpha)$ is an efficient solution for f_1 and f_3.

If $\alpha = 1$ and there exist more then one optimal solution for $f_1(x)$,then the solutions for which $f_3(x)$ is minimum are efficient solutions (for f_1, f_3).

If α=0 and there exist more then one optimal solution for f_1, then the solutions for which $f_1(x)$ is largest are efficient solutions. Note that if $\lambda = \frac{-\alpha}{1-\alpha}$, then $f_4(\lambda) = \frac{-Q(x)}{1-\alpha}$ which proves the equivalence between the present method and that using Lagrange multipliers.

We give below several theorems characterizing the efficient solutions.

THEOREM 3.11. *Let x^* be an efficient point for (f_1, f_2). Then*

$$f_2(x^*) = \min_x f_2(x),$$

where the minimization is taken over all $x \in D$ such that $f_1(x) = f_1(x^*)$.

Proof. Assume that $f_2(x^*) \neq \min_x f_2(x)$, and hence there exists an $\hat{x} \neq x^*$ such that $f_1(\hat{x}) = f_1(x^*)$ and $f_2(\hat{x}) < f_2(x^*)$. This contradicts the fact that x^* is efficient with respect to (f_1, f_2).

THEOREM 3.12. *Assume that the minimum values in problems I and II are strictly increasing with respect to $\bar{f}_1$. Then x^* is efficient for (f_1, f_2) if and only if it is efficient for (f_1, f_3).*

Proof. Let x^* be efficient for (f_1, f_2). From Theorem 3.11,

$$f_2(x^*) = \min_{\substack{x \in D \\ f_1(x) = f_1(x^*)}} f_2(x).$$

But $f_2 = f_3 - f_1^2$ and hence

$$f_3(x^*) = \min_{\substack{x \in D \\ f_1(x) = f_1(x^*)}} f_3(x).$$

Assume that x^* is not efficient for (f_1, f_3). Hence, there exists an $\hat{x}$ efficient for (f_1, f_3) such that

$$(3.25) \quad \begin{aligned} f_1(\hat{x}) &\geqslant f_1(x^*), \\ f_3(\hat{x}) &\leqslant f_3(x^*), \end{aligned}$$

and at least one of the inequalities is strict.

Then,

$$f_3(x^*) = \min_{\substack{x \in D \\ f_1(x) = f_1(x^*)}} f_3(x) < \min_{\substack{x \in D \\ f_1(x) = f_1(\hat{x})}} f_3(x) = f_3(\hat{x}),$$

i.e. $f_3(x^*) < f_3(\hat{x})$ which contradicts (3.25). Hence x^* is efficient with respect to (f_1, f_3). The reciprocal relation can be proved in a similar manner.

THEOREM 3.13. *If the minimum values of problems* I *and* II *are strictly increasing with respect to* $\bar{f}_1$ *and if* $f_2(x^*) = \min f_2(x)$ *for* $x \in D$ *and* $f_1(x) = f_1(x^*)$, *then* x^* *is efficient with respect to* (f_1, f_2) *and* (f_1, f_3).

The proof is straightforward.

Now let us state and prove Theorem 3.12 under different conditions.

THEOREM 3.14. *Assume that* $f_1(x) \geqslant 0$ *for any* $x \in D$. *If* x^* *is efficient with respect to* (f_1, f_2), *then* x^* *is also efficient for* (f_1, f_3).

Proof. Assume that x^* is not efficient for (f_1, f_3), and hence there exists a point $\bar{x}$ such that

$$(3.26) \quad \begin{aligned} 0 \leqslant f_1(x^*) &\leqslant f_1(\bar{x}) \\ f_3(x^*) &\geqslant f_3(\bar{x}) \end{aligned}$$

and at least one of the inequalities is strict.

But

$$f_2(x^*) = f_3(x^*) - f_1^2(x^*) > f_3(\bar{x}) - f_1^2(\bar{x}) = f_2(\bar{x}).$$

Relations (3.26) imply that x^* is not efficient for (f_1, f_2).

The above statements, especially Theorem 3.13, show that the solution of the stochastic programming problem with two associated deterministic functions (i.e. finding its efficient points) is equivalent to the solution obtained by minimizing the variance subject to constraints on the mean value.

Consider now the following stochastic programming problem with several objective functions:

$$\max c_i'(\omega)x, \qquad i = 1, \ldots, r,$$

subject to

$$x \in D.$$

Assume that $F_i(z)$ is the cumulative distribution function of $c_i'(\omega)x$ and $z_i = z_i(x)$ ranges over $c_i'(\omega)x$. (In particular, $z_i(x)$ can be $\mathbf{E}(c_i'(\omega)x)$). Let $0 < \alpha < 1$ such that $P[c_i'(\omega)x \geqslant$ $\geqslant z_i(x)] = 1 - F_i(z(x)) = 1 - \alpha$.

The *set of stochastic, efficient solutions at level* α (denoted by $\mathbf{S}_\alpha$) is defined as follows [197].

$\mathbf{S}_\alpha = \{x \mid x \in D,\ F_i(z_i(x)) = \alpha,\ i = 1, \ldots, r,$ there exists no other $x' \in D$ such that $F_i(z_i(x')) = \alpha$, for all $i = 1, \ldots, r$ $z_i(x') \geqslant z_i(x)$ for all $i \in \{1, \ldots, r\}$ and at least for one i_0, $z_{i_0}(x') \geqslant z_{i_0}(x)\}$.

It is interesting to note that this definition is essentially that given in Chapter 1, Section 1.1, provided α remains constant.

3.6. THE PROTRADE METHOD

The PROTRADE (Probabilistic Tradeoff Development) method was introduced by Goicoechea, Duckstein and Bulfin [76] to solve stochastic programming problems with multiple objective functions. The method is also presented in ref. [197] and was successfully applied to the land reclamation and management problem of Black Mesa region, northern Arizona, USA.

This method can be considered to be the stochastic analog to the deterministic POP and STEM methods presented in Chapter 1. It is an interative method providing a so-called "dialogue" between the user and the model, the decision maker having the option of conducting the search process for the efficient solutions by modifying the initial conditions according to the partial results obtained.

During the iterations the decision-maker can improve upon the value already attained by one objective function as well as upon the probability of reaching the corresponding objective (or both).

Consider the mathematical programming problem with multiple objective functions

$$\max_{x \in D_0} Z(x) = (Z_1(x), Z_2(x), \ldots, Z_r(x)),$$

where

$$D_0 = \{x \mid x \in R^n, g_i(x) \leqslant 0, \; x > 0, \; i = 1, \ldots, m\}$$

$$Z_i' = \sum_{j=1}^{n} c_{ij} x_j,$$

$$Z_i(x) = \mathbf{E}(Z_i'(x))$$

$$c_{ij} = N[\mathbf{E}(c_{ij}), \; \mathrm{Var}(c_{ij})].$$

The functions $g_i(x)$ are convex and differentiable.

The PROTRADE method is carried out in the following steps:

I. For each function Z_i determine the maximum and minimum value, i.e. the co-domain of each function

$$Z_{i\max} = Z_i(x^{i*}) = \max_{x \in D_0} Z_i(x), \qquad (i = 1, \ldots, r),$$

and

$$Z_{i\min} = Z_i(y^{i*}) = \min_{x \in D_0} Z_i(x), \qquad (i = 1, \ldots, r).$$

Let

$$U_1 = (Z_1(x^{1*}), \ldots, Z_r(x^{r*}))$$

be the vector whose components are the maximum values of the objective functions.

II. Transform the functions Z_i such that their values belong to the interval $[0, 1]$:

$$G_i(x) = \frac{Z_i(x) - Z_{i\min}}{Z_{i\max} - Z_{i\min}}$$

and construct the auxiliary function

$$F(x) = \sum_{i=1}^{r} G_i(x).$$

III. Solve

$$\max_{x \in D_0} F(x).$$

Let X^1 be the optimal solution. Construct the vector

$$G_1 = (G_1(X^1), G_2(X^1), \ldots, G_r(X^1)),$$

whose components are the values of G_i for the solution X^1.

Note that in the first 3 steps we have applied the maximum global utility method to the mean-value functions.

IV. Estimate the multi-dimensional utility function $u(G)$ which represents the decision-maker's options.

For example, one can use the multiplicative utility function given by Fishburn [66] and Keeney [93]

$$1 + ku(G) = \prod_{i=1}^{r} [1 + kk_i u_i(G_i)],$$

where u and u_i are utility functions scaled from zero to one, k_i are the scaling factors, $0 < k_i < 1$, and $k > -1$ is another nonzero scaling factor.

V. Re-define the auxiliary function

$$S_1(x) = \sum_{i=1}^{r} w_i G_i(x),$$

where

$$w_i = 1 + \frac{r}{G_i(X^1)} \left[\frac{\partial u(G)}{\partial G_i}\right]_{G_1},$$

r is the step size which induces an increment $\Delta u(G)$ of the utility function.

In practice, one calculates $u(G_1)$, decides upon $0 \leqslant \Delta u(G) \leqslant 1$ and then determines r from

$$\Delta u(G) = u[G_1 + r\nabla u(G_1)] - u(G_1).$$

VI. Generate a new solution X^2 by solving

$$\max_{x \in D_0} S_1(x)$$

and construct the vectors

$$G_2 = (G_1(X^2), G_2(X^2), \ldots, G_r(X^2))$$

and

$$U_2 = (Z_1(X^2), Z_2(X^2), \ldots, Z_r(X^2)).$$

VII. Construct the vector V_1, the elements of which reflect the link between the values of the objective functions and the probabilities of attaining the respective values.

$$V_1 = \begin{pmatrix} G_1(X^2) & G_2(X^2) & \ldots & G_r(X^2) \\ 1 - \alpha_1 & 1 - \alpha_2 & \ldots & 1 - \alpha_r \end{pmatrix}.$$

The second row in V_1 is defined such that

$$P(Z_i'(x) \geqslant Z_i(X^2)) \geqslant 1 - \alpha_i.$$

As is known, this inequality has a deterministic equivalent

$$\sum_{j=1}^{n} \mathbf{E}(c_{ij})x_j + K_{\alpha_i}(x'Vx)^{1/2} \geqslant Z_i(X^2),$$

where K_{α_i} is such that

$$\Phi(K_{\alpha_i}) = \alpha_i$$

(Φ is the Laplace function and V is the covariance matrix).

VIII. At this stage the decision-maker has to answer to whether X^2 leads to satisfactory values for the functions $Z_1, \ldots, Z_r$ or not. If all values $Z_1(X^2), \ldots, Z_r(X^2)$ are satisfactory, then X^2 is the optimal solution. Else, one goes to the next step.

IX. Choose that $Z_k(x)$ having the least satisfactory pair $\begin{pmatrix} G_k(X^2) \\ 1 - \alpha_k \end{pmatrix}$. The decision-maker will have to establish the values of $e_k \in R^+$ and $\alpha_k^0 (0 \leqslant \alpha_k^0 \leqslant 1)$ such that

$$P(Z_k'(x) \geqslant e_k) \geqslant 1 - \alpha_k^0.$$

This inequality adds to the set of constraints defining the domain D_0.

The decision-maker has to specify the above items if he is not satisfied with the level reached by the function G_k or by the probability with which this value is reached or by both.

X. The feasible solutions domain is redefined by adding the previous restriction:

$$D_1 = \left\{ x \mid x \in D_0, \sum_{j=1}^{n} \mathbf{E}(c_{kj}) x_j + K_{\alpha_k^0} (x' V x)^{1/2} \geqslant \right.$$

$$\left. \geqslant e_k, \quad x > 0 \right\}.$$

XI. Construct a new auxiliary function

$$S_2(x) = \sum_{i \neq k}^{r} w_i G_i(x)$$

and restart from step VI.

Solve the mathematical programming problem

$$\max_{x \in D_1} S_2(x)$$

to find a new solution X^3, etc.

Finally, we shall find the vector

$$V_2 = \begin{pmatrix} e_1 & e_2 \ldots & e_r \\ 1 - \alpha_1^0 & 1 - \alpha_2^0 \ldots & 1 - \alpha_r^0 \end{pmatrix},$$

i.e. an acceptable value and acceptable probability for each function.

The decision-maker may content with the solution found or may further search for new solutions.

Note that each time one constructs the functions $S_i(x)$, the summation does not include the functions already transformed into restrictions in the feasible solution domain.

Example. Consider [76], [197]

$$\max f_1(x) = c_{11}x_1 + c_{12}x_2$$

$$\min f_2(x) = c_{21}x_1^2 + c_{22}x_2^2$$

$$\max f_3(x) = c_{31}x_1 + c_{32}x_2$$

on the domain

$$D_0 = \{(x_1, x_2) \mid x_1^2 + x_2^2 \leqslant 25, \quad 3x_1 + x_2 \leqslant 12,$$

$$x_2 \geqslant 1, \; x_1, x_2 \geqslant 0\},$$

where the coefficients c_{ij} are normal random variables :

$$c_{11} \sim N(2, 16), \; c_{12} \sim N(3, 4),$$

$$c_{21} \sim N(1, 1), \; c_{22} \sim N(1, 1),$$

$$c_{31} \sim N(4, 9), \; c_{32} \sim N(1, 1).$$

Choose

$$Z_1(x) = \bar{f}_1(x) = 2x_1 + 3x_2$$

$$Z_2(x) = -\bar{f}_2(x) = -x_1^2 - x_2^2$$

$$Z_3(x) = \bar{f}_3(x) = 4x_1 + x_2.$$

I. For each function Z_i determine its maximum value

$$Z_{1\max} = 17.94\,;\ Z_{2\max} = -1\,;\ Z_{3\max} = 15.64,$$

corresponding to points $x^1 = (2.58\,;\ 4.26)$; $x^2 = (0\,;1)$; $x^3 =$ $= (3.66\,;1)$ (fig. 3.4).

Also, find the minimum values

$$Z_{1\min} = 3\,;\ Z_{2\min} = -25\,;\ Z_{3\min} = 1 \text{ hence}$$

$$U_1 = (17.94\,;\quad -1\,;\ 15.64).$$

II. Transform the functions Z_i such that their values always range between zero and one.

$$G_1(x) = \frac{Z_1(x) - 3}{17.94 - 3} = 0.133x_1 + 0.2x_2 - 0.2,$$

$$G_2(x) = \frac{Z_2(x) + 25}{-1 + 25} = -0.041x_1^2 - 0.041x_2^2 + 1.04,$$

$$G_3(x) = \frac{Z_3(x) - 1}{15.64 - 1} = 0.273x_1 + 0.068x_2 - 0.06.$$

Construct the auxiliary function $F(x)$ as a sum of the three functions G_1, G_2, G_3:

$$F(x) = G_1(x) + G_2(x) + G_3(x) = 0.406x_1 - 0.041x_2^2 +$$

$$+ 0.268x_2 - 0.041x_1^2 + 0.78.$$

III. Maximize $F(x)$ on D_0 and obtain

$$X^1 = (3.09\,;\ 2.73).$$

Hence,

$$G_1 = (0.756\ ;\ 0.344\ ;\ 0.968).$$

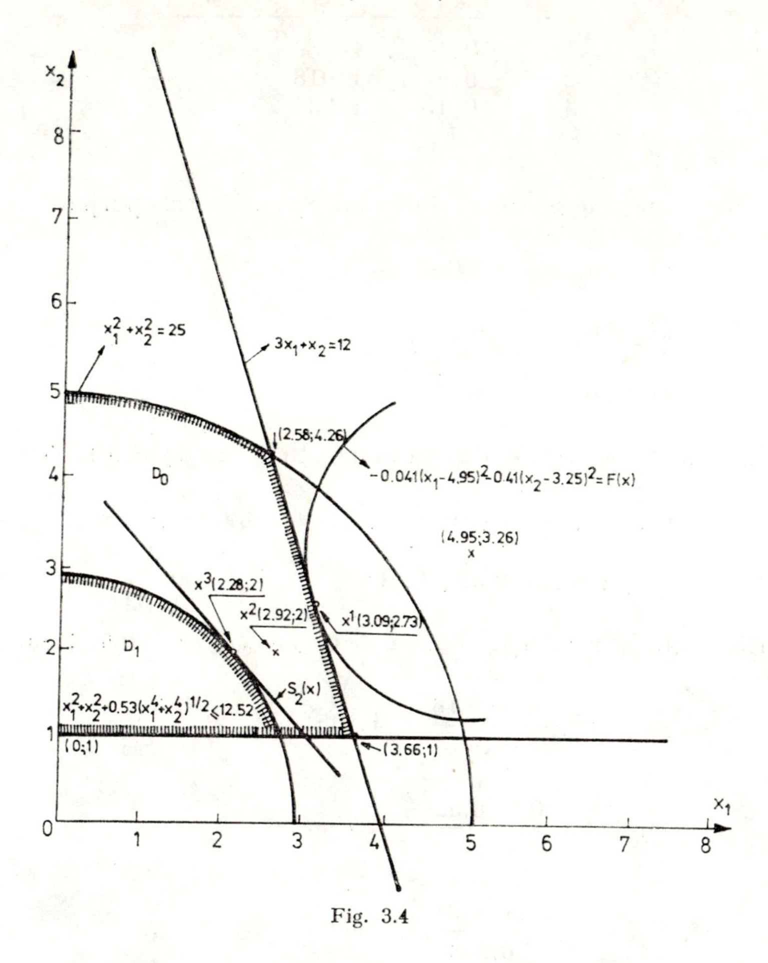

Fig. 3.4

IV. Estimate the multi-dimensional utility function $u(G)$ as

$$ku(G) = \prod_{i=1}^{3} (1 + kk_i c_i(1 - e^{b_i G_i})) - 1$$

with the following parameters

i	k_i	c_i	b_i
1	0.4	1.156	−2
2	0.6	1.018	−4
3	0.15	1.03	−3.5

and $k = -0.4$.

V. Re-define the auxiliary function $S_1(x)$. To this purpose use

$$kk_1u_1(G_1) = -0.144,$$

$$kk_2u_2(G_2) = -0.182,$$

$$kk_3u_3(G_3) = -0.059,$$

and obtain $u(G_1) = 0.852$. Assume that $\Delta u(G) = 0.1$. Then

$$G_1 + r\nabla u(G_1) = (0.756\,;\ 0.344\,;\ 0.968) + \\ + r(0.16\,;\ 0.52\,;\ 0.01).$$

From $u(G_i + r\nabla u(G_1)) - 0.952 = 0$ we get $r = 0.5$. Hence,

$$w_1 = 1 + \frac{0.5 \cdot 0.16}{0.756} = 1.105,$$

$$w_2 = 1 + \frac{0.5 \cdot 0.52}{0.344} = 1.755,$$

$$w_3 = 1 + \frac{0.5 \cdot 0.01}{0.968} = 1.005.$$

Thus,

$$S_1(x) = 1.105G_1(x) + 1.755G_2(x) + 1.005G_3(x) = \\ = 0.421x_1 - 0.072x_1^2 + 0.289x_2 - 0.072x_2^2 + 1.544.$$

VI. Solve

$$\max_{x \in D_0} S_1(x)$$

and find the solution

$$X^2 = (2.92\,;\,2),$$

and use the latter to construct the vectors

$$G_2 = (0.588\,;\ 0.527\,;\ 0.873)$$

and

$$U_2 = (11.84\,;\ -12.52\,;\ 13.68).$$

VII. Generate the vector V_1

$$V_1 = \begin{pmatrix} 0.588 & 0.527 & 0.873 \\ 0.5 & 0.5 & 0.5 \end{pmatrix}.$$

VIII. If the decision-maker is satisfied with the values of the components of the vector U_2, then the solution X^2 is optimal. Suppose that this is not the case.

IX. Assume that the decision-maker is not satisfied with the function $Z_2(x)$ which for X^2 is -12.52 with a probability of at least 0.5. Choose $\alpha_2^0 = 0.7$, and hence consider the constraint

$$P(Z_2'(x) \geqslant -12.52) \geqslant 0.7$$

or

$$-x_1^2 - x_2^2 - 0.53(x_1^4 + x_2^4)^{1/2} \geqslant -12.52.$$

X. The previous constraint is added to those defining the domain D_0 and hence

$$D_1 = \left\{ x \;\middle|\; \begin{aligned} & x_1^2 + x_2^2 \leqslant 25 \\ & 3x_1 + x_2 \leqslant 12 \\ & x_2 \geqslant 1\,;\ x_1, x_2 \geqslant 0 \\ & -x_1^2 - x_2^2 - 0.53(x_1^4 + x_1^4)^{1/2} \geqslant -12.52 \end{aligned} \right\}.$$

Construct a new auxiliary function

$$S_2(x) = w_1G_1(x)+w_3G_3(x) = 0.421x_1+0.289x_2 - 0.281$$

and maximize it on D_1 to yield the optimal solution

$$X^3 = (2.28\,;\ 2).$$

Restart from step VI for the function $S_2(x)$:

$$G_3 = (0.503\,;\ 0.663\,;\ 0.698).$$

Next construct the vector V_2. First calculate the values $1 - \alpha_1$ and $1 - \alpha_3$. If one takes $e_1 = 11.84$, then

$$2 \cdot (2.28) + 3(2) + K_{\alpha_1}[16(2.28)^2 + 4(2)^2]^{1/2} \geqslant e_1$$

yields

$$K_{\alpha_1} \geqslant 0.128 \text{ hence } 1 - \alpha_1 = 0.450 \text{ and}$$

hence the value $Z_1 = 11.84$ is attained with a probability of at least 0.450.

One computes $1 - \alpha_3$ in a similar manner. If $e_3 = 13.68$, then inequality

$$4(2.28) + 2 + K_{\alpha_3}[9(2.28)^2 + 2^2]^{1/2} \geqslant e_3$$

yields $K_{\alpha_3} \geqslant 0.639$, i.e. $1 - \alpha_3 = 0.361$. Hence, $Z_3 = 13.68$ is obtained with a probability of at least 0.361.
Thus,

$$V_2 = \begin{pmatrix} 0.588 & 0.527 & 0.873 \\ 0.450 & 0.7 & 0.361 \end{pmatrix}.$$

Suppose the decision-maker is satisfied with the data given by the vector V_2. The algorithm stops here and solution X^3 is considered optimal for the initial problem.

3.7. THE CASE OF DISCRETE DISTRIBUTIONS

Consider now a particular case of problem (3.1)–(3.2). One takes deterministic constraints

$$Ax \leqslant b\,;\; x \geqslant 0$$

and r efficiency functions

$$(\max)\, Z_k = \sum_{j=1}^{n} c_{kj}\, x_j, \qquad (k = 1, \ldots, r)$$

where the coefficients c_{kj} are random variables with s discrete values

$$c_{kj} \in \{c_{kj}^1, c_{kj}^2, \ldots, c_{kj}^s\}$$

corresponding to s natural states $N_1, N_2, \ldots, N_s$.

In the most general case, neither the efficiency function to be chosen for the optimisation process nor the natural state (i.e. the effective values of c_{kj}) are known.

If the natural state (say N_l) and the objective function to be maximized (say Z_u) are known, then we have a deterministic mathematical programming problem with one objective function.

$$\max_{x \in D} Z_{lu}(x) = \sum_{j=1}^{n} c_{uj}^l\, x_j.$$

However, at the moment of decision, neither the function to be chosen nor the natural state are known. A provision must be made that the solution vector x is calculated with consideration of all the objective functions and all the natural states.

Let p_k $(k = 1, \ldots, r)$ be the probability that function Z_k is chosen as an objective function and let q_h $(h = 1, \ldots, s)$ be the probability that the natural state N_h occurs.

If we consider simultaneously the r objective functions and s natural states, then we obtain $r \cdot s$ efficiency functions

$$\max Z_{hk} = \sum_{j=1}^{n} c_{kj}^h x_j, \qquad (h = 1, \ldots, s\,;\; k = 1, \ldots, r)$$

For a given natural state N_h, calculate a synthesis function of the r functions Z_{hk}

$$(3.27)\quad Z_h^* = \sum_{k=1}^{r} p_k Z_{hk} = \sum_{k=1}^{r} p_k \sum_{j=1}^{n} c_{kj}^h x_j = \sum_{j=1}^{n} \left(\sum_{k=1}^{r} p_k c_{kj}^h \right) x_j =$$

$$= \sum_{j=1}^{n} c_j x_j \, ,$$

where $c_j = \sum_{k=1}^{r} p_k c_{kj}^h$.

If the functions Z_{hk} express some distinct economical quantities that cannot be reduced to a common yardstick, then they will be replaced by utilities Z'_{hk} (in the von Neumann-Morgenstern sense) and then the function Z_h^* is calculated with the aid of formula (3.27).

According to the available information about the natural states one can have:

a) *Certainty conditions.* In this case, a natural state N_h is certain to occur ($q_h = 1, q_t = 0, t \neq h$). The stochastic programming problem with multiple objective function reduces to an ordinary one-objective function stochastic programming problem (3.27) with a random vector c.

b) *Risk conditions* (the probabilities q_h that the natural states occur are known). For each natural state N_h one determines a synthesis function Z_h^* ($h = 1, 2, \ldots, s$), then solves s linear programming problems and obtains the solutions $S_1, S_2, \ldots$ $\ldots, S_s$. Then calculate Z_{uv} ($u = 1, \ldots, s$; $v = 1, \ldots, s$), the value of the function Z_v^* associated to the solutions S_u and to the natural states N_v, and obtain the decision matrix shown in Table 3.1.

Table 3.1

Solutions \ Natural states	N_1	N_2	. . .	N_s
S_1	Z_{11}	Z_{12}	. . .	Z_{1s}
S_2	Z_{21}	Z_{22}	. . .	Z_{2s}
. . .	. . .	. . .	. . .	. . .
S_s	Z_{s1}	Z_{s2}	. . .	Z_{ss}

Some of the solutions S_i can be eliminated. For instance, if for S_i and S_j

$$Z_{iv} \leqslant Z_{jv}, \qquad (v = 1, \ldots, s),$$

then S_i is eliminated since S_i is dominated by S_j. Thus, suppose now that the solutions $S_1, \ldots, S_s$ do not contain dominated solutions. Then choose that solution $S^* \in \{S_1, \ldots, S_s\}$ which maximizes the weighted mean of the values Z_{ij}, assuming that the weights are the probabilities p_v, i.e.

$$\max \sum_{v=1}^{s} p_v Z_{uv}. \qquad (u = 1, \ldots, s)$$

c) *Uncertainty conditions* (the probabilities that the natural states occur are not known). In this case use various nonanalytical techniques and rules to estimate the probabilities that the natural states take place. Several criteria can be adopted to characterize the best solution : Wald's, Laplace's, Hurwicz's and Savage's [100]. We give below a brief presentation of these criteria.

Wald's criterion (Prudent's Law) is a minimax criterion. For each solution (criterion) S_i choose

$$Z_i = \min_j Z_{ij}$$

and consider the optimal solution S_k such that

$$\min_j Z_{kj} = \max_i \min_j Z_{ij}$$

is the solution of the problem.

Laplace's criterion (Equally probable criterion) consists in assuming that all the natural states are equally probable. This means that for each natural state, there exists the same probability of occurrence $q'_j = \frac{1}{s}$, i.e. we deal with risk conditions b). Then one chooses the solution for which

$$\max_i \left[\frac{1}{s} \sum_{j=1}^{s} Z_{ij} \right].$$

Savage's criterion. (Regrets Criterion). Calculate the matrix of regrets

$$r_{ij} = Z_{ij} - \max_i Z_{ij}$$

to which one applies Wald's criterion, i.e. choose that solution S_k which realizes

$$\max_i \min_j r_{ij}.$$

Hurwicz's criterion. Estimate for each situation a probability α_i that the most advantageous value $(\max_j Z_{ij})$ takes place and a probability $1 - \alpha_i$ that the most disadvantageous value $(\min_j Z_{ij})$ occurs and then consider

$$\alpha_i \max_j Z_{ij} + (1 - \alpha_i) \min_j Z_{ij}$$

and choose as solution the most advantageous solution S_k which makes

$$\max_i \left[\alpha_i \max_j Z_{ij} + (1 - \alpha_i) \min_j Z_{ij}\right].$$

Obviously, we have no preference in using one or another criterion. Any of them can assist us in finding a starting solution to be subsequently improved during the iterations as soon as data about the occurrence of the natural states can be gathered.

Chapter 4

Some generalizations of the minimum-risk problem and Kataoka's problem

The minimum-risk and Kataoka's problems were defined in Chapter 2. Here we shall reconsider these problems and give some generalizations. Thus, Section 4.1 presents the two problems developed by Bereanu, Bergthaller, Charnes, Cooper, Dragomirescu, Geoffrion and Kataoka and Section 4.2 gives some generalizations of these problems when the objective function appears as a product or a ratio of two linear functions. Deterministic equivalents for these two problems are also given. Section 4.3 furnishes a minimum-risk approach to the Chebyshev problem.

4.1. THE MINIMUM-RISK PROBLEM AND KATAOKA'S PROBLEM

Consider the linear stochastic programming problem

$$\text{(4.1)} \quad \min c'x$$

subject to

$$\text{(4.2)} \quad Ax = b; \quad x \geqslant 0,$$

where $A = (a_{ij})$, $(i = 1, \ldots, m; \ j = 1, \ldots, n)$ is an $m \times n$ $(m < n)$ matrix of rank m; 0 and x are n-vectors; b is an m-vector and c is an n-dimensional random vector.

In connection with problem (4.1)—(4.2), the notion of Minimum-Risk Solution is independently introduced in stochastic programming by Bereanu [16], [17], [18] and Charnes and Cooper [38]. Thus, one considers a level k of resources and expenditures and then one is satisfied if the objective function takes at most the value k.

DEFINITION 4.1 $\bar{x}(k)$ is said to be a *minimum-risk solution* of level k for problem (4.1)—(4.2) if

$$\bar{x}(k) \in D = \{| Ax = b, \ x \geqslant 0\}$$

(4.3) and

$$P\{\omega | c'(\omega)\bar{x}(k) \leqslant k\} = \max_{x \in D} \ P\{\omega | c'(\omega)x \leqslant k\}.$$

If the set D is regular (i.e. non-empty and bounded) and does not contain the origin, then there exists at least one solution $\bar{x}(k)$ to (4.3). One assumes that $0 \notin D$ only to rule out some trivial cases. Thus, if $0 \in D$ then, for $x = 0$, $P\{\omega | c'(\omega)x \leqslant k\} = 1$ if $k \geqslant 0$ (i.e. $x = 0$ is trivially optimal), while for $k \leqslant 0$ $P\{\omega | c'(\omega)x \leqslant k\} = 0$ so that $x = 0$ cannot be optimal.

Related to the minimum-risk problem is Kataoka's problem [91], [92]

$$\min f$$

subject to

$$P\{\omega | c'(\omega) \leqslant f\} \geqslant \alpha$$

$$x \in D.$$

Our aim is to obtain a value as small as possible for the objective function such that the probability of obtaining such a result exceeds a given threshold α.

Several problems consider the maximization of f such that

$$P\{\omega | c'(\omega)x \leqslant f\} = \alpha.$$

There are also cases when the minimum-risk problem appears as

$$(4.3') \quad \max_{x \in D} \ P\{\omega | c'(\omega)x \geqslant k\}.$$

These situations will be tackled in a similar manner. For the particular case presented below, the solution of the minimum-risk problem does not depend on the distribution function of c and is obtained by solving a hyperbolic (fractional) programming problem. It is obvious that the probability in (4.3′) will depend on the distribution of c.

THEOREM 4.1. *If* $c(\omega) = c^0 + c^1\tau(\omega)$, $c^1 > 0$ *and the distribution* $F_\tau(z)$ *of* $\tau(\omega)$ *is continuous, then the minimum-risk solution of* (4.3′) *does not depend on* $F_\tau(z)$ *and is obtained by solving the following fractional programming problem:*

$$\max_{x\in D}\left[\frac{-k + c'^0 x}{c'^1 x}\right].$$

Proof. Indeed,

$$\{\omega\,|(c^0 + c^1\tau(\omega))'x \geqslant k\} = \left\{\omega\,|\,\tau(\omega) \geqslant \frac{k - c'^0 x}{c'^1 x}\right\}.$$

Hence,

$$P\{\omega\,|(c^0 + c^1\tau(\omega))'x \geqslant k\} =$$

$$= P\left\{\omega\,|\,\tau(\omega) \geqslant \frac{k - c'^0 x}{c'^1 x}\right\} = 1 - F_\tau\left(\frac{k - c'^0 x}{c'^1 x}\right)$$

and

$$\max_{x\in D}\ P\{\omega\,|((c^0 + c^1\tau(\omega))'x \geqslant k\} =$$

$$= 1 - F_\tau\left(\max \frac{c'^0 x - k}{c'^1 x}\right).$$

The proof is now obvious considering that F_τ is continuous and non-decreasing.

Since deterministic equivalents are known only for stable probability laws, of which the normal distribution is widely used, the costs c_i are ordinarily treated as normal independent random variables. If c has the mean value vector $\bar{c}$ and the covariance matrix V, then $c'x$ is of type $N(\bar{c}'x,\ x'Vx)$. Then one can state

THEOREM 4.2. *The minimum-risk solution of problem* (4.3′) *is given by the solution of the following hyperbolic programming problem*

$$\max_x \left\{ \frac{\bar{c}'x - k}{\sqrt{x'Vx}} \,\middle|\, Ax \leqslant b, \quad x \geqslant 0 \right\}. \tag{4.4}$$

Proof. Indeed,

$$P\{\omega \,|\, c'(\omega)x \geqslant k\} = P\left\{\omega \,\middle|\, \frac{c'(\omega)x - \bar{c}'x}{\sqrt{x'Vx}} \geqslant \frac{k - \bar{c}'x}{\sqrt{x'Vx}}\right\} = 1 - \Phi\left(\frac{-\bar{c}'x + k}{\sqrt{x'Vx}}\right)$$

and

$$\max_{x \in D} P\{\omega \,|\, c'(\omega)x \geqslant k\} = 1 - \Phi\left(\max_{x \in D} \frac{\bar{c}'x - k}{\sqrt{x'Vx}}\right).$$

Note that if x^* is the solution of problem (4.4) and $z(x^*) = \dfrac{\bar{c}'x^* - k}{\sqrt{x^{*\prime}Vx^*}}$ is the value of the objective function, then we obtain $P\{\omega \,|\, c'(\omega)x^* \geqslant k\}$ since $z(x) \sim N(0,1)$.

Bereanu [16] uses the Zoutendijk's [171] method of feasible directions to solve (4.4). Bergthaller [25] adopts the transformation method to change (4.4) into a quadratic programming problem.

THEOREM 4.3 (Bergthaller [25]). *Let* $P \subset R^p$, $Q \subset R^q$ *and* $f : P \to R$, $g : Q \to R$. *If there exists a transformation* $T : Q \to P$ *such that* $T(Q) = P$ *and if* $\max_{x \in P} f(x)$ *exists, then* $\max_{y \in Q} g(y)$ *also exists and*

$$T(Q_0) = P_0,$$

where

$$P_0 = \{x \in P \mid \max_{z \in P} f(z) = f(x)\},$$

$$Q_0 = \{y \in Q \mid \max_{z \in Q} g(z) = g(y)\},$$

$$g = f \circ T.$$

Proof. Let $x_0 \in P_0$, $y_0 \in Q$ such that $T(y_0) = x_0$. Consider $y \in Q$ and $x = Ty$. Obviously $f(x) \leqslant f(x_0)$. From the definition of g, $g(x) \leqslant g(x_0)$, i.e. $P_0 \subset T(Q_0)$. Similarly, $P_0 \supset T(Q_0)$, i.e.

$$T(Q_0) = P_0.$$

The proof is complete now.
By virtue of this theorem, one can prove

THEOREM 4.4 (Bergthaller [25]). *If* $0 \neq \alpha < Z^* = \max \{c'x \mid Ax \leqslant b,\ x \geqslant 0\}$ *and if* x^* *is an optimal solution of problem* (4.4), *then* x^* *is unique and*

$$x_i^* = \frac{\alpha \hat{y}_i}{c'\hat{y} - \dfrac{1}{|\alpha|}}, \tag{4.5}$$

where $\hat{y}$ *is an optimal solution of the quadratic programming problem*

$$\min y'Vy$$

subject to

$$\begin{aligned} A^*y &\leqslant b^* \\ y &\geqslant 0 \end{aligned} \tag{4.6}$$

where

$$a_{ij}^* = |\alpha| a_{ij} - (\operatorname{sgn} \alpha) c_j b_i$$

$$b_i^* = -\frac{1}{\alpha} b_i.$$

Proof. Consider

$$P = \{x \mid Ax \leqslant b, \ \ x \geqslant 0, \ c'x > \alpha\} \subset R^n,$$

$$Q = \left\{y \mid A^*y \leqslant b^*, \ \ y \geqslant 0, \ (\text{sgn}) \ c'y > \frac{1}{\alpha}\right\} \subset R^n,$$

$$f(x) = \frac{c'x - \alpha}{\sqrt{x'Vx}},$$

$$g(y) = \frac{1}{|\alpha| \sqrt{y'Vy}}.$$

Define the mapping $T : Q \to R^n$; $(Ty = x)$ as

$$x_i = \frac{\alpha y_i}{c'y - \dfrac{1}{|\alpha|}}. \tag{4.7}$$

Let us prove that this mapping fulfils the requirements of Theorem 4.3.

If $y \in Q$, then $x \geqslant 0$. Indeed, if $\alpha > 0$, then $\alpha y_i > 0$, $c'y > \dfrac{1}{\alpha}$, i.e. $x \geqslant 0$ and if $\alpha > 0$, then $\alpha y_i \leqslant 0$, $c'y < \dfrac{1}{|\alpha|}$, i.e. $x \geqslant 0$. Since $A^*y \leqslant b^*$, then $A \dfrac{\alpha y}{c'y - \dfrac{1}{|\alpha|}} \leqslant b$, i.e. $Ax \leqslant b$. But

$$c'x = \frac{\alpha c'y}{c'y - \dfrac{1}{|\alpha|}} > \alpha \quad \text{implies that } T(Q) \subset P.$$

T is a one-to-one mapping as one can easily prove by writing (4.7) as a linear set

$$\sum_{j=1}^{n} (c_j x_i - \delta_i^j \alpha) y_j = \frac{x_i}{|\alpha|}, \qquad (i = 1, \ldots, n) \tag{4.8}$$

(δ_i^j are Kronecker's symbols) having non-zero determinant:

$$\Delta = (-\alpha)^{n-1}(c'x - \alpha) \neq 0.$$

Consider now that $x \in P$. The solution y of (4.8) satisfies $x = Ty$. One cannot have $c'y = \frac{1}{|\alpha|}$ since then $x = 0$, $y = 0$ but $0 \notin P$. It is impossible that $(\mathrm{sgn})c'y < \frac{1}{\alpha}$ since then $c'x < \alpha$, i.e. $x \notin P$. Thus, $(\mathrm{sgn})\, c'y > \frac{1}{\alpha}\cdot$ Since also $y \geqslant 0$ and $A^*y \leqslant \leqslant b^*$, then $y \in Q$. Hence,

$$g(y) = (f \circ T)(y) = \frac{1}{|\alpha|\sqrt{y'Vy}}\quad .$$

According to Theorem 4.3, the problem

$$\max_{y \in Q}\; g(y)$$

has at least one optimal solution $y^* = T^{-1}(x^*)$. Each optimal solution is the optimal solution of problem

$$\min\; \{y'Vy\,|y \in Q\}.$$

i.e. of problem (4.6). Since x^* is the only solution of (4.4), then $y^*(x^* = Ty^*)$ is the unique solution for (4.6). The proof is complete now.

Problem (4.4) can be solved by assimilating it with a parametric quadratic programming problem

$$\min \left\{\frac{1}{2}\, x'Vx + \lambda c'x\,|Ax \;\leqslant\; b,\;\; x \;\geqslant\; 0\right\}\cdot$$

In this respect, see Dragomirescu [56], Geoffrion [73] and Kataoka [91], [92].

The above model was also obtained by Kataoka using an equivalent problem. Let us consider Kataoka's problem in the form

$$\max f$$

subject to

$$P(c'x \leqslant f) = \alpha$$

$$Ax \leqslant b; \quad x \geqslant 0.$$

As is well known [247], if the vector c is normal, the expression $P(c'x \leqslant f) = \alpha$ is equivalent with $f = \bar{c}'x - \Phi^{-1}(\alpha)\sqrt{x'Vx}$. Hence, Kataoka's problem becomes:

Problem 1

$$\max f_{\mathrm{I}} = \bar{c}'x - q\sqrt{x'Vx},$$

subject to

$$Ax \leqslant b; \quad x \geqslant 0,$$

where $q = \Phi^{-1}(\alpha)$.

To solve this problem, Kataoka [91] uses the auxiliary problem

Problem 2

$$\max f_{\mathrm{II}} = \bar{c}'x - \frac{q}{2R}\, x'Vx,$$

subject to

$$Ax \leqslant b, \quad x \geqslant 0,$$

where R is a positive parameter. The equivalence of the two problems is proved by:

THEOREM 4.5. *If $\hat{x}(R)$ is an optimal solution of Problem 2 and satisfies*

$$R = \sqrt{\hat{x}(R)'\, V\hat{x}(R)},$$

then $\hat{x}(R)$ is an optimal solution for Problem 1 too, and conversely.

Proof. Lagrange's function corresponding to Problem 1 is

$$F_{\mathrm{I}}(x,u) = \sum_{j=1}^{n} \bar{c}_j x_j - q\sqrt{\sum_{i,j} v_{ij}x_i x_j} + \sum_i u_i(b_i - \sum_j a_{ij}x_j).$$

If $\hat{x} = (\hat{x}_1, \hat{x}_2, \ldots, \hat{x}_n)$, $\hat{u} = (\hat{u}_1, \hat{u}_2, \ldots, \hat{u}_m)$ is an optimal solution of Problem 1, then the Kuhn-Tucker conditions are satisfied:

$$(4.9) \quad \left(\frac{\partial F_{\mathrm{I}}}{\partial x_j}\right)_{\hat{x},\hat{u}} = \bar{c}_j - \frac{q \sum_i v_{ij}\hat{x}_i}{\sqrt{\sum_{i,j} v_{ij}\hat{x}_i\hat{x}_j}} - \sum_i \hat{u}_i a_{ij} \begin{cases} \leqslant 0 \text{ if } \hat{x}_j = 0, \\ = 0 \text{ if } \hat{x}_j > 0. \end{cases}$$

$$(4.10) \quad \left(\frac{\partial F_{\mathrm{I}}}{\partial u_i}\right)_{\hat{x}} = b_i - \sum_j a_{ij}\hat{x}_j \begin{cases} \geqslant 0 \text{ if } \hat{u}_i = 0, \\ = 0 \text{ if } \hat{u}_i > 0. \end{cases}$$

Lagrange's function corresponding to Problem 2 is

$$F_{\mathrm{II}}(x, u) = \sum_j \bar{c}_j x_j - \frac{q}{2R} \sum_{i,j} v_{ij}x_i x_j + \sum_i u_i(b_i - \sum_j a_{ij}x_j)$$

If $\hat{x}(R)$ is the optimal solution of Problem 2, then the Kuhn-Tucker conditions are satisfied:

$$(4.11) \quad \left(\frac{\partial F_{\mathrm{II}}}{\partial x_j}\right)_{\hat{x},\hat{u}} = \bar{c}_j - \frac{q}{R} \sum_i v_{ij}\hat{x}_i - \sum_i \hat{u}_i a_{ij} \begin{cases} \leqslant 0 \text{ if } \hat{x}_j = 0 \\ = 0 \text{ if } \hat{x}_j > 0 \end{cases}$$

$$(4.12) \quad \left(\frac{\partial F_{\mathrm{II}}}{\partial u_i}\right)_{\hat{x}} = b_i - \sum_j a_{ij}\hat{x}_j \begin{cases} \geqslant 0 \text{ if } \hat{u}_i = 0 \\ = 0 \text{ if } \hat{u}_i > 0. \end{cases}$$

Let $\hat{x}(R)$ be the optimal solution to Problem 2. If the condition of the theorem is satisfied, then (4.9)–(4.10) and (4.11)–(4.12) are equivalent and hence $\hat{x}(R)$ is also a solution of Problem 1. Conversely, if $\hat{x}$ is an optimal solution of Problem 1, then denoting

$$R = \sqrt{\hat{x}' V \hat{x}}$$

(4.9)—(4.10) is equivalent to (4.11)—(4.12), and hence $\hat{x}$ is optimal for Problem 2.

Dragomirescu [56] gave an alternate solution of (4.4), i.e of the minimum-risk problem. He considers the following. parametric quadratic programming problem :

$$(4.13)\quad \max \left\{ \lambda \bar{c}'x - \frac{1}{2} x'Vx \right\} = \Psi_\lambda(x)$$

and then gives the necessary and sufficient condition for the solution x_λ of problem (4.13) to be a solution of the minimum-risk problem. Since V is positive definite the objective function of (4.13) is strictly concave, and hence for every $\lambda \in R$ the quadratic problem has an unique optimal solution.

THEOREM 4.6. *If x^* is a minimum-risk solution corresponding to level $k(k < \max_x \{\bar{c}'x\})$, then there exists a positive solution λ^* of*

$$(4.14)\quad \bar{c}'x - (1/\lambda) x_\lambda' V x_\lambda = k,$$

such that

$$x^* = x_{\lambda^*}.$$

Proof. Consider

$$\lambda^* = \frac{x^{*\prime}Vx^*}{\bar{c}'x^* - k}, \qquad \varphi_k(x) = \frac{\bar{c}'x - k}{(x'Vx)^{1/2}}.$$

Since $k < \max_x \{\bar{c}'x\}$, $\bar{c}'x^* - k > 0$, and hence $\lambda^* > 0$. But

$$\varphi_k(x^*) = \frac{\bar{c}'x^* - k}{(x'^*Vx^*)^{1/2}} \geqslant \frac{\bar{c}'x - k}{(x'Vx)^{1/2}} = \varphi_k(x).$$

If account is taken of the obvious inequality

$$u^{1/2} \leqslant (u+1)/2, \qquad u \geqslant 0,$$

then

$$\frac{\bar{c}'x - k}{\bar{c}'x^* - k} \leqslant \frac{(x'Vx)^{1/2}}{(x'^*Vx^*)^{1/2}} \leqslant \frac{1}{2}\left(\frac{x'Vx}{x'^*Vx^*} + 1\right).$$

Multiply both sides by $x^{*\prime}Vx^*$, add $\lambda^*k - \frac{1}{2}x'Vx$ and use the definition of λ^*; then, for all $x \in D$,

$$\lambda^*\bar{c}'x - \frac{1}{2}x'Vx \leqslant \lambda^*k + \frac{1}{2}x^{*\prime}Vx^* = \lambda^*\bar{c}'x^* -$$

$$- \frac{1}{2}x^{*\prime}Vx^*.$$

This last relation shows that for $\lambda = \lambda^*$, x^* is an optimal solution for (4.13), i.e. $x^* = x_{\lambda^*}$. Then, the definition of λ^* implies that

$$\bar{c}'x^* - (1/\lambda)x'_{\lambda^*}Vx_{\lambda^*} = k$$

The proof is complete now.

The above theorem also implies a method of solving the minimum-risk problem:

1) Solve the quadratic parametric programming problem (4.13) for $\lambda \geqslant 0$. Using the long version of Wolfe's algorithm, we obtain in a finite number of steps both the optimal solution x_λ and the corresponding Lagrange multipliers y_λ for $\lambda \geqslant 0$. These are continuous and piecewise linear functions of λ. Hence, there exist critical values $0 \equiv \lambda_0 < \lambda_1 < \ldots < \lambda_s \equiv +\infty$ such that $x_\lambda = x_1^i + \lambda x_2^i$, $y_\lambda = y_1^i + \lambda y_2^i$ for $\lambda_i \leqslant \lambda \leqslant \lambda_{i+1}$ $(i = 0,1, \ldots, s-1)$, where x_1^i, x_2^i, y_1^i, y_2^i $(i = 0,1, \ldots, s-1)$ are constant vectors.

2) Determine those values λ^* satisfying equation (4.14). Then x_{λ^*} will be the minimum-risk solution corresponding to level k.

Geoffrion [73] considers the minimum-risk problem and Kataoka's problem as applications of the results obtained for the mathematical programming problem with two objective functions.

Using the notation of Chapter 1, Theorem 1.10, one can write

Kataoka's Problem. Consider

$$f_1(x) = \bar{c}'x$$
$$f_2(x) = -\, x'Vx,$$
$$u(f_1(x)) = f_1,$$
$$u(f_2(x)) = -(-f_2)^{1/2}.$$

Then

$$\Psi(x) = \varphi(u_1(f_1(x)), u_2(f_2(x))) = u(f_1(x)) -$$
$$- \Phi^{-1}(\alpha)u(f_2(x)) = \bar{c}'x + \Phi^{-1}(\alpha)\,(x'Vx)^{1/2}.$$

Thus the results of Theorem 1.10, Chapter 1, are directly applicable. To determine that value of x which maximizes the function $\Psi(x)$ consider the parametric program

$$\max_{x \in D} \,[\lambda\bar{c}'x - (1-\lambda)x'Vx],$$

which can be easily solved (e.g. by Wolfe's algorithm) to get the solution $x^*(\lambda)$. In solving the parametric problem one assumes that $\lambda = 1$ and then decreases the value of λ until the above function reaches its maximum in [0,1]. The parametric program will give the solution $x^*(\lambda\alpha)$, which, according to Theorem 1.10, is the optimal solution for Kataoka's problem.

The minimum-risk problem. Consider

$$D = \{x \mid Ax \leqslant b,\;\; x \geqslant 0,\;\; \bar{c}'x - k \geqslant 0\}$$

$$f_1(x) = \bar{c}'x$$
$$f_2(x) = -\, x'Vx$$
$$u(f_1) = f_1 - k$$
$$u(f_2) = -(-f_2)^{1/2}$$

$$\varphi(u(f_1),\; u(f_2)) = \frac{u(f_1)}{-\,u(f_2)}\,.$$

Use again the parametric program

$$\max_{x \in D} \; [\lambda \bar{c}'x - (1 - \lambda)x'Vx].$$

Assume that $\lambda = 1$ and decrease λ until the unimodal function $\dfrac{\bar{c}'x^*(\lambda) - k}{[x^{*\prime}(\lambda)\, Vx^*(\lambda)]^{1/2}}$ reaches its maximum value on [0,1]. If this happens for $\lambda = \lambda_k$ then, according to Theorem 1.10, $x^*(\lambda_k)$ is an optimum solution of the minimum-risk problem.

Related to problem (4.1)—(4.2) is the minimization of the variance of the linear function $c'x$.

(4.15) Minimize $\{\text{Var}\ (c'x) = \mathbf{E}[(c'x - \bar{c}'x)^2] = x'Vx\}$

subject to

$$Ax = b\,; \quad x \geqslant 0.$$

Recently [198], Gupta and Swarup considered the minimization of the upper permissible limit "f" of $(c'x - \bar{c}'x)^2$ for a given probability level p.

(4.16) Minimize f

subject to

$$P[(c'x - \bar{c}'x)^2 \leqslant f] = p$$

$$Ax = b\,; \quad x \geqslant 0,$$

where $\bar{c}$ is the mean vector of c.

The optimal solution of this problem would tend to minimize the variance $c'x$. This model might be preferred to model (4.15) since an optimal solution x_0 of (4.15) may be more risky in the sense that the probability of obtaining a very high value of $\xi = (c'x - \bar{c}'x)^2$ for x_0 is greater than for some other value of x because of the random variable ξ.

The model (4.16) can be reduced to a standard quadratic problem if the vector c is normally distributed.

Then the random variable

$$\chi^2 = \frac{(c'x - \bar{c}'x)^2}{(x'Vx)}$$

follows a Chi-square distribution with one degree of freedom. Thus, the chance constraint

$$P[(c'x - \bar{c}'x)^2 \leqslant f] = p$$

is equivalent to

$$P\left[\chi^2 \leqslant \frac{f}{x'Vx}\right] = p,$$

or

$$f = q(x'Vx),$$

where q is the p-fractile of the probability distribution of a χ^2-variate with one degree of freedom, i.e.

$$\int_0^q \frac{2^{-1/2}}{\Gamma(1/2)} e^{-1/2\chi^2}(\chi^2)^{1/2-1} d\chi^2 = p.$$

The value of q for a given p can be obtained from the tables of χ^2-probability integral or from the tables of percentage points of χ^2-distribution (for example, Pearson and Hartley (eds.), *Biometrika Tables for Statisticians*, Vol. I, Tables 7 and 8, Cambridge Univ. Press, U.K. (1966)).

Whence the deterministic equivalent of problem (4.16)

$$\text{Minimize } f = q(x'Vx)$$

subject to

$$Ax = b; \quad x \geqslant 0.$$

4.2. GENERALIZATIONS

We shall now present the deterministic equivalents for the minimum-risk problem and Kataoka's problem when the objective function is a product, or a ratio, of linear functions.

4.2.1. Consider the minimum-risk problem for a product of linear functions

$$(4.17)\quad \max\ P\{\omega\,|\,c'(\omega)x\cdot d'x \leqslant k\}$$

subject to

$$Ax = b;\quad x \geqslant 0.$$

Note that $d = (d_1, \ldots, d_n)'$ is an n vector.

Assume that only the vector c is random and that it is normally distributed, with mean value $\bar{c} = (\bar{c}_1, \ldots, \bar{c}_n)$ and covariance matrix V. The vector d is constant.

To avoid trivial solutions assume that $0 \notin D = \{x\,|\,Ax = b, x \geqslant 0\}$. Also note that

$$\{\omega\,|\,c'(\omega)x\cdot d'x \leqslant k\} = \left\{\omega\,|\,\frac{c'(\omega)x\cdot d'x - \bar{c}'x\cdot \mathrm{d}x'}{d'x\cdot\sqrt{x'Vx}} \leqslant\right.$$

$$\left.\leqslant \frac{k - \bar{c}'x\cdot d'x}{d'x\cdot\sqrt{x'Vx}}\right\}.$$

Hence,

$$P\{\omega\,|\,c'(\omega)x\cdot d'x \leqslant k\} =$$

$$= P\left\{\omega\,|\,\frac{c'(\omega)\cdot d'x - \bar{c}'x\cdot d'x}{d'x\sqrt{x'Vx}} \leqslant \frac{k - \bar{c}'x\cdot d'x}{d'x\sqrt{x'Vx}}\right\}.$$

For each x fixed, $z(\omega) = c'(\omega)x\cdot d'x$ follows a normal distribution law with mean value

$$\mathbf{E}z(\omega)) = \bar{c}'x\cdot d'x$$

and standard deviation

$$\mathbf{D}(z(\omega)) = d'x\sqrt{x'Vx}.$$

Hence, $\dfrac{c'(\omega)x \cdot d'x - \bar{c}'x \cdot d'x}{d'x\sqrt{x'Vx}}$ follows also a normal distribution law, namely $N(0,1)$. Thus,

$$P\{\omega | c'(\omega)x \cdot d'x \leqslant k\} = \Phi\left(\frac{k - \bar{c}'x \cdot d'x}{d'x\sqrt{x'Vx}}\right),$$

where

$$\Phi(t) = \frac{1}{\sqrt{2\pi}} \int_{-\infty}^{t} e^{-\frac{s^2}{2}} ds \quad \text{is the Laplace function.}$$

Then

$$\max_{x \in D} P\{\omega | c'(\omega)x \cdot d'x \leqslant k\} = \Phi\left(\max_{x \in D} \frac{k - \bar{c}'x \cdot d'x}{d'x\sqrt{x'Vx}}\right).$$

Since Φ is non-decreasing and continuous, the deterministic equivalent of the efficiency function (4.17) is the function

$$\max_{x \in D} \frac{k - \bar{c}'x \cdot d'x}{d'x\sqrt{x'Vx}}.$$

Hence, one can state

THEOREM 4.7. *The minimum-risk solution of the stochastic programming problem* (4.17) *is the solution of the following nonlinear fractional programming problem*

$$\max_{x \in D} \frac{k - \bar{c}'x \cdot d'x}{d'x\sqrt{x'Vx}}.$$

If d is also a random variable (say normal with mean d and covariance matrix W), then the problem becomes considerably more complicated. If c and d are independent variables, then for a given x,

$$\mathbf{E}(z(\omega)) = \bar{c}'x \cdot \bar{d}'x$$

and

$$\mathbf{D}^2(z(\omega)) = (\bar{c}'x)^2\mathbf{D}^2(d'(\omega)x) + (\bar{d}'x)^2\mathbf{D}^2(c'(\omega)x) +$$

$$+ \mathbf{D}^2(c'(\omega)x) \cdot \mathbf{D}^2(d'(\omega)x) = (\bar{c}'x)^2(xWx) +$$

$$+ (x'Vx) \cdot (\bar{d}'x)^2 + (x'Vx)(x'Wx).$$

As in the previous case,

$$P\{\omega \,|\, c'(\omega)x \cdot d'(\omega)x \leqslant k\} =$$

$$= \Phi\left(\frac{k - \bar{c}'x \cdot \bar{d}'x}{\sqrt{(\bar{c}'x)^2(x'Wx) + (x'Vx)(\bar{d}'x)^2 + (x'Vx)(x'Wx)}}\right).$$

Using a reasoning similar to the one above yields the following

THEOREM 4.7′. *If c and d are independent random variables, then the solution of the stochastic programming problem* (4.17) *is the same with the solution of the fractional programming problem*:

$$\max_{x \in D} \frac{k - \bar{c}'x \cdot \bar{d}'x}{\sqrt{(\bar{c}'x)^2(x'Wx) + (x'Vx)(\bar{d}'x)^2 + (x'Vx)(x'Wx)}}.$$

4.2.2. Consider now the minimum-risk problem for the ratio of two linear functions whose numerator has random coefficients:

$$(4.18) \quad \max P\left\{\omega \,|\, \frac{c'(\omega)x}{d'x} \leqslant k\right\},$$

subject to

$$Ax = b;\; x \geqslant 0.$$

Let $z(x, \omega) = \dfrac{c'(\omega)x}{d'x}$. If the coefficients follow a normal distribution then, for a given x,

$$\mathbf{E}(z(\omega)) = \frac{\bar{c}'x}{d'x}$$

and

$$\mathbf{D}(z(\omega)) = \frac{\sqrt{x'Vx}}{d'x}.$$

Hence,

$$\left\{\omega \mid \frac{c'(\omega)x}{d'x} \leqslant k\right\} = \left\{\omega \mid \frac{\dfrac{c'(\omega)x}{d'x} - \dfrac{\bar{c}'x}{d'x}}{\dfrac{\sqrt{x'Vx}}{d'x}} \leqslant \frac{k - \dfrac{\bar{c}'x}{d'x}}{\dfrac{\sqrt{x'Vx}}{d'x}}\right\} =$$

$$= \left\{\omega \mid \frac{c'(\omega)x - \bar{c}'x}{\sqrt{x'Vx}} \leqslant \frac{kd'x - \bar{c}'x}{\sqrt{x'Vx}}\right\},$$

and thus

$$P\left\{\omega \mid \frac{c'(\omega)x}{d'x} \leqslant k\right\} = \Phi\left(\frac{k \cdot d'x - \bar{c}'x}{\sqrt{x'Vx}}\right).$$

Hence, one can state

THEOREM 4.8. *The minimum-risk solution of the stochastic programming problem* (4.18) *is given by the solution of the nonlinear fractional programming problem*

$$\max_{x \in D} \frac{k \cdot d'x - \bar{c}'x}{\sqrt{x'Vx}}.$$

4.2.3. Kataoka's problem for the product of two linear functions is

(4.19) $\max f,$

subject to

$$Ax = b$$
$$(4.20) \quad x \geqslant 0$$
$$P\{\omega \mid c'x \cdot d'x \leqslant f\} = \alpha.$$

If the vector c has normal distribution (mean value $\bar{c}$, covariance matrix V) and d is constant, then the previous argument yields

$$P\{\omega \mid c'(\omega)x \cdot d'x \leqslant f\} = \Phi\left(\frac{f - \bar{c}'x\, d'x}{d'x\sqrt{x'Vx}}\right) = \alpha,$$

i.e.

$$\frac{f - \bar{c}'x \cdot d'x}{d'x\sqrt{x'Vx}} = \Phi^{-1}(\alpha)$$

or

$$f = [\bar{c}'x + \Phi^{-1}(\alpha)\sqrt{x'Vx}]\, d'x.$$

If $\alpha \leqslant 0.5$, then $\Phi^{-1}(\alpha) = -q \leqslant 0$ and the above problem becomes

$$(4.21) \quad \max_{x \in D} f(x) = (\bar{c}'x - q\sqrt{x'Vx})\, d'x.$$

THEOREM 4.9. *The solution of Kataoka's problem* (4.19)—(4.20) *is given by the solution of the mathematical programming problem having* (4.21) *as objective function.*

Two methods for solving such problems are given in ref. [151].

If the vector d is also random (c and d are assumed to be independent variables), then the following deterministic equivalent to problem (4.19)—(4.20) is obtained.

THEOREM 4.9′. *The solution of Kataoka's problem for a product of linear functions whose coefficients are independent random variables is given by the solution of the mathematical programming problem with objective function*

$$\max_{x \in D} [\bar{c}'x \cdot \bar{d}'x + \Phi^{-1}(\alpha) \cdot$$
$$\cdot \sqrt{(\bar{c}'x)^2(x'Wx) + (x'Vx)(\bar{d}'x)^2 + (x'Vx)(x'Wx)}].$$

4.2.4. Kataoka's problem for the ratio of two linear functions is

$$\max\ k$$

subject to

$$Ax = b\,;\ x \geqslant 0$$

$$P\left\{\omega \mid \frac{c'x}{d'x} \leqslant k\right\} = \alpha,$$

where c is a random vector with n-dimensional normal distribution and d is constant.

Let us find a deterministic equivalent to this problem. From the previous calculations,

$$P\left\{\omega \mid \frac{c'x}{d'x} \leqslant k\right\} = \Phi\left(\frac{k \cdot d'x - \bar{c}'x}{\sqrt{x'Vx}}\right) = \alpha,$$

i.e.

$$\frac{k \cdot d'x - \bar{c}'x}{\sqrt{x'Vx}} = \Phi^{-1}(\alpha)$$

or

$$k = \frac{\bar{c}'x + \Phi^{-1}(\alpha)\sqrt{x'Vx}}{d'x}.$$

Let now $\alpha \leqslant 0.5$. Hence $\Phi^{-1}(\alpha) = -q \leqslant 0$. Then one can state

THEOREM 4.10. *The solution to Kataoka's problem for the ratio of two linear functions whose numerator has random coefficients, is given by the solution of the following fractional programming problem:*

$$\max_{x \in D} \frac{\bar{c}'x - q\sqrt{x'Vx}}{d'x}.$$

4.2.5. We give below two possible methods for solving Kataoka's problem for a ratio of two linear functions.

Problem I:

$$\max F(x) = \frac{\bar{c}'x - q\sqrt{x'Vx}}{d'x},$$

subject to

$$Ax = b; \quad x \geqslant 0.$$

Suppose that
(i) $d'x > 0$ for any $x \in D$;
(ii) the problem $\max\limits_{x \in D} \dfrac{\bar{c}'x}{d'x}$ has a finite optimal solution $\hat{x}_0$.

THEOREM 4.11. *Under assumptions* (i) *and* (ii) *the optimal solution of Problem* I *is finite.*

Proof. Show that for each feasible solution x, the value of the efficiency function for Problem I does not exceed $\dfrac{\bar{c}'\hat{x}_0}{d'\hat{x}_0}$, i.e.

$$\frac{\bar{c}'x - q\sqrt{x'Vx}}{d'x} \leqslant \frac{\bar{c}\hat{x}_0}{d'\hat{x}_0}. \tag{4.22}$$

Assume that the contrary is true, i.e. there exists an $\bar{x}$ such that

$$\frac{(\bar{c}'\bar{x} - q\sqrt{\bar{x}'V\bar{x}})}{d'\bar{x}} > \frac{\bar{c}'\hat{x}_0}{d'\hat{x}_0}.$$

But $-q \leqslant 0$ and hence

$$\frac{\bar{c}'\bar{x}}{d'\bar{x}} > \frac{\bar{c}'\hat{x}_0}{d'\hat{x}_0},$$

which contradicts (ii). Hence, (4.22) is true for any $x \in D$ and the proof of the theorem springs from (ii).

In the most general case, when V is positive semi-definite, the numerator of $F(x)$ is no longer differentiable and $F(x)$ is explicitly quasi-concave, as is implicit in the following

THEOREM 4.12. *The function* $F(x) = \dfrac{\bar{c}'x - q\sqrt{x'Vx}}{d'x}$ *is explicitly quasi-concave on the convex domain D.*

Proof. Let $x_1 \neq x_2$, $x_1, x_2 \in D$ and $x_0 = \lambda x_1 + (1-\lambda)x_2$, $0 < \lambda < 1$. Assume, for instance, that $F(x_1) < F(x_2)$, i.e.

$$\frac{\bar{c}'x_1 - q\sqrt{x_1'Vx_1}}{d'x_1} < \frac{\bar{c}'x_2 - q\sqrt{x_2'Vx_2}}{d'x_2}.$$

Then

(4.23)

$$\frac{\bar{c}'x_2 \cdot \bar{d}'x_1 - \bar{c}'x_1 \cdot d'x_2 - q \cdot d'x_1\sqrt{x_2'Vx_2} + q \cdot d'x_2\sqrt{x_1'Vx_1}}{d'x_1 \cdot d'x_2} > 0.$$

Let us show that

(4.24)

$$F(x_0) = \frac{\bar{c}'[\lambda x_1 + (1-\lambda)x_2] - q\sqrt{[\lambda x_1 + (1-\lambda)x_2]'V[\lambda x_1 + (1-\lambda)x_2]}}{d'[\lambda x_1 + (1-\lambda)x_2]} >$$

$$> \frac{\bar{c}'x_1 - q\sqrt{x_1'Vx_1}}{d'x_1} = F(x_1).$$

Indeed, the function $h(x) = \sqrt{x'Vx}$ is convex by virtue of Theorem 2.6, Chapter 2 and hence

(4.25)

$$\sqrt{[\lambda x_1 + (1-\lambda)x_2]'V[\lambda x_1 + (1-\lambda)x_2]} \leqslant \lambda\sqrt{x_1'Vx_1} +$$

$$+ (1-\lambda)\sqrt{x_2'Vx_2}.$$

According to (4.25),

(4.26)

$$F(x_0) = \frac{\bar{c}'[\lambda x_1 + (1-\lambda)x_2] - q\sqrt{[\lambda x_1 + (1-\lambda)x_2]'V[\lambda x_1 + (1-\lambda)x_2]}}{d'[\lambda x_1 + (1-\lambda)x_2]} \geqslant$$

$$\geqslant \frac{\lambda\bar{c}'x_1 + (1-\lambda)\bar{c}'x_2 - q[\lambda\sqrt{x_1'Vx_1} + (1-\lambda)\sqrt{x_2'Vx_2}]}{\lambda d'x_1 + (1-\lambda)d'x_2}.$$

Consider the expression

$$\frac{\lambda\bar{c}'x_1+(1-\lambda)\bar{c}'x_2-q[\lambda\sqrt{x_1'Vx_1}+(1-\lambda)\sqrt{x_2'Vx_2}]}{\lambda d'x_1+(1-\lambda)d'x_2}-\frac{\bar{c}'x_1-q\sqrt{x_1'Vx_1}}{d'x_1}=$$

$$=\frac{\lambda\bar{c}'x_1\cdot d'x_1+(1-\lambda)\bar{c}'x_2\cdot d'x_1-q\lambda d'x_1\sqrt{x_1'Vx_1}-q(1-\lambda)d'x_1\sqrt{x_2'Vx_2}-}{d'x_1[\lambda d'x_1+(1-\lambda)d'x_2]}$$

$$\frac{-\lambda\bar{c}'x_1\cdot d'x_1-(1-\lambda)\bar{c}'x_1\cdot d'x_2+\lambda qd'x_1\sqrt{x_1'Vx_1}+(1-\lambda)qd'x_2\sqrt{x_1'Vx_1}}{d'x_1[\lambda d'x_1+(1-\lambda)d'x_2]}=$$

$$=\frac{(1-\lambda)[-\bar{c}'x_1\cdot d'x_2+\bar{c}'x_2d'x_1-qd'x_1\sqrt{x_2'Vx_2}+qd'x_2\sqrt{x_1'Vx_1}]}{d'x_1[\lambda d'x_1+(1-\lambda)d'x_2]}>0$$

which is a direct consequence of the fact that $0<\lambda<1$ (i.e. $1-\lambda>0$), the numerator is positive for any $x \in D$, and of (4.23).Combining (4.26) with the above inequality yields (4.24). The proof is complete now.

Note that the domain of feasible solutions D is convex.

Since an explicit quasi-convex function has the property that each local maximum is a global maximum [150], Theorem 4.12 leads to

THEOREM 4.13. *Each local maximum of function $F(x)$ is a global maximum.*

If V is positive definite, the numerator of $F(x)$ is differentiable and since each explicit quasi-concave differentiable function is also pseudoconcave [103], it follows that $F(x)$ is pseudoconcave.

Denote $q=t^2$.

Then

$$F(x)=\frac{\bar{c}'x-q\sqrt{x'Vx}}{d'x}=\frac{\bar{c}'x-\sqrt{x'tVx}}{d'x}=$$

$$=\frac{\bar{c}'x-\sqrt{x'Ux}}{d'x},$$

where $U = tV$ is a positive semi-definire matrix. One can consider a parametric form of Problem I [53].

Problem II :

$$\max_{x \in D} F^*(x) = (\bar{c} - Fd)'x - \sqrt{x'Ux}.$$

Dinkelbach's theorem [53] of fractional programming theory states that $\max F^*(x) = 0$, and if one gets the zero value, then the solution of the parametric Problem II is the solution of the original Problem I.

The dual to Problem II is easily written using the duality theorem derived by Sinha [129]

Problem II′ :

$$\min G = b'y,$$

subject to

$$A'y + Uz + dF \geqslant \bar{c},$$

$$z'Uz \leqslant 1,$$

$$y \geqslant 0.$$

Such problems can be generally solved with reasonable efficiency by the Sequential Unconstrained Minimization Technique (SUMT) [64]. This technique consists in transforming the constrained problem into an unconstrained one. To this purpose, the constraints are brought to the form "$\geqslant$", and then introduced into the objective function. One of the forms used is

$$\min[b'y - r_k \ln(A'y + Uz + dF - \bar{c}) - r_k \ln(1 - z'Uz)], \tag{4.27}$$

where r_k has the following properties :

(1) $r_k \geqslant 0$, $(k = 1,2, \ldots)$,

(2) r_k is a decreasing sequence,

(3) $\lim_{k \to \infty} r_k = 0$.

One gets a sequence of feasible solutions $x(r_k)$ which minimizes the function (4.27). Under certain (convexity) conditions,

$$\lim_{k\to\infty} x(r_k) = \bar{x},$$

where $\bar{x}$ is the optimum solution of Problem II′.

In solving Problem II′, we have actually solved Problem II and, hence, Problem I.

The unconstrained minimization problem can be solved with algorithms of metric variable [64].

When U is positive semidefinite, the numerator of $F(x)$ is not differentiable, and consequently one cannot use convex programming methods to solve Problem I. Several techniques can be used in this respect.

First we shall consider two computation procedures for Problem I. We assume that $\bar{c}'x - q\sqrt{x'Vx} > 0$, for any $x \in D$.

In order to solve Problem I, use the transformation $z = tx$ with $t \geqslant 0$, which was used by Charnes-Cooper [36] in fractional linear programming. Using this transformation, Problem I reduces to one of convex mathematical programming. Whence,

Problem III:

$$\max H(z, t) = \bar{c}'z - q\sqrt{z'Vz}$$

subject to

$$Az \leqslant bt,$$

$$d'z = 1,$$

$$z, t \geqslant 0.$$

Let $Y = \{(z, t) \in R^n \times R^1 |\ Az - bt \leqslant 0,\ d'z = 1,\ z \geqslant 0, t \geqslant 0\}$ be the domain of feasible solutions for Problem III.

Recall a result obtained by Charnes and Cooper [36]:

LEMMA 4.1. *For any feasible solution* $(z, t) \in Y$ *of Problem* III $t > 0$.

Proof. If $t = 0$, then $Az \leqslant 0$, which contradicts the regularity condition for D. Indeed, the set $\{z | Az \leqslant 0\}$ is a cone, i.e. an unbounded set, which contradicts the assumption that D is bounded. The proof is complete now.

The mapping $z = tx$ sets a one-to-one correspondence between the domains D and Y. The relation between the optimal solutions of Problems I and III is given by

THEOREM 4.14. *If $(\bar{z}, \bar{t}) \in Y$ is an optimal solution of Problem* III, *then* $x = \frac{\bar{z}}{\bar{t}}$ *is an optimal solution of Problem* I.

Proof. Since $(\bar{z}, \bar{t})$ is an optimal solution of Problem III,

$$\bar{c}'\bar{z} - q\sqrt{\bar{z}'V\bar{z}} \geqslant \bar{c}'z - q\sqrt{z'Vz} \tag{4.28}$$

for any $(z, t) \in Y = \{(z, t) \mid Az \leqslant bt,\ d'z = 1,\ z,\ t \geqslant 0\}$. Assume that $x = \frac{\bar{z}}{\bar{t}}$ is not an optimal solution of Problem I, i.e. there exists another point x^0 with the property that

$$\frac{\bar{c}'x^0 - q\sqrt{x^{0\prime}Vx^0}}{d'x^0} > \frac{\bar{c}'\frac{\bar{z}}{\bar{t}} - q\sqrt{\left(\frac{\bar{z}}{\bar{t}}\right)'V\left(\frac{\bar{z}}{\bar{t}}\right)}}{d'\frac{\bar{z}}{\bar{t}}}.$$

But

$$\frac{\bar{c}'\frac{\bar{z}}{\bar{t}} - q\sqrt{\left(\frac{\bar{z}}{\bar{t}}\right)'V\left(\frac{\bar{z}}{\bar{t}}\right)}}{d'\frac{\bar{z}}{\bar{t}}} = \frac{\bar{c}'\bar{z} - q\sqrt{\bar{z}'V\bar{z}}}{d'\bar{z}} =$$

$$= \bar{c}'\bar{z} - q\sqrt{\bar{z}'V\bar{z}},$$

since $d'\bar{z} = 1$, where $(\bar{z}, \bar{t})$ is the solution of Problem III. Hence,

$$\frac{\bar{c}'x^0 - q\sqrt{x^{0\prime}Vx^0}}{d'x^0} > \bar{c}'\bar{z} - q\sqrt{\bar{z}'V\bar{z}}.$$

Let $d'x^0 = \rho > 0$, $\hat{z} = \frac{x^0}{\rho}$ and $\hat{t} = \frac{1}{\rho}$. Then $(\hat{z}, \hat{t})$ is a feasible solution of Problem III since

$$\hat{z}, \hat{t} \geqslant 0$$

$$d'\hat{z} = \frac{d'x^0}{\rho} = \frac{\rho}{\rho} = 1$$

$$A\hat{z} = A\frac{x^0}{\rho} = \frac{Ax^0}{\rho} \leqslant b\hat{t}.$$

One has

$$\text{(4.30)} \quad \frac{\bar{c}'x^0 - q\sqrt{x^{0\prime}Vx^0}}{d'x^0} = \frac{\bar{c}'\frac{x_0}{\rho} - q\sqrt{\left(\frac{x^0}{\rho}\right)'V\left(\frac{x^0}{\rho}\right)}}{d'\frac{x^0}{\rho}} =$$

$$= \frac{\bar{c}'\hat{x} - q\sqrt{\hat{x}'V\hat{x}}}{d'\hat{x}} = \bar{c}'\hat{x} - q\sqrt{\hat{x}'V\hat{x}}.$$

Then from (4.29) and (4.30) it results that

$$\bar{c}'\hat{x} - q\sqrt{\hat{x}'V\hat{x}} > \bar{c}'\bar{x} - q\sqrt{\bar{z}'V\bar{z}},$$

which contradicts (4.28). The proof is complete now.

Since the function $H(x, t)$ is concave, the solution of Problem I, where the efficiency function is neither concave nor convex, reduces to the solution of a concave programming problem. Such a problem can be solved by any of the algorithms given in the literature.

An alternative solution may be obtained by reducing Problem I to a more simple problem. A problem equivalent to Problem I is

Problem IV:

$$\min K(x, t) = d'x$$

subject to

$$Ax \leqslant bt,$$
$$\bar{c}'x - q\sqrt{x'Vx} \geqslant 1,$$
$$x \geqslant 0,\ t \geqslant 0.$$

We assume that (*i*) $d'x > 0$ and (*ii*) $\bar{c}'x - q\sqrt{x'Vx} > 0$ are satisfied for any $x \in D$. Let

$$Z = \{(x, t) \in R^n \times R^1 | Ax \leqslant bt,\ \bar{c}'x - q\sqrt{x'Vx} \geqslant 1,$$
$$x \geqslant 0,\quad t \geqslant 0\},$$

be the domain of feasible solution for Problem IV. The equivalence between Problems I and IV is given by

THEOREM 4.15. *If* $(x^*, t^*) \in Z$ *is an optimal solution for Problem* IV, *then* $\dfrac{x^*}{t^*}$ *is an optimal solution for Problem* I.

Proof. Assume *ad absurdum* that the optimal solution of Problem II is $\hat{x} \neq \dfrac{x^*}{t^*}$ Let

$$\hat{t} = \frac{1}{\bar{c}'\hat{x} - q\sqrt{\hat{x}'V\hat{x}}}.$$

Then $(\hat{t}\hat{x}, \hat{t})$ is a feasible solution of Problem IV. Indeed,

$$\hat{t}\hat{x} \geqslant 0,\ \hat{t} \geqslant 0,$$
$$A\hat{t}\hat{x} - b\hat{t} = \hat{t}(A\hat{x} - b) \leqslant 0,$$
$$\bar{c}'(\hat{t}\hat{x} - q\sqrt{(\hat{t}\hat{x})'V(\hat{t}\hat{x})}) = \hat{t}(\bar{c}'\hat{x} - q\sqrt{\hat{x}'V\hat{x}}) = 1.$$

Since (x^*, t^*) is an optimal solution for Problem IV,

$$(4.31)\quad d'x^* \leqslant d'\hat{x}.$$

But $\frac{x^*}{t^*}$ is a feasible solution for Problem I, as one can easily verify. Then,

$$F\left(\frac{x^*}{t^*}\right) = \frac{\bar{c}'\frac{x^*}{t^*} - q\sqrt{\left(\frac{x^*}{t^*}\right)' V \left(\frac{x^*}{t^*}\right)}}{d'\frac{x^*}{t^*}} =$$

$$= \frac{\bar{c}'x - q\sqrt{x^{*\prime}Vx^*}}{d'x^*} > \frac{1}{d'x^*} > \frac{1}{\hat{t}d'\hat{x}} =$$

$$= \frac{\bar{c}'\hat{x} - q\sqrt{\hat{x}'V\hat{x}}}{d'\hat{x}} = F(\hat{x}),$$

and thus the point $x = \frac{x^*}{t^*}$ is optimal for the function $F(x)$. Thus, Problem I has been reduced to a concave programming problem with a convex feasible solutions domain.

Example [145]. Let

$$\max \frac{c_1x_1 + c_2x_2}{5x_1 + 3x_2 + 1},$$

subject to

$$\begin{aligned} &3x_1 + 5x_2 \leqslant 15,\\ (4.32)\quad &5x_1 + 2x_2 \leqslant 10,\\ &x_1, x_2 \geqslant 0. \end{aligned}$$

Assume that c_1 and c_2 are normally distributed random variables with mean value vector $\bar{c}$ and covariance matrix V:

$$\bar{c} = \begin{pmatrix} 5 \\ 3 \end{pmatrix}, \quad V = \begin{pmatrix} 1 & 1 \\ 1 & 2 \end{pmatrix}.$$

The corresponding Kataoka's problem is

$$\max k,$$

subject to

$$
(4.33)\quad
\begin{aligned}
&3x_1 + 5x_2 \leqslant 15,\\
&5x_1 + 2x_2 \leqslant 10,\\
&P\left(\omega \,\middle|\, \frac{c_1x_1 + c_2x_2}{5x_1 + 2x_2 + 1} \leqslant k\right) = \alpha,\\
&x_1, x_2 \geqslant 0.
\end{aligned}
$$

The deterministic equivalent to this problem is

$$
(4.34)\quad \max F(x) = \frac{5x_1 + 3x_2 - \Phi^{-1}(\alpha)\sqrt{x_1^2 + 2x_1x_2 + 2x_2^2}}{5x_1 + 2x_2 + 1},
$$

subject to

$$
\begin{aligned}
&3x_1 + 5x_2 \leqslant 15,\\
&5x_1 + 2x_2 \leqslant 10,\\
&x_1, x_2 \geqslant 0.
\end{aligned}
$$

Note that requirements (i) and (ii) are satisfied.
Indeed, $5x_1 + 2x_2 + 1 > 0$ on the domain $D = \{(x_1, x_2) | 3x_1 + 5x_2 \leqslant 15, 5x_1 + 2x_2 \leqslant 10, x_1, x_2 \geqslant 0\}$, and the solution of problem

$$
T = \max_{x \in D} \frac{\bar{c}'x}{d'x} = \max_{x \in D} \frac{5x_1 + 3x_2}{5x_1 + 2x_2 + 1}
$$

is finite [123]: $x_1 = 0,\ x_2 = 3,\ T = \dfrac{9}{7} = 1.285714.$

Let us make the variable transformation $x_1^* = tx_1;\ x_2^* = tx_2$. Problem IV corresponding to problem (4.34) is

$$
\min K(x, t) = 5x_1^* + 2x_2^* + t,
$$

Table 4.1

No.	α	$\Phi^{-1}(\alpha)$	Solution (x_1^*, x_2^*, t) of problem (4.35)	Solution $x_1 = \frac{x_1^*}{t}$, $x_2 = \frac{x_2^*}{t}$ of Kataoka's problem	Maximum value of k	$P\left(\frac{c_1x_1 + c_2x_2}{5x_1+2x_2+1} > k\right) = 1 - \alpha$
1	0.5	0		$x_1 = 0$; $x_2 = 3$	1.285714	0.5
2	0.48	−0.05	$x_1^* = 0$; $x_2^* = 0.3414$; $t = 0.1138$	$x_1 = 0$; $x_2 = 3$	1.2554	0.52
3	0.44	−0.15	$x_1^* = 0$; $x_2^* = 0.3587$; $t = 0.1196$	$x_1 = 0$; $x_2 = 2.9991$	1.1947	0.56
4	0.40	−0.25	$x_1^* = 0$; $x_2^* = 0.3779$; $t = 0.126$	$x_1 = 0$; $x_2 = 2.99921$	1.134148	0.60
5	0.20	−0.84	$x_1^* = 0.1186$; $x_2^* = 0.2669$; $t = 0.1127$	$x_1 = 1.05235$ $x_2 = 2.368234$	0.806655	0.80
6	0.158	−1	$x_1^* = 0.1678$; $x_2^* = 0.1877$; $t = 0.1214$	$x_1 = 1.3822$; $x_2 = 1.5461$	0.7486	0.842
7	0.138	−1.09	$x_1^* = 0.2013$; $x_2^* = 0.1246$; $t = 0.1256$	$x_1 = 1.6027$; $x_2 = 0.992$	0.7012	0.862
8	0.105	−1.25	$x_1^* = 0.2375$; $x_2^* = 0.0678$; $t = 0.1323$	$x_1 = 1.79516$; $x_2 = 0.512471$	0.687081	0.895
9	0.096	−1.3	$x_1^* = 0.2463$; $x_2^* = 0.056$; $t = 0.1344$	$x_1 = 1.8325$; $x_2 = 0.4166$	0.6756	0.904
10	0.074	−1.45	$x_1^* = 0.2701$; $x_2^* = 0.0278$; $t = 0.1406$	$x_1 = 1.921$; $x_2 = 0.197724$	0.6465	0.926
11	0.023	−2	$x_1^* = 0.333$; $x_2^* = 0$; $t = 0.1667$	$x_1 = 1.9976$; $x_2 = 0$	0.54539	0.977
12	0.010	−2.33	$x_1^* = 0.3745$ $x_2^* = 0$; $t = 0.1873$	$x_1 = 1.9994$; $x_2 = 0$	0.4854	0.990

subject to

$$3x_1^* + 5x_2^* \leqslant 15t,$$

$$(4.35)\quad 5x_1^* + 2x_2^* \leqslant 10t,$$

$$5x_1^* + 3x_2^* - 1 \geqslant \Phi^{-1}(\alpha)\sqrt{x_1^{*2} + 2x_1^* x_2^* + 2x_2^{*2}},$$

$$x_1, x_2, t \geqslant 0.$$

Problem (4.35) was solved for various values of α. The results obtained are given in Table 4.1.

The level α varies between 0 and 0.5 and the value of k for which $P\left(\omega \mid \dfrac{c_1x_1 + c_2x_2}{5x_1 + 2x_2 + 1} \leqslant k\right) = \alpha$ varies from $-\infty$ to $T = \dfrac{9}{7}$.

The value of α increases as k decreases. For instance, for $\alpha = 0.44$, $P\left(\omega \mid \dfrac{c_1x_1 + c_2x_2}{5x_1 + 2x_2 + 1} > 1.1947\right) = 0.56$ and for $\alpha = 0.074$, $P\left(\omega \mid \dfrac{c_1x_1 + c_2x_2}{5x_1 + 2x_2 + 1} > 0.6465\right) = 0.926$.

Another method for solving Problem I consists in using the dual problem. Thus, we reconsider Problem I in a more general case *).

Problem I* :

$$\max F(x) = \frac{\bar{c}'x - \sqrt{x'Ux} + \alpha}{d'x + \beta},$$

subject to

$$Ax = b$$

$$x \geqslant 0,$$

where α, β are positive numbers and U is an $n \times n$ positive semidefinite matrix. Assume that the denominator does not

---

*) An even more general case is treated by Vivek Patkar and Stancu-Minasian in ref. [226].

change sign, i.e. it is positive over the whole domain of feasible solutions. The dual to Problem I* is

Problem V :

$$\min L(v, z, w) = z,$$

subject to

$$A'v + dz + Uw \geqslant \bar{c},$$
$$-b'v + \beta z \geqslant \alpha,$$
$$w'Uw \leqslant 1,$$
$$v \geqslant 0,$$

where v is an m-vector and w is an n-vector.

Denote by D_D and D_P the feasible solutions domain for Problems I* and V, respectively.

THEOREM 4.16 [129]. $\max_{x \in D_p} F(x) \leqslant \min_{(v, z, w) \in D_D} L(v, z, w).$

Proof. Let $x \in D_P$ and $(v, z, w) \in D_D$ be two feasible solutions to the two problems.

Multiplying $A'v + dz + Uw \geqslant \bar{c}$ by $x \geqslant 0$ yields

$$x'A'v + x'dz + x'Uw \geqslant x'\bar{c}. \tag{4.36}$$

Since $x'Av = b'v$, relation (4.36) becomes

$$b'v + x'dz + x'Uw \geqslant x'\bar{c}.$$

Adding $-b'v + \beta z \geqslant \alpha$, yields

$$(x'd + \beta)z + x'Uw \geqslant x'\bar{c} + \alpha. \tag{4.37}$$

In the sequel, we shall use the following result :

LEMMA 4.2. *If $x, y \in R^m$ and C is a positive semidefinite matrix,*

$$(x'Cy) \leqslant (x'Cx)^{1/2}(y'Cy)^{1/2}.$$

The equality holds for $\lambda \geqslant 0$ *such that* $Cx = \lambda Cy$.
Using Lemma 4.2., expression (4.37) yields

$$(x'd + \beta)z + (x'Ux)^{1/2}(w'Uw)^{1/2} \geqslant x'\bar{c} + \alpha$$

or

$$(x'd + \beta)z + (x'Ux)^{1/2} \geqslant x'\bar{c} + \alpha, \text{ since } w'Uw \leqslant 1.$$

Hence,

$$z \geqslant \frac{\bar{c}'x - \sqrt{x'Ux} + \alpha}{x'd + \beta}$$

and the proof is immediate.

THEOREM 4.17. *If Problem* I* *has an optimal solution* x_0, *then there exists an optimal solution* (v_0, z_0, w_0) *for Problem* V *and the two extreme values coincide*:

$$F(x_0) = L(v_0, z_0, w_0).$$

Proof. Let x_0 be an optimal solution for Problem I*. In compliance with what was stated in Sub-section 4.2.5, Problem I* can be transformed into an equivalent problem using the transformation $y = tx (t \geqslant 0)$.

Problem I**

$$\max H(y, t) = \bar{c}'y - \sqrt{y'Uy} + \alpha,$$

subject to

$$Ay \leqslant bt,$$

$$d'y + \beta t = 1,$$

$$y, t \geqslant 0.$$

According to Theorem 4.14, between the optimal solutions x_0 and (y_0, t_0) of Problems I* and I** there exists the relation $x_0 = \dfrac{y_0}{t_0}$ and then the duality principle implies that

$$F(x_0) = H(y_0, t_0). \tag{4.38}$$

The dual to Problem I** is Problem V [129].

Then, if (y_0, t_0) and (v_0, z_0, w_0) are optimal solutions to Problems I** and V, respectively, then

$$(4.39) \quad H(y_0, t_0) = L(v_0, z_0, w_0).$$

Relations (4.38) and (4.39) yield

$$F(x_0) = L(v_0, z_0, w_0).$$

The proof is complete now.

Let us now prove the duality theorem conversely.

THEOREM 4.18 [3]. *If (v_0, z_0, w_0) is an optimal solution of Problem V, then there exists an optimal solution x_1 to Problem I*, and the corresponding optimum values of the objective functions are equal.*

Proof. Assume (v_0, z_0, w_0) to be the optimal solution for Problem V. The Kuhn-Tucker conditions [96] are satisfied, and hence there exists a vector $(x_0, y_0) \geqslant 0$ and $\mu \geqslant 0$ such that

$$\begin{aligned}
&\text{a)}\ x_0'd + y_0\beta = 1,\\
&\text{b)}\ Ax_0 - by_0 \leqslant 0,\\
&\text{c)}\ v_0'(Ax_0 - by_0) = 0,\\
(4.40)\quad &\text{d)}\ Ux_0 = 2\mu Uw_0,\\
&\text{e)}\ \mu(1 - w_0'Uw_0) = 0,\\
&\text{f)}\ x_0'(A'u_0 + dz_0 + Uw_0 - \bar{c}) = 0,\\
&\text{g)}\ - b'v_0 + \beta z_0 - \alpha \geqslant 0,\\
&\text{h)}\ y_0(- b'v_0 + \beta z_0 - \alpha) = 0.
\end{aligned}$$

Let $x_1 = \dfrac{x_0}{y_0}$. According to Lemma 4.1, $y_0 > 0$. From this, and from (4.40a), x_1 is a feasible solution for Problem I*.

Expressions (4.40c) and (4.40h) give

$$v_0'Ax_0 = y_0(\beta z_0 - \alpha).$$

Substituting this in (4.40f) yields

$$(4.41)\quad \bar{c}'x_0 - x_0'Uw_0 = (d'x_0 + \beta y_0)z_0 - \alpha y_0.$$

Relation (4.40 d) and Lemma (4.3) imply that

$$x_0'Uw_0 = (x_0'Ux_0)^{1/2}(w_0'Uw_0)^{1/2}$$

or

$$(4.42)\quad x_0'Uw_0 = (x_0'Ux_0)^{1/2}$$

since relation (4.40e) leads to either $\mu = 0$ or $w_0'Uw = 1$. Then (4.41) and (4.42) yield

$$\bar{c}'x_0 - (x_0'Ux_0)^{1/2} = (d'x_0 + \beta y_0)z_0 - \alpha y_0$$

or

$$\frac{\bar{c}'x_0 - (x_0'Ux_0)^{1/2} + \alpha y_0}{d'x_0 + \beta y_0} = z_0$$

and

$$\frac{\bar{c}'x_1 - (x_1'Ux_1)^{1/2} + \alpha}{d'x_1 + \beta} = z_0.$$

The proof is complete now.

Thus, if the matrix U is positive semidefinite, then the solution of Problem I* cannot be obtained by convex programming techniques since the numerator function is not differentiable. For this reason, one has to solve the dual problem V, which is a convex programming problem.

For the minimum-risk approach to special classes of mathematical programming problems (e.g. the Chebyshev problem, linear plus indefinite (or fractional) programming), see ref. [236]. We give below the minimum-risk approach to the Chebyshev problem.

4.3. THE MINIMUM-RISK APPROACH TO THE CHEBYSHEV PROBLEM

Consider the Chebyshev problem :

$$(4.43) \quad \min_{x} \max_{1 \leqslant \ \leqslant r} \{z_i(x)\},$$

subject to

$$(4.44) \quad Ax = b; \qquad x \geqslant 0,$$

where $z_i(x) = (c^i)'x + \alpha^i (i = 1, \ldots, r)$, $c^i (i = 1, \ldots, r)$, are constant vectors and $\alpha^i (i = 1, \ldots, r)$ are scalar constants.

Let now c^i and α^i assume random values :

$$c^i(\omega) = c_1^i + t(\omega) c_2^i$$

$$\alpha^i(\omega) = \alpha_1^i + t(\omega) \alpha_2^i,$$

where c_1^i, $c_2^i (i = 1, \ldots, r)$ are constant vectors, α_1^i, $\alpha_2^i (i = 1, \ldots$ $\ldots, r)$ are scalar constants and $t(\omega)$ is a random variable.

The distribution function of the optimum value for the stochastic Chebyshev problem was determined in Section 3.1, Chapter 3. Let us now use the minimum-risk approach to obtain the solution of this problem.

We assume that

$$(4.45) \quad c_2^{i\prime} x + \alpha_2^i \neq 0 \text{ for all } x \in D, \text{ and } i \in M = \{1, 2, \ldots, r\}.$$

The minimum-risk approach to the Chebyshev problem (4.43)—(4.44) consists in finding the optimal solution of the following programming problem :

$$(4.46) \quad v(w) = \max_{x \in D} P\{\omega \,|\max_{i \in M} [(c^i(\omega))'x + \alpha^i(\omega)] \leqslant w\}.$$

$$\text{Let } z_i''(x) = c_2^{i\prime} x + \alpha_2^i, \quad g_i(x, w) = \frac{w - c_1^{i\prime} x - \alpha_1^i}{c_2^{i\prime} x + \alpha_2^i},$$

for all $x \in D$ and $i \in M$.

Then, we have

$$F(x, w) = P\{\omega \mid \max_{i \in M}[(c_1^i + t(\omega)c_2^i)'x + \alpha_1^i + t(\omega)\alpha_2^i] \leqslant$$

$$\leqslant w\} = P\{\omega \mid (c_1^i + t(\omega)c_2^i)'x + \alpha_1^i + t(\omega)\alpha_2^i \leqslant w, \ \forall i \in M\} =$$

$$= \begin{cases} P\{\omega \mid t(\omega) \leqslant g_i(x, w), \ \forall \ i \in M\} & \text{if} \quad z_i''(x) > 0 \\ P\{\omega \mid t(\omega) \geqslant g_i(x, w), \ \forall \ i \in M\} & \text{if} \quad z_i''(x) < 0 \end{cases} =$$

$$= \begin{cases} P\{\omega \mid t(\omega) \leqslant \min_{i \in M} g_i(x, w)\} & \text{if} \quad z_i''(x) > 0 \\ P\{\omega \mid t(\omega) \geqslant \max_{i \in M} g_i(x, w)\} & \text{if} \quad z_i''(x) < 0 \end{cases} =$$

$$= \begin{cases} T(\min_{i \in M} g_i(x, w)) & \text{if} \quad z_i''(x) > 0 \\ 1 - T(\max_{i \in M} g_i(x, w)) & \text{if} \quad z_i''(x) < 0. \end{cases}$$

Hence, the problem becomes

$$\max_x F(x, w) = \begin{cases} \max_x T(\min_{i \in M} g_i(x, w)) & \text{if } z_i''(x) > 0 \\ 1 - \min_x T(\max_{i \in M} g_i(x, w)) & \text{if } z_i''(x) < 0. \end{cases}$$

If the distribution function $T(z)$ of the random variable $t(\omega)$ is continuous and strictly increasing, then

$$v(w) = \max_x F(x, w) = \begin{cases} T(\max_x \min_{i \in M} g_i(x, w)), & \text{if } z_i''(x) > 0 \\ 1 - T(\min_x \max_{i \in M} g_i(x, w)), & \text{if } z_i''(x) < 0. \end{cases}$$

The following theorem is immediate.

THEOREM 4.19 [246]. *If assumption* (4.45) *holds and the distribution function* $T(z)$ *of* $t(\omega)$ *is continuous and strictly increasing, then the minimum-risk solution of problem* (4.46) *does not depend on*

$T(z)$ and can be obtained by solving the deterministic piecewise linear fractional programming problems:

$$(4.47)\quad \max_x \min_{i \in M} \frac{w - c_1^{i\prime}x - \alpha_1^i}{c_2^{i\prime}x + \alpha_2^i} \quad \text{if} \quad z_i''(x) > 0$$

or

$$(4.48)\quad \min_x \max_{i \in M} \frac{w - c_1^{i\prime}x - \alpha_1^i}{c_2^{i\prime}x + \alpha_2^i} \quad \text{if} \quad z_i''(x) < 0.$$

Remarks. 1) The piecewise linear fractional programming problems (4.47) or (4.48) can be solved by use of a parametric algorithm similar to Dinkelbach's algorithm for fractional programming;
2) If

$$c^i(\omega) = c_1^i + t^i(\omega)c_2^i, \quad i \in M,$$
$$\alpha^i = 0, \quad i \in M,$$

and if $t^i(\omega)$ $(i \in M)$ are independent random variables whose distribution function T_i is continuous and strictly increasing, then the minimum-risk solution of problem (4.46) depends on T_i as follows:

$$\max_x F(x, w) = \max_x \prod_{i=1}^{r} T_i\left(\frac{w - c_1^{i\prime}x}{c_2^{i\prime}x}\right);$$

3) When the functions $z_i(i \in M)$ defining the objective function z are nonlinear (linear fractional, linear-plus-fractional or linear-plus-indefinite functions), the minimum-risk approach to the Chebyshev problem can be restated and solved under appropriate hypothesis in a manner similar to the linear case.

Chapter 5

Multiple minimum-risk solutions in stochastic programming

We shall introduce the concept of *multiple minimum-risk solution in stochastic programming*. An algorithm used for obtaining such solutions is given [138], [140], [141]. Section 5.5. gives a method for solving a multiple criteria problem. The last section deals with interactive approach to stochastic programming with multiple objective functions.

5.1. PROBLEM DEFINITION

Consider a stochastic programming problem with r objective functions:

$$(5.1) \qquad \gamma_k(\omega) = \max_x \{Z_k = c_k'(\omega)x\}, \qquad (k = 1, \ldots, r)$$

subject to

$$(5.2) \qquad \begin{aligned} Ax &\leqslant b \\ x &\geqslant 0, \end{aligned}$$

where: A and b are $m \times n$ and $m \times 1$ matrices, $x = (x_1, \ldots, x_n)'$ is an unknown n-dimensional column vector, and $c_k(\omega) = (c_{k1}(\omega), \ldots, c_{kn}(\omega))$ $(k = 1, \ldots, r)$ are n-dimensional vectors whose elements are random variables on the probability space $\{\Omega, K, P\}$.

Denote by

$$D = \{x | x \in R^n, \ Ax \leqslant b, \ x \geqslant 0\}$$

the set of feasible solutions of (5.2). Assume that D is bounded and nonempty.

Suppose that the random vectors $c_k(\omega)$ $(k = 1, \ldots, r)$ follow a non-degenerate n-dimensional normal distribution. Denote their mean values by

$$\bar{c}_k = (\bar{c}_{k1}, \ldots, \bar{c}_{kn}), \qquad (k = 1, \ldots, r)$$

and the covariance matrix by V_k

$$\bar{c}_{ki} = \mathbf{E}(c_{ki}), \qquad (k = 1, \ldots, r; i = 1, \ldots, n)$$

$$v_{kij} = \mathbf{E}[(c_{ki} - \bar{c}_{ki})(c_{kj} - \bar{c}_{kj})], \quad (k = 1, \ldots, r;$$

$$i, j = 1, \ldots, n).$$

Then $c_k'(\omega)x$ $(k = 1, \ldots, r)$ is a normal random variable with mean value

$$M_k = \bar{c}_k'x$$

and variance

$$\mathbf{D}^2(c_k'(\omega)x) = \mathbf{E}\left(\sum_{j=1}^{n} (c_{kj} - \bar{c}_{kj})x_j\right)^2 =$$

$$= \sum_{i,j} \mathbf{E}[(c_{ki} - \bar{c}_{ki})(c_{kj} - \bar{c}_{kj})]x_i x_j = x'V_k x.$$

Our aim is to transform the problem such that instead of the r random objective functions we obtain only one deterministic objective function. Several solution variants exist in this respect.

1. For each Z_k function, one maximizes its mean value

$$\max F_k = \max_{x \in D} \bar{c}_k'x, \quad (k = 1, \ldots, r).$$

Thus, we obtain a linear programming problem with r deterministic objective functions. Next, we can use the solution methods described by Maruşciac [109].

For example, one can construct a unique function

$$F^* = \sum_{i=1}^{r} k_i \bar{c}_i'x$$

and then maximize it. This is tantamount to using in (5.1)—(5.2) a function like

$$\widetilde{F} = \sum_{i=1}^{r} k_i c_i' x$$

and maximizing its mean value. Since "the mean of a sum of random variables is equal to the sum of the mean values",

$$\mathbf{E}(\widetilde{F}) = \mathbf{E}\left(\sum_{i=1}^{r} k_i c_i' x \right) = \sum_{i=1}^{r} k_i \mathbf{E}(c_i' x) = \sum_{i=1}^{r} k_i \bar{c}_i' x = F^*.$$

Observation 1. For $r = 1$, one gets the E-model considered by Charnes and Cooper [38].

2. For each $k = 1, \ldots, r$, one minimizes the variance

$$\min F_k = \min x' V_k x, \qquad (k = 1, \ldots, r).$$

Hence, problem (5.1)—(5.2) becomes a deterministic problem with r quadratic objective functions, which can be solved by use of the method presented in ref. [109].

Observation 2. For $r = 1$, one gets the V-model expounded in ref. [38].

3. For each Z_k, one chooses a minimum level u_k to be attained and maximizes the probability that Z_k exceeds the value u_k. In this manner we obtain the following mathematical programming problem [138]:

$$(5.3) \qquad \begin{aligned} Ax &\leqslant b \\ x &\geqslant 0 \end{aligned}$$

with r functions to be maximized:

$$(5.4) \qquad \begin{aligned} &\max P\{\omega \mid c_1'(\omega)x \geqslant u_1\} = Z_1 \\ &\max P\{\omega \mid c_2'(\omega)x \geqslant u_2\} = Z_2 \\ &\ldots\ldots\ldots\ldots \\ &\max P\{\omega \mid c_r'(\omega)x \geqslant u_r\} = Z_r. \end{aligned}$$

$u_1, \ldots, u_r$ above are given positive values.

Such a problem is called [138] a "multiple minimum-risk problem with levels $u_1, u_2, \ldots, u_r$".

As was already shown in Chapter 1, the optimum solution for an objective function is not necessarily optimum for the other functions and, hence, one introduces the notion of the "best compromise solution". For problem (5.3)—(5.4), a "compromise" can be reached by using a synthesis function for the r functions

$$(5.5)\qquad F = F(P\{\omega \mid c_1'(\omega)x \geqslant u_1\}, \ldots, P\{\omega \mid c_r'(\omega)x \geqslant u_r\}).$$

This function must be increasing with respect to each of its components. A possible expression for such a function is:

$$\max \sum_{i=1}^{r} k_i P\{\omega \mid c_i'(\omega)x \geqslant u_i\},$$

i.e. the maximization of the weighted sum of the r original functions (The weight of function Z_i is k_i).

Another possible objective is to maximize the joint probability [204]:

$$P\{\omega \mid c_1'(\omega)x \geqslant u_1, \ldots, c_r'(\omega)x \geqslant u_r\}.$$

In order to solve problems of type (5.3)—(5.4), we suggest a method consisting in the sequential solving of some minimum-risk problems which include also quadratic constraints. The following hypotheses are necessary:

i) the programs of the mean values are bounded, i.e.

$$T_i = \max \bar{c}_i'x < \infty, \qquad (i = 1, \ldots, r),$$

ii) $u_i < T_i$,

iii) $0 \notin D$.

Then one proceeds as follows:

First order the criteria according to their importance, i.e. if Z_1 is more important than Z_2, Z_2, than Z_3 a.s.o., then one gets the sequence $Z_1, Z_2, \ldots, Z_r$.

Solve the minimum-risk problem

$$(5.6) \qquad \max_{x \in D} P\{\omega \,|\, c_1'(\omega)x \geqslant u_1\},$$

which corresponds to function Z_1 and level u_1.

Let this solution be $x_{u_1}^1 \in D$ and let p_1 be the optimum value of (5.6).

Next, define the domain

$$U_1 = \{x \,|\, x \in D,\ P\{\omega \,|\, c_1'(\omega)x \geqslant u_1\} \geqslant$$

$$\geqslant p_1 - \varepsilon_1\},\ (\varepsilon_1 > 0, \text{ known})$$

on which we solve the minimum-risk problem

$$\max P\{\omega \,|\, c_2'(\omega)x \geqslant u_2\}.$$

Let now the maximum value be p_2 and let $x_{u_2}^2 \in U_1$ be the corresponding optimum solution.
Apply the same procedure on U_2, where

$$U_2 = \{x \,|\, x \in U_1,\ P\{\omega \,|\, c_2'(\omega)x \geqslant u_2\} \geqslant$$

$$\geqslant p_2 - \varepsilon_2\}(\varepsilon_2 > 0, \text{ known}),$$

and solve the minimum-risk problem

$$\max P\{\omega \,|\, c_3'(\omega)x \geqslant u_3\}, \text{ etc.}$$

Assume now that for the first $r - 1$ optimum values $p_1, p_2, \ldots, p_{r-1}$ the corresponding domains $U_1, U_2, \ldots, U_{r-2}$ have been identified. Then define

$$U_{r-1} = \{x \,|\, x \in U_{r-2},\ P\{\omega \,|\, c_{r-1}'(\omega)x \geqslant u_{r-1}\} \geqslant p_{r-1} -$$

$$- \varepsilon_{r-1}\}\ (\varepsilon_{r-1} > 0, \text{ known})$$

and solve the minimum-risk problem

$$\max P\{\omega \,|\, c_r'(\omega)x \geqslant u_r\}$$

whose solution, x^r_{ur}, is called [138], [140], [141] "the multiple minimum-risk solution corresponding to levels $u_1, u_2, \ldots, u_r$ and to the relaxation values $\varepsilon_1, \varepsilon_2, \ldots, \varepsilon_{r-1}$.

The solution x^r_{ur} is "the best compromise solution" for (5.3)—(5.4) and the ordering scheme described above. Obviously, for a different ordering scheme, one may get other solutions. Hence, one can formulate the following

DEFINITION 5.1 (Stancu-Minasian [138]). A *multiple minimum-risk solution* corresponding to levels $u_1, \ldots, u_r$ and relaxation values $\varepsilon_1, \varepsilon_2, \ldots, \varepsilon_{r-1}$ is the vector $x=(x_1, \ldots, x_n)'$ which maximizes

$$P\{\omega \mid c_r'(\omega)x \geqslant u_r\},$$

subject to

$$Ax \leqslant b$$

$$x \geqslant 0$$

$$P\{\omega \mid (c_1'\omega)x \geqslant u_1\} \geqslant p_1 - \varepsilon_1$$

$$P\{\omega \mid c_2'(\omega)x \geqslant u_2\} \geqslant p_2 - \varepsilon_2$$

$$\cdots\cdots\cdots\cdots$$

$$P\{\omega \mid c_{r-1}'(\omega)x \geqslant u_{r-1}\} \geqslant p_{r-1} - \varepsilon_{r-1},$$

where

$$p_i = \max_{x \in U_{i-1}} P\{\omega \mid c_i'(\omega)x \geqslant u_i\}, \quad (i = 1, 2, \ldots, r-1)$$

$$U_0 = D = \{x \mid Ax \leqslant b,\ x \geqslant 0\}$$

$$U_{i-1} = \{x \mid x \in U_{i-2}, P\{\omega \mid c_{i-1}'(\omega)x \geqslant u_{i-1}\} \geqslant p_{i-1} - \varepsilon_{i-1}\}, \quad (i = 1, \ldots, r-1).$$

Observation 3. For $r = 1$, one gets the minimum risk model due to Bereanu [17] and the P-model due to Charnes and Cooper [38].

Observation 4. The constraints $P(\sum_j c_j x_j \geqslant u) \geqslant \alpha$, where c_j are normally distributed with means $\bar{c}_j$ and covariance matrix V, are equivalent with the following deterministic constraints:

$$\sum_j \bar{c}_j x_j + \tau(x'Vx)^{1/2} \geqslant u, \tag{5.7}$$

where τ is given by

$$\frac{1}{\sqrt{2\pi}} \int_{-\infty}^{\tau} \exp\left(-\frac{1}{2}x^2\right) \mathrm{d}x = 1 - \alpha.$$

Numerical example. Consider the following stochastic programming problem with two objective functions:

$$\max F_1 = (c_1^{(1)} x_1 + c_2^{(1)} x_2)$$

$$\max F_2 = (c_1^{(2)} x_1 + c_2^{(2)} x_2)$$

subject to

$$x_1 + x_2 \leqslant 2$$

$$3x_1 + 2x_2 \geqslant 3$$

$$x_1, x_2 \geqslant 0,$$

where $c^{(1)} = (c_1^{(1)}, c_2^{(1)})$ and $c^{(2)} = (c_1^{(2)}, c_2^{(2)})$ have normal distributions. Their means and covariances are

$$m^{(1)} = \begin{pmatrix} 2 \\ 3 \end{pmatrix}, \quad V_1 = \begin{pmatrix} 1 & 1 \\ 1 & 2 \end{pmatrix},$$

$$m^{(2)} = \begin{pmatrix} 9 \\ 1 \end{pmatrix}, \quad V_2 = \begin{pmatrix} 2 & -1 \\ -1 & 1 \end{pmatrix}.$$

The separate solutions of the two problems ($c^{(1)}$ and $c^{(2)}$ are replaced by their means) are $S_1(x_1 = 0\,;\, x = 2)$ and $S_2\ (x_1 = 2,$ $x_2 = 0)$.

The value of the functions is 6 and 18, respectively.

Since $0 \notin D$, u_1 can take values only in the interval $(-\infty, 6)$, whereas u_2 lies in $(-\infty, 18)$. Choose $u_1 = \frac{5}{4}$ and $u_2 = 6$.

According to their importance, the two functions can be ordered as F_1, F_2.

For the minimum-risk problem

$$\max_{x \in D} P\left(c_1^{(1)} x_1 + c_2^{(1)} x_2 \geqslant \frac{5}{4}\right),$$

one gets [25] the solution $x_1^* = \frac{6}{11} = 0.54545$, $x_2^* = \frac{16}{11} = 1.45544$. The value of the objective function is then

$$P\left(c_1^{(1)} x_1^* + c_2^{(1)} x_2^* \geqslant \frac{5}{4}\right) =$$

$$= 1 - \Phi\left(\frac{5/4 - 2x_1^* - 3x_2^*}{x_1^{*2} + 2x_1^* x_2^* + 2x_2^{*2})}\right) =$$

$$= 1 - \Phi(-1.75677) = 1 - 0.0392 = 0.9608.$$

To the initial system of constraints one adds the constraint

$$P\left(c_1^{(1)} x_1 + c_2^{(1)} x_2 \geqslant \frac{5}{4}\right) \geqslant 0.9608 - 0.005 =$$

$$= 0.9558 \ (\varepsilon_1 = (0.005)$$

which, by virtue of (5.7), is equivalent with

$$2x_1 + 3x_2 - 1.75 \sqrt{x_1^2 + 2x_1 x_2 + 2x_2^2} \geqslant 5/4,$$

i.e.

$$1.11x_1^2 + 3.22x_2^2 - 5x_1 - 7.5x_2 + 6.22x_1 x_2 +$$

$$+ 1.5625 \geqslant 0.$$

Next, one solves the mathematical programming problem defined by the constraints :

$$x_1 + x_2 \leqslant 2$$

$$3x_1 + 2x_2 \geqslant 3$$

$$1.11x_1^2 + 3.22x_2^2 - 5x_1 - 7.5x_2 + 6.22x_1x_2 + 1.5625 \geqslant 0$$

and the objective function

$$\max P(c_1^{(2)}x_1 + c_2^{(2)}x_2 \geqslant 6).$$

The objective function is equivalent with

$$\min \frac{6 - (9x_1 + x_2)}{\sqrt{2x_1^2 - 2x_1x_2 + x_2^2}}.$$

The solution is $x_1 = 0.572335$, $x_2 = 1.427665$, and corresponds to the best compromise between F_1 and F_2, which take the values 0.955 and 0.7123, respectively.

Next, we shall describe a method for obtaining the multiple minimum-risk solution which is derived from Aggarwal and Swarup's method [175] for maximizing linear fractional functionals subject to one quadratic constraint and a number of linear constraints. (Aggarwal also applied it [2] for maximizing an indefinite quadratic fractional function subject to a quadratic constraint.) Here, this method will be introduced gradually, first for the case of two objective functions, then for three, and finally for r objective functions.

5.2. THE CASE OF TWO OBJECTIVE FUNCTIONS [140]

Consider the problem

$$\begin{aligned} &\max P\{\omega \,|\, c_1'(\omega)x \geqslant u_1\} = Z_1, \\ (5.8)\quad &\\ &\max P\{\omega \,|\, c_2'(\omega)x \geqslant u_2\} = Z_2, \end{aligned}$$

subject to

$$Ax \leqslant b, \tag{5.9}$$
$$x \geqslant 0.$$

Assume that $c_k(\omega)$ $(k = 1, 2)$ is an n-dimensional random variable of normal distribution with mean value $\bar{c}_k = (\bar{c}_{k1}, \ldots, \bar{c}_{kn})$ $(k = 1, 2)$ and covariance matrix V_k, $(k = 1, 2)$. Also suppose that the domain $U_0 = \{x | Ax \leqslant b, \ x \geqslant 0\}$ is bounded and non-empty.

Consider first the minimum-risk problem

$$\max_{x \in U_0} P\{\omega | c_1'(\omega)x \geqslant u_1\}$$

and let p_1 be the optimum value. Decrease p_1 by a value ε_1 $(\varepsilon_1 > 0)$ and solve

Problem 1:

$$\max P\{\omega | c_2'(\omega)x \geqslant u_2\},$$

subject to

$$Ax \leqslant b,$$
$$x \geqslant 0,$$
$$P\{\omega | c_1'(\omega)x \geqslant u_1\} \geqslant p_1 - \varepsilon_1.$$

The vector c_k follows a normal distribution, and hence

$$P\{\omega | c_1'(\omega)x \geqslant u_1\} = P\left\{\omega | \frac{c_1'(\omega)x - \bar{c}_1'x}{\sqrt{x'V_1x}} \geqslant \frac{u_1 - \bar{c}_1'x}{\sqrt{x'V_1x}}\right\},$$

$$P\left\{\omega | \frac{c_1'(\omega)x - \bar{c}_1'x}{\sqrt{x'V_1x}} < \frac{u_1 - \bar{c}_1'x}{\sqrt{x'V_1x}}\right\} \leqslant 1 - p_1 + \varepsilon_1$$

or

$$\Phi\left(\frac{u_1 - \bar{c}_1'x}{\sqrt{x'V_1x}}\right) \leqslant 1 - p_1 + \varepsilon_1,$$

where Φ is the Laplace function. Then,

$$\bar{c}_1'x + \Phi^{-1}(1 - p_1 + \varepsilon_1)(x'V_1x)^{1/2} \geqslant u_1.$$

According to [25], the efficiency function for Problem 1 is equivalent with

$$\max \frac{\bar{c}_2'x - u_2}{(x'V_2x)^{1/2}}$$

Hence, instead of Problem 1 one can solve

Problem 1′:

$$\max \frac{\bar{c}_2'x - u_2}{(x'V_2x)^{1/2}},$$

subject to

$$Ax \leqslant b, \; x \geqslant 0,$$

$$\bar{c}_1'x + \Phi^{-1}(1 - p_1 + \varepsilon_1)(x'V_1x)^{1/2} \geqslant u_1.$$

The solution of Problem 1′ will be the multiple minimum-risk solution corresponding to levels u_1 and u_2, and to the relaxation value ε_1. This solution also corresponds to a global-maxim, since according to assumption (ii), Section 5.1., the objective function is strictly quasi-concave. A method for solving Problem 1′ is described in the sequel. This consists in two phases :

Phase I. Solve the problem

$$\max \frac{\bar{c}_2'x - u_2}{(x'V_2x)^{1/2}},$$

subject to

$$Ax \leqslant b, \; x \geqslant 0.$$

This problem is obtained from Problem 1′ if the quadratic constraint is neglected.

A solution $\hat{x}$ to this problem can be obtained with the methods given in refs. [17], and [18]. If $\hat{x}$ also satisfies the constraint

$$\bar{c}_1'x + \Phi^{-1}(1 - p_1 + \varepsilon_1)(x'V_1x)^{1/2} \geqslant u_1,$$

then it is the solution of Problem 1′ and, hence, the multiple minimum-risk solution. If not, then for $x = \hat{x}$,

$$\bar{c}_1'x + \Phi^{-1}(1 - p_1 + \varepsilon_1)(x^1V_1x)^{1/2} < u_1,$$

and assuming that

$$\hat{\mu} = \frac{\bar{c}_2'\hat{x} - u_2}{(\hat{x}'V_2\hat{x})^{1/2}},$$

one can pass to Phase II.

Phase II. Solve the problem

$$\max[\bar{c}_1'x + \Phi^{-1}(1 - p_1 + \varepsilon_1)(x'V_1x)^{1/2}],$$

subject to

$$Ax \leqslant b, \ x \geqslant 0,$$

$$\frac{\bar{c}_2'x - u_2}{(x'V_2x)^{1/2}} \geqslant \hat{\mu}.$$

The last constraint can be also written as

$$\bar{c}_2'x - \hat{\mu}(x'V_2x)^{1/2} \geqslant u_2.$$

Obviously, a possible solution to this problem is $\hat{x}$, i.e. the optimum solution of Problem 1′.

The value of $\hat{\mu}$ is then gradually decreased. The value of the efficiency function may then rise or remain unchanged. If for $\mu = \bar{\mu}$ the maximum value of the efficiency function reaches u_1, then the optimum solution thus found will be also optimum for the Problem 1′. (This is justified by virtue of Theorem 5.1 below).

If there exists no value μ such that

$$\bar{c}_1'x + \Phi^{-1}(1 - p_1 + \varepsilon_1)(x'V_1x)^{1/2} \geqslant u_1,$$

then Problem 1′ has no solution for given p_1 and ε_1 and one has to resume the process with a larger value of ε_1.

THEOREM 5.1. *Consider*

Problem 2 :

$$\max[\bar{c}_1'x + \Phi^{-1}(1 - p_1 + \varepsilon_1)(x'V_1x)^{1/2}],$$

subject to

$$Ax \leqslant b,\ x \geqslant 0,$$

$$\bar{c}_2'x - \mu(x'V_2x)^{1/2} \geqslant u_2.$$

An optimum solution of Problem 2 *such that the objective function is equal to* u_1 *is also an optimum solution of Problem* 1′.

Proof. Write the Kuhn-Tucker conditions for an optimal solution to Problem 1′

$$\frac{\bar{c}_2'(x'V_2x) - (V_2x)(\bar{c}_2'x - u_2)}{(x'V_2x)^{3/2}} + \Bigg[\bar{c}_1' + \Phi^{-1}(1 -$$

$$- p_1 + \varepsilon_1)\frac{V_1x}{(x'V_1x)^{1/2}}\Bigg]v_\alpha - A'v + u = 0,$$

$$\bar{c}_1'x + \Phi^{-1}(1 - p_1 + \varepsilon_1)(x'V_1x)^{1/2} - y_1 = u_1,$$

$$Ax + y = b,$$

$$u'x + v'y + v_\alpha y_1 = 0,$$

$$x,\ y,\ u,\ v,\ v_\alpha,\ y_1 \geqslant 0,$$

where u and v are the vectors of the Lagrange multipliers associated with the variables x and y having n and m elements, respectively, y_1 is the slack variable and v_α is the Lagrange multiplier associated with y_1.

The Kuhn-Tucker conditions for Problem 2 are

$$\bar{c}_1' + \frac{\Phi^{-1}(1 - p_1 + \varepsilon_1)(V_1 x)}{(x'V_1 x)^{1/2}} - A'v + u +$$

$$+ \left[\bar{c}_2' - \frac{\mu V_2 x}{(x'V_2 x)^{1/2}}\right] v_\beta = 0$$

$$Ax + y = b$$

$$\bar{c}_2' x - \mu(x'V_2 x)^{1/2} - u_2 - y_2 = 0$$

$$u'x + v'y + v_\beta y_2 = 0$$

$$x, y, u, v, v_\beta, y_2 \geqslant 0,$$

where u and v are defined above, y_2 is the slack variable and v_β is the Lagrange multiplier associated with y_2.

If for $\mu = \bar{\mu}$ one gets

$$\bar{c}_1' x + \Phi^{-1}(1 - p_1 + \varepsilon_1)(x'V_1 x)^{1/2} = u_1$$

then $y_1 = 0$.

Suppose that $\hat{x}, \hat{y}, \hat{u}, \hat{v}, \hat{y}_2 = 0, \hat{v}_\beta$ is an optimum solution of Problem 2, i.e. it satisfies the Kuhn-Tucker conditions:

(5.10.1)

$$\bar{c}_1' + \frac{\Phi^{-1}(1 - p_1 + \varepsilon_1) V_1 \hat{x}}{(\hat{x}'V_1 \hat{x})^{1/2}} - A'\hat{v} + \hat{u} +$$

$$+ \left[\bar{c}_2' - \frac{\bar{\mu} V_2 \hat{x}}{(\hat{x}'V_2 \hat{x})^{1/2}}\right] \hat{v}_\beta = 0,$$

(5.10.2) $A\hat{x} + \hat{y} = b,$

(5.10.3) $\bar{c}_2' x - \bar{\mu}(\hat{x}'V_2 \hat{x})^{1/2} - u_2 = 0,$

(5.10.4) $\bar{c}_1'\hat{x} + \Phi^{-1}(1 - p_1 + \varepsilon_1)(\hat{x}'V_1\hat{x})^{1/2} = u_1,$

(5.10.5) $\hat{u}'\hat{x} + \hat{v}'\hat{y} = 0.$

(5.10.6) $\hat{x}, \hat{y}, \hat{u}, \hat{v}, \hat{v}_\beta \geqslant 0.$

From (5.10.3) it follows that

$$\bar{\mu} = \frac{\bar{c}_2'x - u_2}{(\hat{x}'V_2\hat{x})^{1/2}}.$$

Substituting $\bar{\mu}$ in (5.10.1) yields

$$\bar{c}_1' + \frac{\Phi^{-1}(1 - p_1 + \varepsilon_1)V_1\hat{x}}{(\hat{x}'V_1\hat{x})^{1/2}} - A'\hat{v} + \hat{u} +$$

$$+\left[\bar{c}_2' - \frac{V_2\hat{x}(\bar{c}_2'\hat{x} - u_2)}{(\hat{x}'V_2\hat{x})}\right]\hat{v}_\beta = 0$$

Divide by $\hat{v}_\beta(\hat{x}'V_2\hat{x})^{1/2}$ ($\hat{v}_\beta > 0$) and get

$$\left[\bar{c}_1' + \frac{\Phi^{-1}(1 - p_1 + \varepsilon_1)V_1\hat{x}}{(\hat{x}^1V_1\hat{x})^{1/2}}\right]\frac{1}{\hat{v}_\beta(\hat{x}'V_2\hat{x})^{1/2}} -$$

(5.11)

$$- A'\frac{\hat{v}}{\hat{v}_\beta(\hat{x}'V_2\hat{x})^{1/2}} + \frac{\bar{c}_2'}{(\hat{x}'V_2\hat{x})^{1/2}} -$$

$$- \frac{(V_2\hat{x})(c_2'\hat{x} - u_2)}{(\hat{x}'V_2\hat{x})^{1/2}} + \frac{\hat{u}}{\hat{v}_\beta(\hat{x}'V_2\hat{x})^{1/2}} = 0.$$

Denote

$$u^* = \frac{\hat{u}}{\hat{v}_\beta(\hat{x}'V_2\hat{x})^{1/2}},$$

$$v^* = \frac{\hat{v}}{\hat{v}_\beta(\hat{x}'V_2\hat{x})^{1/2}},$$

$$v_\alpha^* = \frac{1}{\hat{v}_\beta(\hat{x}'V_2\hat{x})^{1/2}},$$

and hence (5.11) becomes

$$(5.11') \quad \left[\bar{c}_1' + \frac{\Phi^{-1}(1 - p_1 + \varepsilon_1)V_1\hat{x}}{(\hat{x}'V_1\hat{x})^{1/2}}\right]v_\alpha^* - A'v^* + u^* + \frac{\bar{c}_2'}{(\hat{x}V_2\hat{x})^{1/2}} - \frac{V_2\hat{x}(c_2'\hat{x} - u_2)}{(\hat{x}'V_2\hat{x})^{1/2}}.$$

Dividing by $\hat{v}_\beta(\hat{x}'V_2\hat{x})^{1/2}$ relation (5.10.5) becomes (5.10.5') $u^*\hat{x}' + v^{*\prime}\hat{y} = 0$.

Relations (5.11'), (5.10.2), (5.10.4) and (5.10.5') imply that $\hat{x}, \hat{y}, u^*, v^*, \hat{y}_1 = 0, v_\alpha^*$ is an optimal solution of Problem 1' since it verifies the Kuhn-Tucker conditions. The proof is complete now.

5.3. THE CASE OF THREE OBJECTIVE FUNCTIONS [141]

Consider the problem

$$(5.12) \quad \begin{aligned} &\max P\{\omega \,|\, c_1'(\omega)x \geqslant u_1\} \\ &\max P\{\omega \,|\, c_2'(\omega)x \geqslant u_2\} \\ &\max P\{\omega \,|\, c_3'(\omega)x \geqslant u_3\}, \end{aligned}$$

subject to

$$(5.13) \quad \begin{aligned} Ax &\leqslant b \\ x &\geqslant 0. \end{aligned}$$

Assume that $c_k(\omega)$ $(k = 1, 2, 3)$ are n-dimensional random variables of normal distribution with mean value $\bar{c}_k = (\bar{c}_{k1}, \ldots, \bar{c}_{kn})$ $(k = 1, 2, 3)$ and covariance matrix V_k $(k = 1, 2, 3)$.

Then problem (5.12)—(5.13) is equivalent with

Problem 1

$$\max \frac{\bar{c}_1' x - u_1}{(x' V_1 x)^{1/2}}$$

$$(5.14) \quad \max \frac{\bar{c}_2' x - u_2}{(x' V_2 x)^{1/2}}$$

$$\max \frac{\bar{c}_3' x - u_3}{(x' V_3 x)^{1/2}},$$

subject to

$$(5.15) \quad \begin{aligned} Ax &\leqslant b \\ x &\geqslant 0. \end{aligned}$$

Consider also

Problem 1′

$$(5.16) \quad \max \frac{\bar{c}_3' x - u_3}{(x' V_3 x)^{1/2}}$$

subject to

$$Ax \leqslant b; \; x \geqslant 0$$

$$(5.17) \quad \bar{c}_1' x + \Phi^{-1}(1 - p_1 + \varepsilon_1)(x' V_1 x)^{1/2} \geqslant u_1$$

$$\bar{c}_2' x + \Phi^{-1}(1 - p_2 + \varepsilon_2)(x' V_2 x)^{1/2} \geqslant u_2.$$

We shall solve Problem 1, having three objective functions, by solving Problem 1′, which has only one objective function. In order to solve Problem 1′, we shall use the method applied in Section 5.2 herein and in ref. [140].

But let us first solve

Problem 1''

$$(5.18)\quad \max \frac{\bar{c}_3'x - u_3}{(x'V_3x)^{1/2}}$$

subject to

$$Ax \leqslant b; \; x \geqslant 0,$$

which was obtained from Problem 1' by neglecting the nonlinear constraints. If the optimum solution $\hat{x}$ also satisfies

$$(5.19)\quad \bar{c}_1'x + \Phi^{-1}(1 - p_1 + \varepsilon_1)(x'V_1x)^{1/2} \geqslant u_1,$$

$$(5.20)\quad \bar{c}_2'x + \Phi^{-1}(1 - p_2 + \varepsilon_2)(x'V_2x)^{1/2} \geqslant u_2,$$

then $\hat{x}$ is a multiple minimum-risk solution corresponding to levels u_1, u_2, u_3 and relaxation values ε_1 and ε_2.

Assume that $\hat{x}$ does not satisfy any of inequations (5.19) and (5.20). Suppose that

$$\mu^* = \frac{\bar{c}_3'\hat{x} - u_3}{(\hat{x}'V_3\hat{x})^{1/2}},$$

and solve

Problem 1'''

$$\max \, [\bar{c}_1'x + \Phi^{-1}(1 - p_1 + \varepsilon_1)(x'V_1x)^{1/2}],$$

subject to

$$Ax \leqslant b; \; x \geqslant 0$$

$$\bar{c}_3'x - \mu(x'V_3x)^{1/2} \geqslant u_3.$$

Decreasing values of $\mu < \mu^*$ are considered successively until the value $\bar{\mu}$ is reached such that

$$\max[\bar{c}_1'x + \Phi^{-1}(1 - p_1 + \varepsilon_1)(x'V_1x)^{1/2}] = u_1.$$

Let $\bar{x}$ be the optimum solution of Problem 1''' corresponding to $\mu = \bar{\mu}$. If $\bar{x}$ also satisfies the constraint:

$$\bar{c}_2'x + \Phi^{-1}(1 - p_2 + \varepsilon_2)(x'V_2x)^{1/2} \geqslant u_2,$$

then $\bar{x}$ is the multiple minimum-risk solution corresponding to levels u_1, u_2, u_3 and relaxation values ε_1 and ε_2. If not, then one has to solve

Problem 1''''

$$\max[\bar{c}_2'x + \Phi^{-1}(1 - p_2 + \varepsilon_2)(x'V_2x)^{1/2}],$$

subject to

$$Ax \leqslant b, \; x \geqslant 0$$

$$\bar{c}_3'x - \mu(x'V_3x)^{1/2} \geqslant u_3,$$

for various values of $\mu < \bar{\mu}$. If for $\mu = \bar{\bar{\mu}}$,

$$\max[\bar{c}_2'x + \Phi^{-1}(1 - p_2 + \varepsilon_2)(x'V_2x)^{1/2}] = u_2,$$

then the optimum solution $\bar{\bar{x}}$ of Problem 1'''' is the multiple minimum-risk solution corresponding to levels u_1, u_2, u_3 and relaxation values ε_1 and ε_2.

The algorithm for obtaining the multiple minimum-risk solution for a problem with three objective functions is represented by the block diagram shown in fig. 5.1. The idea underlying this algorithm is as follows:

Let C_1 and C_2 (fig. 5.2) be the curves obtained under the assumption that (5.19) and (5.20) are equalities.

In order to simplify the solution of Problem 1', we neglect the nonlinear constraints. If the corresponding optimum solution $\hat{x}$ also satisfies the nonlinear constraints, then the problem is solved. Else, if the optimum solution $\hat{x}$ is outside the hatched area in fig. 5.2, i.e. it does not satisfy (5.19) and (5.20), then one constructs a sequence of values for x until a feasible solution is found. This solution will be optimum.

In obtaining this sequence of values for x, we must bear in mind that the decrease of μ leads to an increase of the value of $\bar{c}_2'x + \Phi^{-1}(1 - p_2 + \varepsilon_2)(x'V_2x)^{1/2}$.

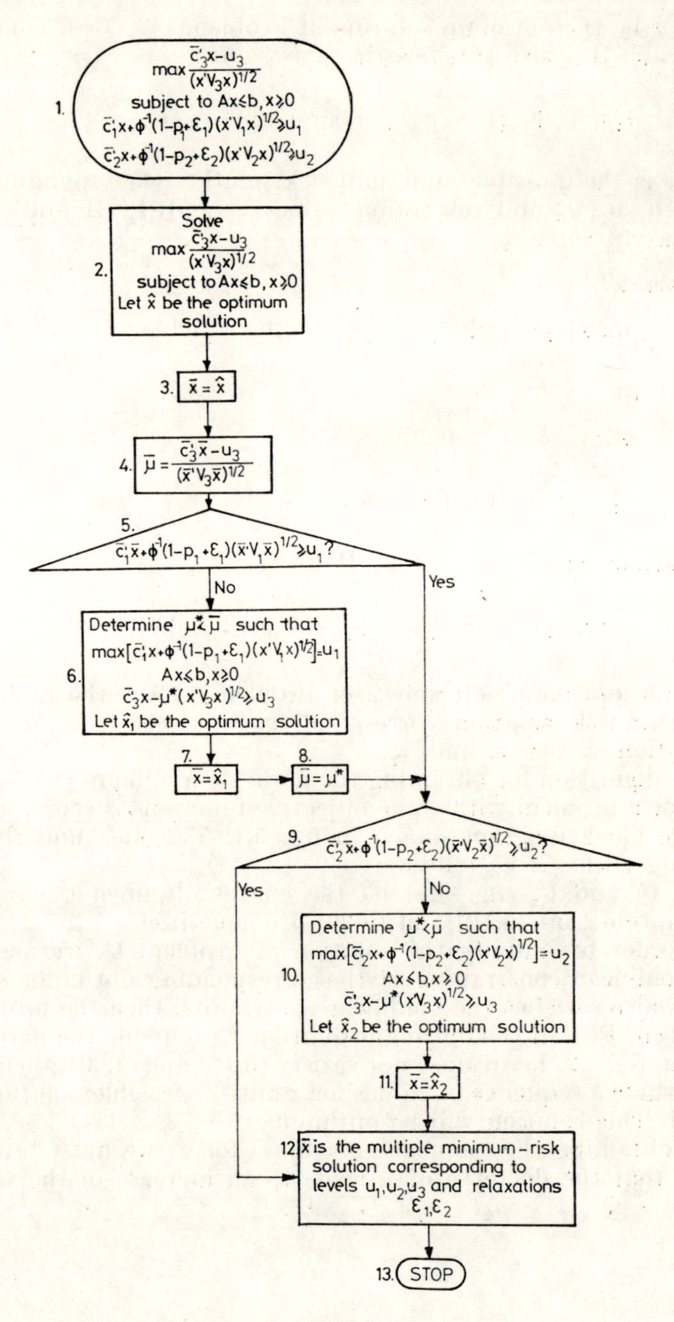
1.
max (c̄'₃x − u₃)/(x'V₃x)^{1/2}
subject to Ax ≤ b, x ≥ 0
c̄'₁x + φ⁻¹(1−p₁+ε₁)(x'V₁x)^{1/2} ≥ u₁
c̄'₂x + φ⁻¹(1−p₂+ε₂)(x'V₂x)^{1/2} ≥ u₂
2.
Solve
max (c̄'₃x − u₃)/(x'V₃x)^{1/2}
subject to Ax ≤ b, x ≥ 0
Let x̂ be the optimum solution
3. x̄ = x̂
4. μ̄ = (c̄'₃x̄ − u₃)/(x̄'V₃x̄)^{1/2}
5.
c̄'₁x̄ + φ⁻¹(1−p₁+ε₁)(x̄'V₁x̄)^{1/2} ≥ u₁ ?
No
Yes
6.
Determine μ* < μ̄ such that
max[c̄'₁x + φ⁻¹(1−p₁+ε₁)(x'V₁x)^{1/2}] = u₁
Ax ≤ b, x ≥ 0
c̄'₃x − μ*(x'V₃x)^{1/2} ≥ u₃
Let x̂₁ be the optimum solution
7. x̄ = x̂₁
8. μ̄ = μ*
9.
c̄'₂x̄ + φ⁻¹(1−p₂+ε₂)(x̄'V₂x̄)^{1/2} ≥ u₂ ?
Yes
No
10.
Determine μ* < μ̄ such that
max[c̄'₂x + φ⁻¹(1−p₂+ε₂)(x'V₂x)^{1/2}] = u₂
Ax ≤ b, x ≥ 0
c̄'₃x − μ*(x'V₃x)^{1/2} ≥ u₃
Let x̂₂ be the optimum solution
11. x̄ = x̂₂
12. x̄ is the multiple minimum-risk solution corresponding to levels u₁, u₂, u₃ and relaxations ε₁, ε₂
13. STOP

Fig. 5.1.

For this reason, if one of the restrictions (5.19) or (5.20) is verified for a given x, then the sequential values of x will also verify it. For instance, if one chooses the branch 5—6—7—8—9—10 (fig. 5.1), then in Step 10 constraint (5.19) was

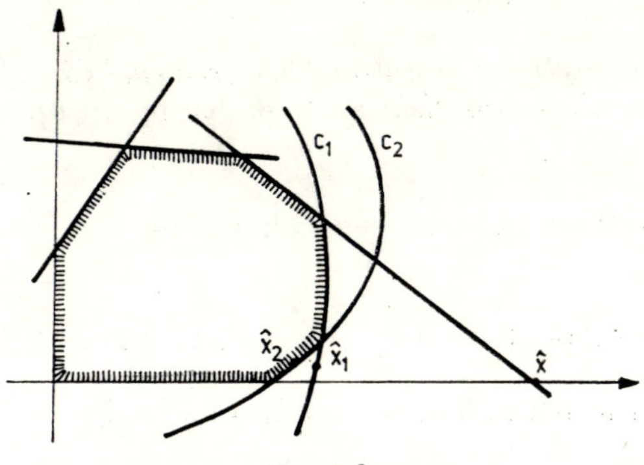

Fig. 5.2.

excluded from the model since it is self-evident. To prove this, one notices that by decreasing μ, the value of

$$\bar{c}_1'x + \Phi^{-1}(1 - p_1 + \varepsilon_1)(x'V_1x)^{1/2}$$

may either be stationary or may increase. Then, for $\mu^* < \bar{\mu}$, its value will be at least u_1 since for $\mu = \bar{\mu}$

$$\max[\bar{c}_1'x + \Phi^{-1}(1 - p_1 + \varepsilon_1)(x'V_1x)^{1/2}] = u_1.$$

The algorithm used to solve problem (5.16)—(5.17) is based on the following

THEOREM 5.2. *Consider*

Problem 2:

$$\max[\bar{c}_i'x + \Phi^{-1}(1 - p_i + \varepsilon_i)(x'V_ix)^{1/2}]. \qquad i \neq 3$$

subject to

$$Ax \leqslant b; \; x \geqslant 0$$

$$\bar{c}_3'x - \mu(x'V_3x)^{1/2} \geqslant u_3.$$

If there exists an index value $i \in \{1, 2\}$ *such that the optimum value of the objective function for this problem is* u_i *and if the optimum solution* $\hat{x}$ *verifies*

$$\bar{c}_j'x + \Phi^{-1}(1 - p_j + \varepsilon_j)(x'V_jx)^{1/2} \geqslant u_j, \; j \in \{1, 2\} - \{i\},$$

then $\hat{x}$ *is the optimal solution of problem (5.16)–(5.17), i.e. the multiple minimum-risk solution for levels* u_1, u_2, u_3 *and relaxation values* ε_1, ε_2.

Proof. Without loss of generality, one can assume that for $i = 2$,

$$\max[\bar{c}_2'x + \Phi^{-1}(1 - p_2 + \varepsilon_2)(x'V_2x)^{1/2}] = u_2.$$

Write now the Kuhn-Tucker conditions for Problem 1′:

$$\frac{\bar{c}_3'(x'V_3x) - (V_3x)(\bar{c}_3'x - u_3)}{(x'V_3x)^{3/2}} + \Big[\bar{c}_1' + \Phi^{-1}(1 -$$

$$- p_1 + \varepsilon_1)\frac{V_1x}{(x'V_1x)^{1/2}}\Big]v_{\alpha_1} - A'v + u +$$

$$+ \Big[\bar{c}_2' + \Phi^{-1}(1 - p_2 + \varepsilon_2)\,\frac{V_2x}{(x'V_2x)^{1/2}}\Big]v_{\alpha_2} = 0,$$

$$Ax + y = b,$$

$$\bar{c}_1'x + \Phi^{-1}(1 - p_1 + \varepsilon_1)(x'V_1x)^{1/2} - y_1 = u_1,$$

$$\bar{c}_2'x + \Phi^{-1}(1 - p_2 + \varepsilon_2)(x'V_2x)^{1/2} - y_2 = u_2,$$

$$u'x + v'y + v_{\alpha_1}y_1 + v_{\alpha_2}y_2 = 0,$$

$$x, y, u, v, v_{\alpha_1}, v_{\alpha_2}, y_1, y_2 \geqslant 0.$$

$y_1, y_2, y = (y^1, \ldots, y^m)$ above are slack variables, and $u, v, v_{\alpha_1}, v_{\alpha_2}$ are Lagrange's multipliers.

The Kuhn-Tucker conditions for Problem 2 are

$$\bar{c}_2' + \Phi^{-1}(1 - p_2 + \varepsilon_2)\frac{V_2x}{(x'V_2x)^{1/2}} - A'v + u +$$

$$+ \left[\bar{c}_3' - \mu\frac{V_3x}{(x'V_3x)^{1/2}}\right]v_{\beta_3} + \left[\bar{c}_1' + \Phi^{-1}(1 - p_1 +\right.$$

$$\left. + \varepsilon_1)\frac{V_1x}{(x'V_1x)^{1/2}}\right]v_{\beta_1} = 0,$$

$$Ax + y = b,$$

$$\bar{c}_3'x - \mu(x'V_3x)^{1/2} - y_3' = u_3,$$

$$\bar{c}_1'x + \Phi^{-1}(1 - p_1 + \varepsilon_1)(x'V_1x)^{1/2} - y_1' = u_1,$$

$$\bar{c}_2'x + \Phi^{-1}(1 - p_2 + \varepsilon_2)(x'V_2x) = u_2,$$

$$u'x + v'y + v_{\beta_1}y_1 + v_{\beta_3}y_3' = 0,$$

$$x, y, u, v, v_{\beta_1}, v_{\beta_3}, y_1', y_3' \geqslant 0.$$

Since for $\mu = \bar{\mu}$, $\bar{c}_2'x + \Phi^{-1}(1 - p_2 + \varepsilon_2)(x'V_2x)^{1/2} = u_2$, $\bar{c}_3'x + \Phi^{-1}(1 - p_3 + \varepsilon_3)(x'V_3x)^{1/2} = u_3$, then $y_2 = 0$ and $y_3' = 0$.

Let $\hat{x}$, $\hat{y}$, $\hat{u}$, $\hat{v}$, $\hat{y}_1'$, $\hat{y}_3' = 0$, $\hat{v}_{\beta_1}$, $\hat{v}_{\beta_3}$ be an optimal solution of Problem 2. It follows that for this solution the Kuhn-Tucker conditions are verified:

$$\bar{c}_2' + \Phi^{-1}(1 - p_2 + \varepsilon_2)\frac{V_2\hat{x}}{(\hat{x}'V_2\hat{x})^{1/2}} - A'\hat{v} + \hat{u} +$$

$$+ \left[\bar{c}_3' - \frac{V_3\hat{x}}{(\hat{x}'V_3\hat{x})^{1/2}}\right]\hat{v}_{\beta_2} + \left[\bar{c}_1' + \Phi^{-1}(1 - p_1 +\right. \tag{5.21.1}$$

$$\left. + \varepsilon_1)\frac{V_1\hat{x}}{(\hat{x}'V_1\hat{x})^{1/2}}\right]\hat{v}_{\beta_1} = 0,$$

(5.21.2) $A\hat{x} + \hat{y} = b,$

(5.21.3) $\bar{c}_1'\hat{x} + \Phi^{-1}(1 - p_1 + \varepsilon_1)(\hat{x}'V_1\hat{x})^{1/2} - \hat{y}_1' - u_1 = 0,$

(5.21.4) $\bar{c}_2'\hat{x} + \Phi^{-1}(1 - p_2 + \varepsilon_2)(\hat{x}'V_2\hat{x})^{1/2} = u_2,$

(5.21.5) $\bar{c}_3'\hat{x} - \bar{\mu}(\hat{x}'V_3\hat{x})^{1/2} - u_3 = 0,$

(5.21.6) $\hat{u}'\hat{x} + \hat{v}'\hat{y} + \hat{v}_{\beta_1}\hat{y}_1 = 0,$

(5.21.7) $\hat{x}, \hat{y}, \hat{u}, \hat{v}, \hat{y}_1', \hat{v}_{\beta_1}, \hat{v}_{\beta_3} \geqslant 0.$

Eliminate $\bar{\mu}$ from (5.21.1)–(5.21.5). From (5.21.5),

$$\bar{\mu} = \frac{\bar{c}_3'\hat{x} - u_3}{(\hat{x}'V_3\hat{x})^{1/2}}.$$

Introduce this value in (5.21.1). One gets:

$$\bar{c}_2' + \Phi^{-1}(1 - p_2 + \varepsilon_2)\frac{V_2\hat{x}}{(\hat{x}'V_2\hat{x})^{1/2}} - A'\hat{v} + \hat{u} +$$

$$+\left[\bar{c}_3' - \frac{(\bar{c}_3'\hat{x} - u_3)V_3\hat{x}}{\hat{x}'V_3\hat{x}}\right]\hat{v}_{\beta_3} + \left[\bar{c}_1' + \Phi^{-1}(1 -\right.$$

$$\left. - p_1 + \varepsilon_1)\frac{V_1\hat{x}}{(\hat{x}'V_1\hat{x})^{1/2}}\right]\hat{v}_{\beta_1} = 0$$

Divide by $\hat{v}_{\beta_3}(\hat{x}'V_3\hat{x})^{1/2}$ and get

$$\left[\bar{c}_2' + \Phi^{-1}(1 - p_2 + \varepsilon_2)\frac{V_2\hat{x}}{(\hat{x}'V_2\hat{x})^{1/2}}\right]\frac{1}{\hat{v}_{\beta_3}(\hat{x}'V_3\hat{x})^{1/2}} -$$

$$- A'\frac{\hat{v}}{\hat{v}_{\beta_3}(\hat{x}'V_3\hat{x})^{1/2}} + \left[\frac{\hat{u}}{\hat{v}_{\beta_3}(\hat{x}'V_3\hat{x})^{1/2}} + \left[\bar{c}_3' -\right.\right.$$

$$\left. - \frac{(\bar{c}_3'\hat{x} - u_3)V_3\hat{x}}{\hat{x}'V_3\hat{x}}\right]\frac{1}{(\hat{x}'V_3\hat{x})^{1/2}} + \left[\bar{c}_1' + \Phi^{-1}(1 - p_1 +\right.$$

$$\left. + \varepsilon_1)\frac{V_1\hat{x}}{(\hat{x}'V_1\hat{x})^{1/2}}\right]\frac{\hat{v}_{\beta_1}}{\hat{v}_{\beta_3}(\hat{x}'V_3\hat{x})^{1/2}} = 0.$$

Denote

$$u^* = \frac{\hat{u}}{\hat{v}_{\beta_3}(\hat{x}'V_3\hat{x})^{1/2}},$$

$$v^* = \frac{\hat{v}}{\hat{v}_{\beta_3}(\hat{x}'V_3\hat{x})^{1/2}},$$

$$v^*_{\alpha_1} = \frac{\hat{v}_{\beta_1}}{\hat{v}_{\beta_3}(\hat{x}'V_3\hat{x})^{1/2}},$$

$$v^*_{\alpha_2} = \frac{1}{\hat{v}_{\beta_3}(\hat{x}'V_3\hat{x})^{1/2}}$$

Then the previous relation becomes

$$\frac{\bar{c}_3'(\hat{x}'V_3\hat{x}) - V_3\hat{x}(\bar{c}_3'\hat{x} - u_3)}{(\hat{x}'V_3\hat{x})^{1/2}} + \Bigg[\bar{c}_1' + \Phi^{-1}(1 -$$

$$(5.22.1) \qquad - p_1 + \varepsilon_1)\frac{V_1\hat{x}}{(\hat{x}'V_1\hat{x})^{1/2}}\Bigg] v^*_{\alpha_1} - A'v^* + u^* +$$

$$+\Bigg[\bar{c}_2' + \Phi^{-1}(1 - p_2 + \varepsilon_2)\frac{V_2\hat{x}}{(\hat{x}'V_2\hat{x})^{1/2}}\Bigg] v^*_{\alpha_2} = 0.$$

Similarly, dividing (5.21.6) by $\hat{v}_{\beta_3}(\hat{x}'V_3\hat{x})^{1/2}$ one gets

$$(5.22.2) \qquad u^{*\prime}\hat{x} + v^{*\prime}\hat{y} + v^*_{\alpha_1}\hat{y}_1 = 0.$$

Relations (5.12.1)—(5.21.5) and (5.22.1), (5.22.2) imply that $\hat{x}$, $\hat{y}$, u^*, v^*, $y_1 = \hat{y}_1$, $\hat{y}_2 = 0$, $v^*_{\alpha_1}$, $v^*_{\alpha_2}$ verify the Kuhn-Tucker conditions for Problem 1′; hence, they are the optimal solution for Problem 1′. The proof is complete now.

5.4. THE CASE OF r OBJECTIVE FUNCTIONS [141]

If one considers r objective functions ($r > 2$), the problem to be solved becomes

$$\max P\{\omega \mid c_1'(\omega)x \geqslant u_1\} = Z_1$$

$$\max P\{\omega \mid c_2'(\omega)x \geqslant u_2\} = Z_2$$

$$\cdots\cdots\cdots\cdots\cdots$$

$$\max P\{\omega \mid c_r'(\omega)x \geqslant u_r\} = Z_r,$$

subject to

$$Ax \leqslant b; \; x \geqslant 0,$$

i.e.

$$\max \frac{\bar{c}_1'x - u_1}{(x'V_1x)^{1/2}}$$

$$\max \frac{\bar{c}_2'x - u_2}{(x'V_2x)^{1/2}}$$

$$\cdots\cdots\cdots\cdots$$

$$\max \frac{\bar{c}_r'x - u_r}{(x'V_rx)^{1/2}}$$

subject to

$$Ax \leqslant b; \; x \geqslant 0.$$

The solution of this problem can be obtained by solving

Problem 3:

$$\max \frac{\bar{c}_r'x - u_r}{(x'V_rx)^{1/2}},$$

subject to

$$Ax \leqslant b,\ x \geqslant 0$$

$$\bar{c}_1'x + \Phi^{-1}(1 - p_1 + \varepsilon_1)(x'V_1x)^{1/2} \geqslant u_1$$

$$\bar{c}_2'x + \Phi^{-1}(1 - p_2 + \varepsilon_2)(x'V_2x)^{1/2} \geqslant u_2$$

$$\cdots\cdots\cdots\cdots\cdots\cdots$$

$$\bar{c}_{r-1}'x + \Phi^{-1}(1 - p_{r-1} + \varepsilon_{r-1})(x'V_{r-1}x)^{1/2} \geqslant u_{r-1}.$$

The algorithm used to solve Problem 3 develops in the following steps:

Step 1. Determine the optimm solution $\hat{x}$ of the problem

$$\max \left\{ \frac{\bar{c}_r'x - u_r}{(x'V_rx)^{1/2}} \,|Ax \leqslant b,\ x \geqslant 0 \right\}.$$

If $\hat{x}$ also verifies the constraints

$$\bar{c}_1'x + \Phi^{-1}(1 - p_1 + \varepsilon_1)(x'V_1x)^{1/2} \geqslant u_1$$

$$\cdots\cdots\cdots\cdots\cdots\cdots$$

$$c_{r-1}'x + \Phi^{-1}(1 - p_{r-1} + \varepsilon_{r-1})(x'V_{r-1}x)^{1/2} \geqslant u_{r-1},$$

then put $\bar{x} = \hat{x}$ and proceed to Step 5. Else, follow Step 2.

Step 2. Choose

$$\bar{\mu} = \frac{\bar{c}_r'\hat{x} - u_r}{(\hat{x}'V_r\hat{x})^{1/2}}$$

$$\bar{x} = \hat{x},$$

and go to Step 3.

Step 3. If there exists an index value i such that

$$\bar{c}_i'\bar{x} + \Phi^{-1}(1 - p_i + \varepsilon_i)(\bar{x}'V_i\bar{x})^{1/2} < u_i,$$

then determine $\mu^* < \bar{\mu}$ such that

$$\max\,[\bar{c}_i'x + \Phi^{-1}(1 - p_i + \varepsilon_i)(x'V_ix)^{1/2}] = u_i,$$

subject to

$$Ax \leqslant b;\ x \geqslant 0,$$

$$\bar{c}_r'x - \mu^*(x'V_rx)^{1/2} \geqslant u_r.$$

Denote the optimal solution by $\bar{x}$ and go to Step 4.

Step 4. If

$$\bar{c}_i'\bar{x} + \Phi^{-1}(1 - p_i + \varepsilon_i)(\bar{x}'V_i\bar{x})^{1/2} \geqslant u_i,\ i = 1, \ldots, r-1,$$

then go to Step 5. Else, put $\bar{\mu} = \mu^*$ and go to Step 3.

Step 5. Consider the multiple minimum-risk solution $\bar{x}$ corresponding to levels $u_1, \ldots, u_r$ and relaxation values $\varepsilon_1, \ldots, \varepsilon_{r-1}$
The flowchart of the above algorithm is given in fig. 5.3. This algorithm is based on the following

THEOREM 5.3. *Consider*

Problem 4

$$\max\,[\bar{c}_i'x + \Phi^{-1}(1 - p_i + \varepsilon_i)(x'V_ix)^{1/2}] \qquad i \neq r$$

subject to

$$Ax \leqslant b;\ x \geqslant 0$$

$$\bar{c}_r'x - \mu(x'V_rx)^{1/2} \geqslant u_r.$$

If there exists an index $i \in \{1, 2, \ldots, r-1\}$ such that the optimum value of the objective function for this problem is equal to u_i and if the optimal solution $\hat{x}$ satisfies

$$\bar{c}_j'x + \Phi^{-1}(1 - p_j + \varepsilon_j)(x'V_jx)^{1/2} \geqslant u_j,\ j \in \{1, \ldots$$

$$\ldots, r-1\} - \{i\},$$

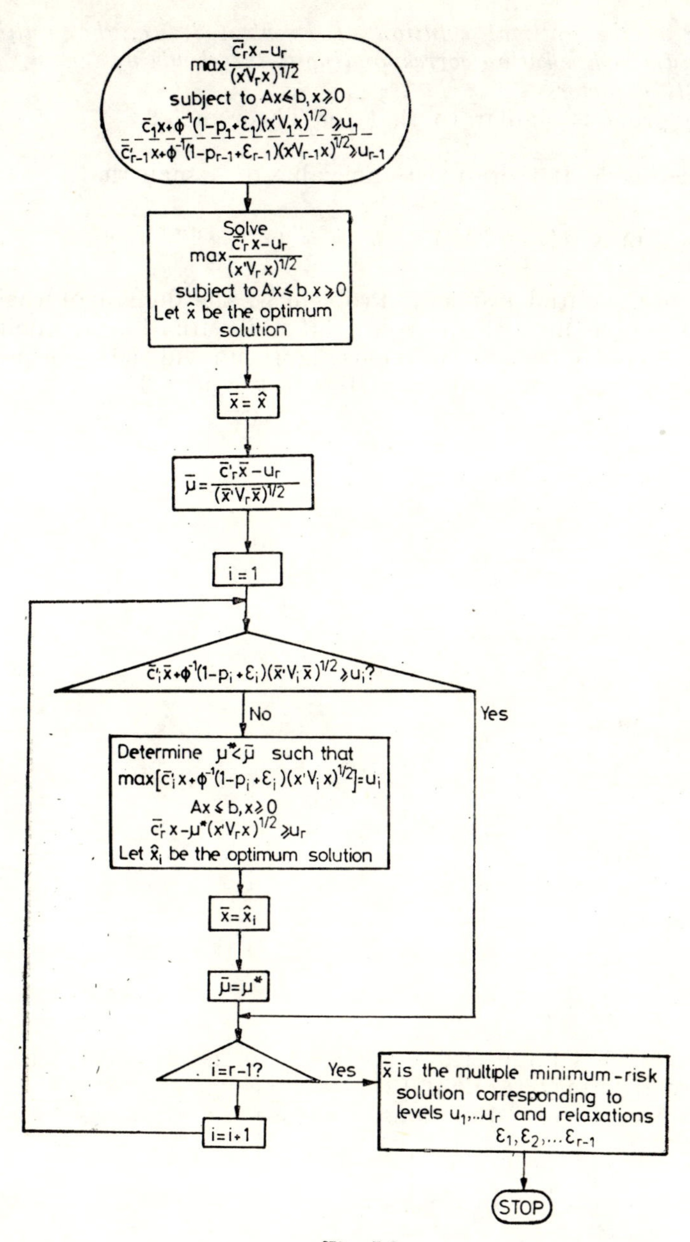

max $\frac{\bar{c}'_r x - u_r}{(x'V_r x)^{1/2}}$
subject to $Ax \leqslant b, x \geqslant 0$
$\bar{c}_1 x + \phi^{-1}(1-p_1+\varepsilon_1)(x'V_1 x)^{1/2} \geqslant u_1$
$\bar{c}_{r-1} x + \phi^{-1}(1-p_{r-1}+\varepsilon_{r-1})(x'V_{r-1} x)^{1/2} \geqslant u_{r-1}$
Solve
max $\frac{\bar{c}'_r x - u_r}{(x'V_r x)^{1/2}}$
subject to $Ax \leqslant b, x \geqslant 0$
Let $\hat{x}$ be the optimum solution
$\bar{x} = \hat{x}$
$\bar{\mu} = \frac{\bar{c}'_r \bar{x} - u_r}{(\bar{x}'V_r \bar{x})^{1/2}}$
$i = 1$
$\bar{c}'_i \bar{x} + \phi^{-1}(1-p_i+\varepsilon_i)(\bar{x}'V_i \bar{x})^{1/2} \geqslant u_i$?
No
Yes
Determine $\mu^* < \bar{\mu}$ such that
$\max[\bar{c}'_i x + \phi^{-1}(1-p_i+\varepsilon_i)(x'V_i x)^{1/2}] = u_i$
$Ax \leqslant b, x \geqslant 0$
$\bar{c}'_r x - \mu^*(x'V_r x)^{1/2} \geqslant u_r$
Let $\hat{x}_i$ be the optimum solution
$\bar{x} = \hat{x}_i$
$\bar{\mu} = \mu^*$
$i = r-1$?
Yes
$\bar{x}$ is the multiple minimum-risk solution corresponding to levels $u_1, \ldots u_r$ and relaxations $\varepsilon_1, \varepsilon_2, \ldots \varepsilon_{r-1}$
$i = i+1$
STOP

Fig. 5.3.

then $\hat{x}$ is the optimal solution of Problem 3, i.e. the multiple minimum-risk solution corresponding to thresholds $u_1, \ldots, u_r$ and relaxation factors $\varepsilon_1, \varepsilon_2, \ldots, \varepsilon_{r-1}$.

The proof is similar to that for Theorem 5.2.

Observation 5. If there exists no value of μ such that

$$\max\,[\bar{c}_i'x + \Phi^{-1}(1 - p_i + \varepsilon_i)(x'V_ix)^{1/2}] = u_i,$$

then, for the initial problem (Problem 3), the domain of feasible solutions is assumed to be void. But this situation was deliberately excluded from the previous algorithm and the corresponding block diagrams plotted in figs. 5.1. and 5.3.

5.5. A DIFFERENT APPROACH

Consider again the problem

$$\max \frac{\bar{c}_1'x - u_1}{(x'V_1x)^{1/2}}$$

$$\max \frac{\bar{c}_2'x - u_2}{(x'V_2x)^{1/2}}$$

$$\cdots\cdots\cdots\cdots$$

$$\max \frac{\bar{c}_r'x - u_r}{(x'V_rx)^{1/2}}$$

subject to

$$Ax \leqslant b$$

$$x \geqslant 0.$$

We give below an alternate approach to the problem, which is largely tributary to Blau [27]:

For the above r functions, consider the synthesis function

$$\min_{j=1,\ldots,r} \left[\frac{\bar{c}_j'x - u_j}{(x'V_jx)^{1/2}}\right]$$

and maximize it. The following model is obtained:

$$(5.23)\quad \max_{x} \min_{j=1,\ldots,r} \left[\frac{\bar{c}_j' x - u_j}{(x' V_j x)^{1/2}}, \right],$$

subject to

$$(5.24)\quad \begin{aligned} Ax &\leqslant b \\ x &\geqslant 0. \end{aligned}$$

Problem (5.23)—(5.24) can be decomposed into r separate problems. For a given $k \in J = \{1, 2, \ldots, r\}$, denote by D_k the subset of $D = \{x \mid Ax \leqslant b;\ x \geqslant 0\}$ defined by

$$D_k = \left\{ x \in D \mid \min_{j \in J} \left[\frac{\bar{c}_j' x - u_j}{(x' V_j x)^{1/2}} \right] = \frac{\bar{c}_k' x - u_k}{(x' V_k x)^{1/2}} \right\}.$$

But D_k is also

$$D_k = \left\{ x \in D \mid \frac{\bar{c}_k' x - u_k}{(x' V_k x)^{1/2}} \leqslant \frac{\bar{c}_j' x - u_j}{(x' V_j x)^{1/2}} \right\}, (j \in J - \{k\}).$$

By virtue of the relation between the objective function and the domains D_k, problem (5.23)—(5.24) has been decomposed into r sub-problems:

$$(5.25)\quad \max_{x} \frac{\bar{c}_k x - u_k}{(x' V_k x)^{1/2}},$$

subject to

$$\begin{aligned} \frac{\bar{c}_k' x - u_k}{(x' V_k x)^{1/2}} &\leqslant \frac{\bar{c}_j' x - u_j}{(x' V_j x)^{1/2}}, \quad j \in J - \{k\}, \\ Ax &\leqslant b, \\ x &\geqslant 0. \end{aligned}$$

Since for some problems of type (5.25), the domain of feasible solutions may be void, define $T = \{k \mid k \in J \text{ and } D_k \neq \emptyset\}$, where $D_k = \{x | x \text{ is feasible for } (5.25)\}$. Hence, Lemma 5.1. can be reformulated as

LEMMA 5.1. *Let $r \in J$ and $r \notin T$. Then*

$$\frac{\bar{c}_k' x - u_r}{(x' V_k x)^{1/2}} < \frac{\bar{c}_r' x - u_r}{(x' V_r x)^{1/2}}, \quad x \in D_k,\, k \in T.$$

According to this lemma, for $k \in T$ one can replace problems (5.25) by the problems

$$(5.25') \quad \max_x \frac{\bar{c}_k' x - u_k}{(x' V_k x)^{1/2}},$$

subject to

$$\frac{\bar{c}_k' x - u_k}{(x' V_k x)^{1/2}} \leqslant \frac{\bar{c}_j' x - u_j}{(x' V_j x)^{1/2}}, \quad j \in T - \{k\}$$

$$Ax \leqslant b,$$

$$x \geqslant 0.$$

Obviously, problems (5.25) and (5.25') have the same domain of feasible solutions.

Problem (5.25'), which is of fractional programming under nonlinear constraints, can be solved by use of a method similar to that presented in Chapter 1, Section 1.8 (linear case), or a method similar to those given in Chapter 4, Section 4.1.

From the above one can state

LEMMA 5.2. *For any $k \in T$, $x \in D_k$ if and only if $x \in D$ and*

$$\min_{j \in J} \left[\frac{\bar{c}_j' x - u_j}{(x' V_j x)^{1/2}} \right] = \frac{\bar{c}_k' x - u_k}{(x' V_k x)^{1/2}}.$$

The solution of problem (5.23)–(5.24) is given by the following

THEOREM 5.4. *Let $X_k \subset D$ be the set of optimal solutions for problem (5.25'). If for a given $r \in T$ one has*

$$(5.26) \quad \max_{k \in T} \left[\frac{\bar{c}_k' x_k - u_k}{(x_k' V_k x_k)^{1/2}} \right] = \frac{\bar{c}_r' x_r - u_r}{(x_r' V_r x_r)^{1/2}},$$

where $x_k \in X_k$, *then* X_r *is the set of optimal solutions for problem* (5.23)—(5.24).

Proof. For $x_k \in X_k$, $k \in T$, Lemma 5.2 implies that $x_k \in D$.

Assume that the optimal solution of problem (5.23)—(5.24) does not belong to the sets X_k. Let $z \in D$, $z \notin \bigcup_{k \in T} X_k$. There exists an index value $s \in T$ such that

$$\min_{j \in J} \left[\frac{\bar{c}_j' z - u_j}{(z' V_j z)^{1/2}} \right] = \frac{\bar{c}_s' z - u_s}{(z' V_s z)^{1/2}} .$$

From Lemma 5.2, it results that $z \in D_s$, but z is not optimal for problem (5.23)—(5.26) since

$$\frac{\bar{c}_s' z - u_s}{(z' V_s z)^{1/2}} < \frac{\bar{c}_s' x_s - u_s}{(x_s' V_s x_s)^{1/2}} .$$

Hence, the set of optimal solutions for problem (5.23)—(5.24) is a subset of $\bigcup_{k \in T} D_k$. According to the assumption in the above theorem, this set is just X_r.

If there are two integers r and s, ($r \neq s$), both of which satisfy the conditions in Theorem 5.4., then $X_r = X_s$.

From Theorem 5.4. one can derive the solution method for problem (5.23)—(5.24):

Solve problems (5.25′) and find the sets of optimum solutions X_k or only an element of each of these sets. Choose an element x_k from each set. The value of r which satisfies (5.26) will give the set X_r of optimal solutions for problem (5.23) — (5.24).

The following questions are implicit in solving problem (5.23)—(5.24):

1) Can one decide that the optimum solution x_r for problem (5.25′) is also optimum for problem (5.23)—(5.24)? If x_r is not optimum for (5.23)—(5.24), how can one of the problems (5.25′) be discarded next?

In order to prove such a result, we use the following [27].

LEMMA 5.3. *Let* $g_j(x)$, $j \in J$ *be continuous functions in the domain* E. *If for* $x \in E$

$$\min_{j \in J} [g_j(x)] = g_l(x), \qquad l \in R \subset J,$$

then there exists a $\zeta > 0$ *such that, for any* $z \in E$ *with* $|z - x| < \zeta$, *there exists a nonempty set* $R_z \subset R$ *for which*

$$\min_{j \in J} [g_j(z)] = g_l(z), \qquad l \in R_z.$$

THEOREM 5.5. *Let* $r \in T$. *If for* x

$$\frac{\bar{c}_r'x - u_r}{(x'V_rx)^{1/2}} = \frac{\bar{c}_l'z - u_l}{(x'V_lx)^{1/2}}, \qquad l \in R \subset T - \{r\},$$

then a necessary and sufficient condition for x *to be an optimum solution of problem* (5.23)—(5.24) *is that* x *is an optimum solution for* (5.25′) *for any* $l \in R \cup \{r\}$.

Proof. We prove only the sufficiency since the necessity derives in a similar manner.

Let x be the optimal solution for (5.25′) ($l \in R \cup \{r\}$). According to Lemma 5.2, $x \in D$ and

$$\min_{j \in J} \left[\frac{\bar{c}_j'x - u_j}{(x'V_jx)^{1/2}}\right] = \frac{\bar{c}_l'x - u_l}{(x'V_lx)^{1/2}} = \frac{\bar{c}_r'x - u_r}{(x'V_rx)^{1/2}}, \quad l \in R \cup \{r\}.$$

By virtue of Lemma 5.3, there exists a $\zeta > 0$ such that for any $z \in D$ with $|z - x| < \zeta$,

$$\min_{j \in J} \left[\frac{\bar{c}_j'x - u_j}{(x'V_jx)^{1/2}}\right] = \frac{\bar{c}_l'x - u_l}{(x'V_lx)^{1/2}}, \quad l \in R_z \subset R \cup \{r\}, \ R_z \neq \emptyset$$

Lemma 5.2. also implies that $z \in D_l$, $l \in R_z$. Since x is the optimal solution for problem (5.25′),

$$\frac{\bar{c}_l'z - u_l}{(x'V_lz)^{1/2}} \leqslant \frac{\bar{c}_l'x - u_l}{(x'V_lx)^{1/2}}, \quad l \in R_z.$$

This means that a neighborhood of x has been found such that for any other point z belonging to this neighbourhood and to the domain of feasible solutions, the value of the objective function of problem (5.23)—(5.24) at point z is not greater than the value at x. The proof is complete now.

We revert now to the questions formulated above. If the solution x_r of problem (5.25′) has been found, 3 cases may exist :

1) $$\frac{\bar{c}_r' x_r - u_r}{(x_r' V_r x_r)^{1/2}} < \frac{\bar{c}_l' x_r - u_l}{(x_r' V_l x_r)^{1/2}}, \quad l \in T - \{r\}$$

2) $$\frac{\bar{c}_r' x_r - u_r}{(x_r' V_r x_r)^{1/2}} = \frac{\bar{c}_l' x_r - u_l}{(x_r' V_l x_r)^{1/2}}, \; l \in R \subset T - \{r\}, \quad R \neq \emptyset$$

and x_r optimal for (5.25′) for any $l \in R$,

3) $$\frac{\bar{c}_r' x_r - u_r}{(x_r' V_r x_r)^{1/2}} = \frac{\bar{c}_l' x_r - u_l}{(x_r' V_l x_r)^{1/2}}, \; l \in R \subset T - \{r\}, \quad R \neq \emptyset$$

but there exists an $s \in R$ such that x_r is not optimal for problem (5.25′) corresponding to $k = s$. In cases 1) and 2), x_r is the optimal solution for problem (5.23)—(5.24). In case (3), one determines the optimum solution x_s for problem (5.25′) and reiterate the reasoning as for x_r.

Thus, in order to solve problem (5.23)—(5.24), one has to solve at most r problems of type (5.25′). Each of these problems gives a global optimum solution and finally the global maximum solution for problem (5.23)—(5.24) is found.

5.6. AN INTERACTIVE APPROACH TO STOCHASTIC PROGRAMMING WITH MULTIPLE OBJECTIVE FUNCTIONS

Leclercq [210], [211], [212], [213] obtained a model somewhat similar to (5.3)—(5.4) starting from a problem with one objective function (which does not exclude the case of more functions), where the coefficients are random variables and some of the constraints contain random variables.

Consider the problem

(5.27) $\max c'x,$

subject to

(5.28) $D_j'x \geqslant d_j \quad j = 1, \ldots, q$

(5.29) $E_k'x \geqslant e_k \quad k = 1, \ldots, r$

(5.30) $x \geqslant 0,$

where the elements of the vectors $c(n \times 1)$, $D_j(n \times 1)$, $d = (d_1, \ldots, d_q)'$ are random variables and those of vectors E_k $(n \times 1)$ and $e = (e_1, \ldots, e_r)'$ are constants.

We must determine the vector x so as to maximize the probability that $c'(\omega)x$ is greater than a given threshold s and the probability that constraints (5.28) are true. The following problem is obtained:

$$(5.31) \quad \begin{array}{l} \max P\{\omega \mid c'(\omega)x \geqslant s\} \\ \max P\{\omega \mid D_j'x \geqslant d_j\} \quad j = 1, \ldots, q, \end{array}$$

subject to

$$x \in D^* = \{x \in R^n \mid E_k'x \geqslant e_k,\ k = 1, \ldots, r;\ x \geqslant 0\}.$$

Since random elements occur also in the constraints, the distinction between the objective function and the constraints fades away.

Denote

$$G = \begin{pmatrix} c' \\ D_1' \\ \ldots \\ D_q' \end{pmatrix}, \quad b = \begin{pmatrix} s \\ d \end{pmatrix},$$

and write (5.31) as

(5.32) $\max P\{\omega \mid G_i x \geqslant b_i\},\ i = 1, \ldots, q + 1,$

subject to

$$x \in D^*,$$

where G_i are the rows in G.

Additionaly, consider the $(q+1)\times(n+1)$ matrix $A = \|G| - b\| = \|a_{ij}\|$, $z = (x, 1)$ and $\bar{D} = \{z \mid x \in D^*\}$. Problem (5.32) becomes

$$(5.33)\quad \max P\{\omega \mid \bigcap_{i=1}^{q+1} A_i z \geqslant 0\},$$

subject to

$$z \in \bar{D},$$

where A_i are the rows of A. Problem (5.33) is a mathematical programming problem with one objective function.

Let us now relax the assumption on the interdependence between the random variables and consider the following possibilities:

I. The rows of A are independent and the elements of row i are linear functions of the same random variable, i.e. $a_{ij} = \alpha_{ij} + T_i\beta_{ij}$ (α_{ij}, β_{ij} are constants and $T_1, \ldots, T_{q+1}$ are independent random variables of known distribution).

If $\alpha_i' = (\alpha_{i1}', \ldots, \alpha_{in}')$ and $\beta_i' = (\beta_{i1}', \ldots, \beta_{in}')$, then Problem (5.33) can be written as

$$(5.34)\quad \max P_i(z) = P(\alpha_i' z + T_i \beta_i' z \geqslant 0) \quad i = 1, \ldots, q+1,$$

subject to

$$z \in \bar{D}.$$

Without any loss of generality, one can suppose that the distributions T_i are symmetric, with zero mean value, i.e.

$$P_i(z) \geqslant 1/2 \text{ if } \alpha_i' z \geqslant 0, \text{ and } P_i(z) < 1/2 \text{ if } \alpha_i' z < 0.$$

Since for an optimal solution z^* we must obtain a value of $P_i(z)$ as large as possible, it only seems reasonable that those feasible solutions for which the value of certain P_i's is less than 1/2 must be eliminated. Thus, one can decrease the domain $\bar{D}$ into $\bar{D}^*$ defined as

$$\bar{D}^* = \bar{D} \cap \{z \mid \alpha_i' z \geqslant 0;\ i = 1, \ldots, q+1\}.$$

By virtue of the symmetry properties of the distributions,

$$P_i(z) = \begin{cases} 1 & \text{if } \beta_i'z = 0 \\ F_{T_i}(|\alpha_i'z|/|\beta_i'z|) & \text{if } \beta_i'z \neq 0, \end{cases}$$

where F_{T_i} is the distribution of the random variable T_i. Problem (5.34) thus becomes

$$(5.35) \quad \max f_i(z) = F_{T_i}(|\alpha_i'z|/|\beta_i'z|) \quad i = 1, \ldots, q + 1$$

subject to

$$z \in \bar{D}^*.$$

II. If each coefficient a_{ij} is a random variable with its own distribution, one assumes that:

(a) a_{ij} are either normal or Cauchy random variables with mean value $\bar{a}_{ij}$ and covariance matrix V^i (here V^i defines the correlation of variables $a_{i1}, \ldots, a_{in}$);

(b) a_{ik} and a_{jl} are independent if $i \neq j$;

(c) $z'V^iz \neq 0$ for all $i = 1, \ldots, q + 1$ and $z \in \bar{D}$.

II.1. a_{ij} are independent random variables with Cauchy distribution of parameters (μ_{ij}, θ_{ij}); then $a_{ij} = \mu_{ij} + t\theta_{ij}$. If $\mu_i' = (\mu_{i1}', \ldots, \mu_{in}')$ and $\theta_i' = (\theta_{i1}', \ldots, \theta_{in}')$, then (5.33) can be written as

$$(5.36) \quad \max P_i(x) = P(\mu_i'z + t(\theta_i'z) \geqslant 0),$$

subject to

$$z \in \bar{D}.$$

Assuming that $\theta_i'z > 0$, then

$$P(\mu_i'z + t(\theta_i'z) \geqslant 0) = P(t \geqslant -\mu_i'z/\theta_i'z),$$

where t is Cauchy-distributed with parameters $(0, 1)$, and hence (5.36) becomes

$$(5.37) \quad \max f_i(z) = \frac{\mu_i'z}{\theta_i'z}, \; i = 1, \ldots, q + 1,$$

subject to

$$z \in \bar{D},$$

which is similar to problem (5.35).

II.2. a_{ij} are normally distributed random variables.
For $z \neq (0, 1)$ and V^i positive definite, one has

$$P(A_i z \geqslant 0) = P\left(\frac{A_i z - \mathbf{E}(A_i)z}{\sqrt{z'V^i z}} \geqslant \right.$$

$$\left. \geqslant \frac{-\mathbf{E}(A_i)z}{\sqrt{z'V^i z}}\right) = P\left(u \geqslant -\frac{\bar{A}_i z}{\sqrt{z'V^i z}}\right),$$

where u is a normal variable obeying $N(0, 1)$ and $\bar{A}_i$ is the i-th row of matrix $\bar{A}$. Thus the deterministic equivalent of problem (5.33) is

$$\text{(5.38)} \quad \max f_i(z) = \frac{\bar{A}_i z}{\sqrt{z'V^i z}}, \quad i = 1, \ldots, q+1,$$

subject to

$$z \in \bar{D}.$$

In order to solve problem (5.38), Leclercq [213] proposes an interactive algorithm similar to the STEM method presented in Chapter 1, Section 1.7. The decision-maker has an active role in searching for the optimum solutions.

First, for each function f_i one considers a threshold B_i and the wheights p_i expressing its importance.

A solution z is satisfactory in terms of the j-th criterion if $f_j(z) \geqslant B_j$. Leclercq [213] defines the notion of non-improvable solution with respect to $B_1, \ldots, B_{q+1}$. This concept is a generalization of that of efficient solution defined in Chapter 1, Section 1.1.

DEFINITION 5.2. The point $x \in \bar{D}$ is said to be *non-improvable* with respect to $B_1, \ldots, B_{q+1}$, if and only if there exists no $y \in \bar{D}$ such that

a) $f_i(y) \geqslant f_i(x)$ if $i \in I = \{i \mid f_i(x) < B_i\}$;

b) $\sum_{i \in I} f_i(y) > \sum_{i \in I} f_i(x)$;

c) $f_j(y) \geqslant B_j$ if $j \in \{j \mid f_j(x) \geqslant B_j\}$.

From this definition it results that x is efficient if and only if it is non-improvable with respect to $\infty, \ldots, \infty$.

In order to present the solution algorithm for problem (5.38) one considers that

$$\bar{\bar{D}} = \bar{D} \cap \{z | z \in \bar{D},\ \bar{A}_i z > 0,\ i = 1, \ldots, q + 1\}.$$

The functions f_i are pseudo-concave and, according to [213], $x \in \bar{\bar{D}}$ is non improvable in $\bar{\bar{D}}$ with respect to $B_1, \ldots, B_{q+1}$ for problem (5.38) if and only if x is non-improvable in $B(x, d)$ (sphere centered in x and of radius d) with respect to $B_1, \ldots$ $\ldots, B_{q+1}$. This means that the property of being "locally" non-improvable (i.e. in $B(x, d)$) implies global non-improvability (i.e. in $\bar{\bar{D}}$). Thus, one will be able to obtain global efficiency using a local test.

The algorithm consists of a series alternation of the computational and decisional stages.

In the decision phase, the decision-maker uses the available information to determine the thresholds $B_1, \ldots, B_{q+1}$, and then, in the computational phase, a non-improvable solution with respect to $B_1, \ldots, B_{q+1}$. The thresholds $B_1, \ldots, B_{q+1}$ differ from one iteration to the next. The decision-maker can choose to decrease a certain B_i in order to improve the values of other functions.

Without insisting on the decision phase, we give below a brief description of the computation phase. One starts with a feasible solution $z_0 \in \bar{\bar{D}}$ and generates a sequence $\{z_k\}_0^N$ of feasible solutions such that:

a) If z_i is a non-improvable point, then the algorithm will find it out and will stop;

b) if z_i is improvable, then the algorithm will compute $z_{k+1} \in K(z_k)$.

Any point belonging to the set $K(z_k)$ is a possible successor to z_k. Hence, the set $K(z_k)$ includes all the points which are better than z_k as described by Definition 5.2. The point z_{k+1} is determined by the composition of two functions, D and M:

– a function $D : \bar{\bar{D}} \to \bar{\bar{D}} \times R^n$ by means of which we can determine a feasible direction d_k in z_k (i.e. a direction for which $\exists\ \varepsilon > 0$ such that $z_k + \varepsilon d_k \in \bar{\bar{D}}$) which leads to points better than z_k.

— a function $M : \bar{\bar{D}} \times R^n \to \bar{\bar{D}}$ which maximizes the function on this direction (determines the best step t^k in this direction).

Then,

$$z_{k+1} = z_k + t^k d_k.$$

One chooses the direction d_k which leads to a local increase of all those objective functions which have not attained the threshold B_i ($i \in I(z_k)$). Thus, one can solve a linear programming problem :

$$\max_{d} G(d) = \sum_{i=1}^{q+1} p_i^k (d' \nabla f_i(z_k)),$$

subject to

$$z_k + d \in Z$$

$$d' \nabla f_i(z_k) \geqslant 0 \qquad i \in I(z_k)$$

$$p_i^k = p_i\{s_i - P[A_i z_k \geqslant 0]\}^+ \quad ([a]^+ : \text{positive part of } a)$$

$$\Phi(B_i) = s_i$$

$$I(z_k) = \{i \mid f_i(z_k) < B_i\}$$

$$J(z_k) = \{j \mid f_j(z_k) \geqslant B_j\}.$$

The coeffcients $\{p_i^k\}$ differ from one iteration to the next, and are determined by considering both the value of p_i and the degree of satisfaction attained in z_k (the error with respect to B_i).

The step t^k is determined as follows :

$$t^k = \min(t_j^k, \bar{t}_i^k, 1) \quad i \in I(z_k), \ j \in J(z_k),$$

where t_j^k is the solution of the problem

$$\max t$$

subject to

$$f_j(z_k + td_k) \geqslant B_j,$$

which is equivalent to solving a 2-nd degree inequality whereas $\bar{t}_i^k$ is obtained by the one-dimensional optimization

$$\max_t \ f_i(z_k + td_k).$$

One can select [210]

$$\bar{t}_i^k = \begin{cases} -\infty & \text{if } B_i(z_k, d_k) \geqslant 0 \\ -\dfrac{C_i(z_k, d_k)}{B_i(z_k, d_k)} & \text{if } B_i(z_k, d_k) < 0, \end{cases}$$

where

$$B_i(z_k, d_k) = (\bar{A}_i h_k)(z_k' V^i h_k) - (\bar{A}_i z_k)(h_k V^i h_k)$$

$$C_i(z_k, d_k) = d_k' \nabla f_i(z_k) = (\bar{A}_i h_k)(z_k' V^i z_k') - (\bar{A}_i z_k)(z_k V^i h_k)$$

$$h_k = (d_k, 0).$$

The stopping conditions are

a) If $(z_k, 0) \in D(z_k)$, then the algorithm stops in z_k.

b) If $0 \in M(z_k, d_k)$, then the algorithm stops in z_k.

The joint use of conditions (a) and (b) as a stopping test avoids cycling but does not ensure non-improvability.

In order to ensure non-improvability, Leclercq suggests a modified algorithm [213] for which condition (a) is used as a stopping condition and ensures non-improvability.

Using Zangwill's theorem [251], it is proved [210], [213] that:

— either the algorithm stops at a point which is a solution of problem (5.38) (i.e. at an efficient and non-improvable solution);

— or one generates a sequence $\{x^k\}_1^\infty$ having a sub-sequence convergent towards the optimum solution of problem (5.38).

Chapter 6

The transportation problem with multiple objective functions

This chapter deals with a single commodity transportation problem with upper bounds on the quantity shipped along each route and for which three efficiency functions are to be treated simultaneously. The functions are the minimization of the total shipping cost and of shipping time along with the maximization of material shipped. A synthesis function of the three efficiency functions is introduced. One shows then that the optimal solutions of a mathematical programming problem involving this synthesis function provide efficient solutions for the original problem. The problem involving the synthesis function is a transportation problem with limited capacities and a fractional objective function. The problem reduces then to a three-dimensional problem without capacity constraints. A simplex-like technique and a variable transformation approach are given for solving the three-dimensional transportation problem. An optimality criterion for a local minimum of this problem is also given. Finally, this problem is treated in stochastic terms.

6.1. THE DETERMINISTIC CASE

In some mathematical programming problems with several objective functions, the consistency of the criteria is not always obvious and often some criteria may be contradictory. We may quote in this respect the choice of a range of commodities [130], several distribution problems [131], scheduling, etc.

We shall consider the transportation problem with several objective functions and contradictory criteria. This problem can be stated as follows: assume that m production centers $A_1, A_2, \ldots, A_m$ contain a certain material in quantities $a_1, a_2, \ldots, a_m$ and that this material is required by n consumption centers $B_1, B_2, \ldots, B_n$ in quantities $b_1, b_2, \ldots, b_n$. For each route (A_i, B_j) let g_{ij} denote the maximum quantity of material that can be transported, c_{ij} the unit cost of the transportation and t_{ij} the time required by the transport.

We must organize the expedition process such that three criteria are simultaneously satisfied: the minimization of total transport-cost, the minimization of total transportation time and the maximization of the quantity transported.

Denote by x_{ij} the transported quantity and construct the following model:

$$(6.1)\quad \min F_1 = \sum_{i=1}^{m} \sum_{j=1}^{n} c_{ij}\, x_{ij},$$

$$(6.2)\quad \min F_2 = \sum_{i=1}^{m} \sum_{j=1}^{n} t_{ij}\, x_{ij},$$

$$(6.3)\quad \max F_3 = \sum_{i=1}^{m} \sum_{j=1}^{n} x_{ij},$$

subject to

$$(6.4)\quad \sum_{j=1}^{n} x_{ij} = a_i, \qquad (i = 1, \ldots, m)$$

$$(6.5)\quad \sum_{i=1}^{m} x_{ij} = b_j, \qquad (j = 1, \ldots, n)$$

$$(6.6)\quad 0 \leqslant x_{ij} \leqslant g_{ij}, \ (i = 1, \ldots, m; j = 1, \ldots, n)$$

$$(6.7)\quad \sum_{i=1}^{m} a_i = \sum_{j=1}^{n} b_j.$$

Condition (6.7) is a necessary condition for the problem to have a feasible (possible) solution. However, this condition is not sufficient because of condition (6.6).

This problem can be solved by use of the utility theory [131]. Thus, functions F_1, F_2, F_3 are transformed into utility functions :

$$F_1' = \sum_{i=1}^{m} \sum_{j=1}^{n} a_1 c_{ij} x_{ij} + b_1$$

$$F_2' = \sum_{i=1}^{m} \sum_{j=1}^{n} a_2 t_{ij} x_{ij} + b_2$$

$$F_3' = \sum_{i=1}^{m} \sum_{j=1}^{n} a_3 x_{ij} + b_3$$

and then we perform the weighted summation using the weight coefficients k_1, k_2, k_3 to get the synthesis function

$$F^* = k_1 F_1' + k_2 F_2' + k_3 F_3'.$$

a_1, a_2, a_3 and b_1, b_2, b_3 above denote the transformation coefficients obtained by using the definition of the utility function in the von Neumann-Morgenstern sense.

Since the determination of the efficient solutions is tantamount to solving a parametric programming problem, it follows that by maximizing the function F^* [131] for a given $k = (k_1, k_2, k_3)$ one obtains efficient solutions.

Another solution to problem (6.1)—(6.7) may be obtained by constructing a synthesis function of the three objective functions :

$$(6.8) \quad F = \frac{F_1 F_2}{F_3^2} = \frac{\left(\sum_{i=1}^{m} \sum_{j=1}^{n} c_{ij} x_{ij}\right)\left(\sum_{i=1}^{m} \sum_{j=1}^{n} t_{ij} x_{ij}\right)}{\left(\sum_{i=1}^{m} \sum_{j=1}^{n} x_{ij}\right)}$$

Recently, Izerman [87] developed an algorithm for solving the transportation problem with multiple objective functions. This algorithm is carried out in three steps :

a) determine an initial efficient-base solution ;

b) determine the other efficient-base solutions ;

c) determine all the efficient solutions.

Gupta [77] considers the minimization of the transportation cost simultaneously with minimization of the maximum transportation time. Thus she considers the problem

$$\min\left(\sum_{i=1}^{m}\sum_{j=1}^{n} c_{ij}x_{ij},\ \max_{\{(i,j)|x_{ij}>0\}} t_{ij}\right)$$

under constraints (6.4)—(6.5) and assuming that $x_{ij} \geqslant 0$.

In order to solve this problem, Gupta develops a multicriterial simplex method using the particular form of the problem matrix. This method is based on the multicriterial simplex method proposed by Yu and Zeleny [165] for solving linear programming problems with multiple objective functions. A similar problem with three indices is considered by Bhatia and Puri [26].

By solving the mathematical programming problem with constraints (6.4)—(6.7) and objective function (6.8), one obtains efficient solutions for problem (6.1)—(6.7) (see Theorem 6.1. below).

Let us recall the definition of efficient solutions with reference to problem (6.1)—(6.7).

DEFINITION 6.1. A point $X^* = (x_{ij}^*)$ is an efficient solution for problem (6.1)—(6.7) if and only if there exists no other feasible solution $X = (x_{ij})$ such that

$$F_1(X) = \sum_{i=1}^{m}\sum_{j=1}^{n} c_{ij}\, x_{ij} \leqslant \sum_{i=1}^{m}\sum_{j=1}^{n} c_{ij}\, x_{ij}^* = F_1(X^*)$$

$$F_2(X) = \sum_{i=1}^{m}\sum_{j=1}^{n} t_{ij}\, x_{ij} \leqslant \sum_{i=1}^{m}\sum_{j=1}^{n} t_{ij}\, x_{ij}^* = F_2(X^*)$$

$$F_3(X) = \sum_{i=1}^{m}\sum_{j=1}^{n} x_{ij} \geqslant \sum_{i=1}^{m}\sum_{j=1}^{n} x_{ij}^* = F_3(X^*),$$

and for at least one subscript $i \in \{1, 2, 3\}$ the inequality is strict.

THEOREM 6.1. *Any feasible solution* $X^* = (x_{ij}^*)$ *which minimizes the function*

$$F(X) = \frac{F_1(X)\, F_2(X)}{F_3^2(X)} = \frac{\left(\sum_{i=1}^{m}\sum_{j=1}^{n} c_{ij}\, x_{ij}\right)\left(\sum_{i=1}^{m}\sum_{j=1}^{n} t_{ij}\, x_{ij}\right)}{\left(\sum_{i=1}^{m}\sum_{j=1}^{n} x_{ij}\right)^2}$$

is an efficient solution for problem (6.1)—(6.7).

Proof. Suppose that $X^* = (x^*_{ij})$ minimizes $F(x)$, i.e. for any feasible solution X,

$$\frac{F_1(X^*)\,F_2(X^*)}{F_3^2(X^*)} \leqslant \frac{F_1(X)\,F_2(X)}{F_3^2(X)} \tag{6.9}$$

but here X^* is not an efficient solution. Then there must exist a feasible solution $\bar{X}$ such that

$$F_1(\bar{X}) \leqslant F_1(X^*)\,;\; F_2(\bar{X}) \leqslant F_2(X^*)\,; F_3(\bar{X}) \geqslant F_3(X^*),$$

and at least one of the inequalities is strict. Hence, using these three inequalities, it results that

$$\frac{F_1(\bar{X})\cdot F_2(\bar{X})}{F_3^2(\bar{X})} < \frac{F_1(X^*)\,F_2(X^*)}{F_3^2(X^*),},$$

which contradicts (6.9). Hence, X^* must be an efficient solution.

The problem defined by constraints (6.4)—(6.7) and the objective function (6.8) is a transportation problem with limited capacities and fractional objective function. Such problems can be stated as three-dimensional problems (but then the capacity condition (6.6) must be removed).

The constraints (6.6) can be written as

$$\sum_{k=1}^{2} y_{ijk} = E_{ij},$$

where $y_{ij1} = x_{ij}$, y_{ij2} are slack variables used to transform the inequalities (6.6) into

$$E_{ij} = g_{ij}.$$

We shall use following notation :

$$l_{ij1} = c_{ij},\; l_{ij2} = 0$$

$$m_{ij1} = t_{ij},\; m_{ij2} = 0$$

$$n_{ij1} = 1,\; n_{ij2} = 0$$

$$A_{i1} = a_i, \; A_{i2} = \sum_{j=1}^{n} g_{ij} - a_i$$

$$B_{1j} = b_j, \; B_{2j} = \sum_{i=1}^{m} g_{ij} - b_j.$$

Then, problem (6.4)–(6.7) and (6.8) can be written as

$$\min F(X) = \frac{\left(\sum_{i=1}^{m}\sum_{j=1}^{n}\sum_{k=1}^{2} l_{ijk}\, x_{ijk}\right)\left(\sum_{i=1}^{m}\sum_{j=1}^{n}\sum_{k=1}^{2} m_{ijk}\, x_{ijk}\right)}{\left(\sum_{i=1}^{m}\sum_{j=1}^{n}\sum_{k=1}^{2} n_{ijk}\, x_{ijk}\right)^2},$$

subject to

$$\sum_{j=1}^{n} x_{ijk} = A_{ik}, \quad (i = 1, \ldots, m;\; k = 1, 2)$$

$$\sum_{i=1}^{m} x_{ijk} = B_{kj}, \quad (j = 1, \ldots, n;\; k = 1, 2)$$

$$\sum_{k=1}^{2} x_{ijk} = E_{ij}, \quad (i = 1, \ldots, m;\; j = 1, \ldots, n)$$

$$x_{ijk} \geqslant 0, \qquad (i = 1, \ldots, m;\; j = 1, \ldots, n;\; k = 1, 2)$$

$$\sum_{i=1}^{m} A_{ik} = \sum_{j=1}^{n} B_{kj};\; \sum_{k=1}^{2} B_{kj} = \sum_{i=1}^{m} E_{ij};\; \sum_{j=1}^{n} E_{ij} = \sum_{k=1}^{2} A_{ik}$$

$$\sum_{i=1}^{m}\sum_{k=1}^{2} A_{ik} = \sum_{k=1}^{2}\sum_{j=1}^{n} B_{kj} = \sum_{i=1}^{m}\sum_{j=1}^{n} E_{ij}.$$

A more general form of this problem is

$$\text{(6.10)} \quad \min F(X) = \frac{\left(\sum_{i=1}^{m}\sum_{j=1}^{n}\sum_{k=1}^{p} l_{ijk}\, x_{ijk}\right)\left(\sum_{i=1}^{m}\sum_{j=1}^{n}\sum_{k=1}^{p} m_{ijk}\, x_{ijk}\right)}{\left(\sum_{i=1}^{m}\sum_{j=1}^{n}\sum_{k=1}^{p} n_{ijk}\, x_{ijk}\right)^2},$$

subject to

$$(6.11)\quad \begin{aligned} &\sum_{j=1}^{n} x_{ijk} = A_{ik}, \ (i = 1, \ldots, m; \ k = 1, \ldots, p), \\ &\sum_{i=1}^{m} x_{ijk} = B_{kj}, \ (j = 1, \ldots, n; \ k = 1, \ldots, p), \\ &\sum_{k=1}^{p} x_{ijk} = E_{ij}, \ (i = 1, \ldots, m; \ j = 1, \ldots, n), \\ &x_{ijk} \geqslant 0, \qquad (i = 1, \ldots, m; \ j = 1, \ldots, n; \ k = \\ &\qquad\qquad = 1, \ldots . p), \end{aligned}$$

$$\sum_{i=1}^{m} A_{ik} = \sum_{j=1}^{n} B_{kj}; \ \sum_{k=1}^{p} B_{kj} = \sum_{i=1}^{m} E_{ij}; \ \sum_{j=1}^{n} E_{ij} = \sum_{k=1}^{p} A_{ik},$$

$$\sum_{i=1}^{m} \sum_{k=1}^{p} A_{ik} = \sum_{k=1}^{p} \sum_{j=1}^{n} B_{kj} = \sum_{i=1}^{m} \sum_{j=2}^{n} E_{ij},$$

DEFINITION 6.2. $X=\{(x_{ijk}), (i, j, k) \in I = (1,\ldots, m)\times(1,\ldots,n)\times \times(1, \ldots, p)\}$ satisfying constraints (6.11) is called a *feasible solution* (or just "a solution").

Definition 6.3. A solution X is called *optimal* if it minimizes the function (6.10).

Assume that for any feasible solution X,

$$\sum_{i=1}^{m} \sum_{j=1}^{n} \sum_{k=1}^{p} l_{ijk} x_{ijk} \geqslant 0; \ \sum_{i=1}^{m} \sum_{j=1}^{n} \sum_{k=1}^{p} m_{ijk} x_{ijk} > 0,$$

$$\sum_{i=1}^{m} \sum_{j=1}^{n} \sum_{k=1}^{p} n_{ijk} x_{ijk} > 0.$$

The function (6.10) is explicitly quasi-concave. For such a function, a local minimum is not necessarily a global minimum.

Martos [105] showed that for such functions the minimum appears at a vertex of the domain of feasible solutions. Thus, problem (6.10)—(6.11) can be solved using a technique similar to the simplex algorithm in linear programming.

We give below a criterion for a local minimum. An initial feasible solution for problem (6.10)—(6.11) can be obtained by any of the known methods of the ordinary transportation problem (e.g. the method of the North-West corner).

Let

$$I_X = \{(i, j, k) \in I \mid x_{ijk} > 0,\ x_{ijk} \in X\}.$$

According to the last two conditions in (6.11), any nondegenerated base solution will contain $mn + pn + mp - m - n - p + 1$ positive components.

Consider the dual variables (the simplex multipliers)

$$u_{ik}^1, u_{ik}^2, u_{ik}^3 \ (i = 1, \ldots, m;\ k = 1, \ldots, p),$$

$$v_{jk}^1, v_{jk}^2, v_{jk}^3 \ (j = 1, \ldots, n;\ k = 1, \ldots, p)$$

and

$$w_{ij}^1, w_{ij}^2, w_{ij}^3 \ (i = 1, \ldots, m; j = 1, \ldots, n),$$

defined as follows :

$$(6.12)\quad u_{ik}^1 + v_{jk}^1 + w_{ij}^1 = l_{ijk}$$

$$(6.13)\quad u_{ik}^2 + v_{jk}^2 + w_{ij}^2 = m_{ijk} \qquad (i, j, k) \in I_X,$$

$$(6.14)\quad u_{ik}^3 + v_{jk}^3 + w_{ij}^3 = n_{ijk}.$$

Let

$$l'_{ijk} = l_{ijk} - u_{ik}^1 - v_{jk}^1 - w_{ij}^1,$$

$$m'_{ijk} = m_{ijk} - u_{ik}^2 - v_{jk}^2 - w_{ij}^2,\ (i, j, k) \in I - I_X,$$

$$n'_{ijk} = n_{ijk} - u_{ik}^3 - v_{jk}^3 - w_{ij}^3.$$

Systems (6.12)—(6.14) can be solved independently. Thus, system (6.12) has $mn + pm + np - m - n - p + 1$ equations with $mn + np + pm$ variables. One can assign arbitrary values (such that the system stays compatible) to $m + n + p - 1$ variables u, v and w (usually the zero value), and then find the other values.

Having determined the values u^1_{ik}, v^1_{jk}, w^1_{ij}, u^2_{ik}, v^2_{jk}, w^2_{ij}, u^3_{ik} v^3_{jk}, w^3_{ij}, one can determine the quantities l'_{ijk}, m'_{ijk} and n'_{ijk} for the variables outside the base.

Let $X^* = (x^*_{ijk})$ be a feasible solution for problem (6.10)—(6.11). In order to establish a criterion for a local minimum we express the function F taking into account only the variables outside the base:

$$\sum_{i=1}^{m} \sum_{j=1}^{n} \sum_{k=1}^{p} l_{ijk}\, x_{ijk} = \sum_{i=1}^{m} \sum_{j=1}^{n} \sum_{k=1}^{p} l_{ijk}\, x_{ijk} +$$

$$+ \sum_{i=1}^{m} \sum_{k=1}^{p} \left(A_{ik} - \sum_{j=1}^{n} x_{ijk} \right) u^1_{ik} +$$

$$+ \sum_{j=1}^{n} \sum_{k=1}^{p} \left(B_{kj} - \sum_{i=1}^{m} x_{ijk} \right) v^1_{jk} + \sum_{i=1}^{m} \sum_{j=1}^{n} \left(E_{ij} - \sum_{k=1}^{p} x_{ijk} \right) w^1_{ij} =$$

$$= \sum_{i=1}^{m} \sum_{j=1}^{n} \sum_{k=1}^{p} (l_{ijk} - u^1_{ik} - v^1_{jk} - w^1_{ij})\, x_{ijk} + \sum_{i=1}^{m} \sum_{k=1}^{p} A_{ik}\, u^1_{ik} +$$

$$+ \sum_{j=1}^{n} \sum_{k=1}^{p} B_{kj}\, v^1_{kj} + \sum_{i=1}^{m} \sum_{j=1}^{n} E_{ij}\, w^1_{ij} = \sum_{(i,j,k) \in I - I_X} l'_{ijk}\, x_{ijk} + V_1,$$

where

$$V_1 = \sum_{i=1}^{m} \sum_{k=1}^{p} A_{ik}\, u^1_{ik} + \sum_{j=1}^{n} \sum_{k=1}^{p} B_{kj}\, v^1_{kj} + \sum_{i=1}^{m} \sum_{j=1}^{n} E_{ij}\, w^1_{ij}.$$

By a similar procedure, one gets

$$\sum_{i=1}^{m} \sum_{j=1}^{n} \sum_{k=1}^{p} m_{ijk}\, x_{ijk} = \sum_{(i,j,k) \in I - I_X} m'_{ijk} x_{ijk} + V_2,$$

$$\sum_{i=1}^{m} \sum_{j=1}^{n} \sum_{k=1}^{p} n_{ijk}\, x_{ijk} = \sum_{(i,j,k) \in I - I_X} n'_{ijk} x_{ijk} + V_3,$$

where

$$V_2 = \sum_{i=1}^{m} \sum_{k=1}^{p} A_{ik} u_{ik}^2 + \sum_{j=1}^{n} \sum_{k=1}^{p} B_{kj} v_{kj}^2 + \sum_{i=1}^{m} \sum_{j=1}^{n} E_{ij} w_{ij}^2,$$

$$V_3 = \sum_{i=1}^{m} \sum_{k=1}^{p} A_{ik} u_{ik}^3 + \sum_{j=1}^{n} \sum_{k=1}^{p} B_{kj} v_{kj}^3 + \sum_{i=1}^{m} \sum_{j=1}^{n} E_{ij} w_{ij}^3.$$

Hence, the function F becomes

$$F(X) = \frac{\left(\sum_{(i,j,k)\in I-I_X} l'_{ijk} x_{ijk} + V_1\right)\left(\sum_{(i,j,k)\in I-I_X} m'_{ijk} x_{ijk} + V_2\right)}{\left(\sum_{(i,j,k)\in I-I_X} n'_{ijk} x_{ijk} + V_3\right)^2}$$

For $(i, j, k) \in I - I_X$, one gets

$$\frac{\partial F}{\partial x_{ijk}} = \frac{l'_{ijk}\left(\sum_{(i,j,k)\in I-I_X} m'_{ijk} x_{ijk} + V_2\right)\left(\sum_{(i,j,k)\in I-I_X} n'_{ijk} x_{ijk} + V_3\right)^2}{\left(\sum_{(i,j,k)\in I-I_X} n'_{ijk} x_{ijk} + V_3\right)^4} +$$

$$+ \frac{m'_{ijk}\left(\sum_{(i,j,k)\in I-I_X} l'_{ijk} x_{ijk} + V_1\right)\left(\sum_{(i,j,k)\in I-I_X} n'_{ijk} x_{ijk} + V_3\right)^2}{\left(\sum_{(i,j,k)\in I-I_X} n'_{ijk} x_{ijk} + V_3\right)^4} -$$

$$- \frac{2n'_{ijk}\left(\sum_{(i,j,k)I\in -I_X} l'_{ijk} x_{ijk} + V_1\right)\left(\sum_{(i,j,k)\in I-I_X} m'_{ijk} x_{ijk} + V_2\right)\left(\sum_{(i,j,k)\in I-I_X} n'_{ijk} x_{ijk} + V_3\right)}{\left(\sum_{(i,j,k)\in I-I_X} n'_{ijk} x_{ijk} + V_3\right)^4}$$

The partial derivatives at the point $X = (x_{ijk}) = (x^*_{ijk})$ are

$$\left.\frac{\partial F}{\partial x_{ijk}}\right|_{x_{ijk}=x^*_{ijk}} = \frac{l'_{ijk} V_2 V_3^2 + m'_{ijk} V_1 V_3^2 - 2n'_{ijk} V_1 V_2 V_3}{V_3^4} =$$

$$= \frac{l'_{ijk} V_2 V_3 + m'_{ijk} V_1 V_3 - 2n'_{ijk} V_1 V_2}{V_3^3}.$$

Denote

$$\Delta_{ijk} = l'_{ijk} V_2 V_3 + m'_{ijk} V_1 V_3 - 2n'_{ijk} V_1 V_2.$$

If the dual variables u^l_{ik}, v^l_{jk}, $w^l_{ij}(l=1, 2, 3;\ i=1, \ldots, m; j=1, \ldots, n; k = 1, \ldots, p)$ are determined, then one can easily estimate Δ_{ijk} for the variables outside the base.

The solution $X^* = (x^*_{ijk})$ can be improved if there exists at least one

$$\Delta_{ijk} < 0.$$

Hence, one can state the following local minimum criterion:

THEOREM 6.2. *A solution $X^* = (x^*_{ijk})$ is a local minimum for problem* (6.10)–(6.11) *if*:

$$\Delta_{ijk} = l'_{ijk} V_2 V_3 + m'_{ijk} V_1 V_3 - 2n'_{ijk} V_1 V_2 \geqslant 0$$

for all the variables outside the base.

If at least one value Δ_{ijk} is negative, then one chooses $\Delta_{i_0,j_0,k_0} = \min \{\Delta_{ijk} | \Delta_{ijk} < 0\}$ and introduces $x_{i_0 j_0 k_0}$ into the base.

The variables leaving the base and the value of the base variables for the new base are established as for an ordinary transportation problem.

In order to solve problem (6.10)–(6.11), one can also apply another method using a change of variables:

$$x_{ijk} = \frac{z_{ijk}}{t},$$

where the parameter t is chosen such that

$$\sum_{i=1}^{m} \sum_{j=1}^{n} \sum_{k=I}^{p} n_{ijk} z_{ijk} = a.$$

Without any loss of generality, one can assume that $a = 1$.

Problem (6.10)—(6.11) then becomes

$$(6.15)\quad \min \widetilde{F} = \left(\sum_{i=1}^{m}\sum_{j=1}^{n}\sum_{k=1}^{p} l_{ijk}\, z_{ijk}\right)\left(\sum_{i=1}^{m}\sum_{j=1}^{n}\sum_{k=1}^{p} m_{ijk}\, z_{ijk}\right),$$

subject to

$$\sum_{j=1}^{n} z_{ijk} = tA_{ik},\ (i = 1, \ldots, m;\ k = 1, \ldots, p)$$

$$(6.16)\quad \sum_{i=1}^{m} z_{ijk} = tB_{kj},\ (j = 1, \ldots, n;\ k = 1, \ldots, p)$$

$$\sum_{k=1}^{p} z_{ijk} = tE_{ij}\ \ (i = 1, \ldots, m;\ j = 1, \ldots, n),$$

$$x_{ijk} \geqslant 0;\ t \geqslant 0,$$

$$\sum_{i=1}^{m}\sum_{j=1}^{n}\sum_{k=1}^{p} n_{ijk}\, z_{ijk} = 1.$$

Using a result due to Charnes and Cooper [36], it results that $t > 0$. The following theorem establishes a connection between the optimal solutions of problems (6.10)—(6.11) and (6.15)—(6.16).

THEOREM 6.3. *If* $(\bar{z}_{ijk}, \bar{t})$ *is an optimum solution for problem* (6.15)—(6.16), *then* $\bar{x}_{ijk} = \dfrac{\bar{z}_{ijk}}{\bar{t}}$ *is an optimum solution for problem* (6.10)—(6.11).

Proof. Let $(\bar{z}_{ijk}, \bar{t})$ be the optimal solution for problem (6.15)—(6.16), i.e.

$$(6.17)\quad \begin{aligned} &\left(\sum_{i=1}^{m}\sum_{j=1}^{n}\sum_{k=1}^{p} l_{ijk}\, \bar{z}_{ijk}\right)\left(\sum_{i=1}^{m}\sum_{j=1}^{n}\sum_{k=1}^{p} m_{ijk}\, \bar{z}_{ijk}\right) < \\ &< \left(\sum_{i=1}^{m}\sum_{j=1}^{n}\sum_{k=1}^{p} l_{ijk}\, z_{ijk}\right)\left(\sum_{i=1}^{m}\sum_{j=1}^{n}\sum_{k=1}^{p} m_{ijk}\, z_{ijk}\right) \end{aligned}$$

for any $Z = (z_{ijk})$ satisfying constraints (6.16). If $\bar{x}_{ijk} = \dfrac{\bar{z}_{ijk}}{\bar{t}}$ were not an optimum solution for problem (6.10)—(6.11), then there exists an optimal $X^0 = (x^0_{ijk})$ such that

$$\frac{\left(\sum_{i=1}^{m}\sum_{j=1}^{n}\sum_{k=1}^{p} l_{ijk}\, x^0_{ijk}\right)\left(\sum_{i=1}^{m}\sum_{j=1}^{n}\sum_{k=1}^{p} m_{ijk}\, x^0_{ijk}\right)}{\left(\sum_{i=1}^{m}\sum_{j=1}^{n}\sum_{k=1}^{p} n_{ijk}\, x^0_{ijk}\right)^2} <$$

$$< \frac{\left(\sum_{i=1}^{m}\sum_{j=1}^{n}\sum_{k=1}^{p} l_{ijk}\, \frac{\bar{z}_{ijk}}{\bar{t}}\right)\left(\sum_{i=1}^{m}\sum_{j=1}^{n}\sum_{k=1}^{p} m_{ijk}\, \frac{\bar{z}_{ijk}}{\bar{t}}\right)}{\left(\sum_{i=1}^{m}\sum_{j=1}^{n}\sum_{k=1}^{p} n_{ijk}\, \frac{\bar{z}_{ijk}}{\bar{t}}\right)^2} =$$

(6.18)

$$= \frac{\left(\sum_{i=1}^{m}\sum_{j=1}^{n}\sum_{k=1}^{p} l_{ijk}\, \bar{z}_{ijk}\right)\left(\sum_{i=1}^{m}\sum_{j=1}^{n}\sum_{k=1}^{p} m_{ijk}\, \bar{z}_{ijk}\right)}{\left(\sum_{i=1}^{m}\sum_{j=1}^{n}\sum_{k=1}^{p} n_{ijk}\, \bar{z}_{ijk}\right)^2} =$$

$$= \left(\sum_{i=1}^{m}\sum_{j=1}^{n}\sum_{k=1}^{p} l_{ijk}\, \bar{z}_{ijk}\right)\left(\sum_{i=1}^{m}\sum_{j=1}^{n}\sum_{k=1}^{p} m_{ijk}\, \bar{z}_{ijk}\right).$$

Let $\sum_{i=1}^{m}\sum_{j=1}^{n}\sum_{k=1}^{p} n_{ijk}\, x^0_{ijk} = \rho > 0$ and $\tilde{z} = \dfrac{x^0}{\rho}$; $\tilde{t} = \dfrac{1}{\rho}\cdot$ One can easily show that $(\tilde{z}, \tilde{t})$ is a feasible solution for problem (6.15) — (6.16). Hence,

$$\frac{\left(\sum_{i=1}^{m}\sum_{j=1}^{n}\sum_{k=1}^{p} l_{ijk}\, x^0_{ijk}\right)\left(\sum_{i=1}^{m}\sum_{j=1}^{n}\sum_{k=1}^{p} m_{ijk}\, x^0_{ijk}\right)}{\left(\sum_{i=1}^{m}\sum_{j=1}^{n}\sum_{k=1}^{p} n_{ijk}\, x^0_{ijk}\right)^2} =$$

$$= \frac{\left(\sum_{i=1}^{m}\sum_{j=1}^{n}\sum_{k=1}^{p} l_{ijk}\, \frac{x^0_{ijk}}{\rho}\right)\left(\sum_{i=1}^{m}\sum_{j=1}^{n}\sum_{k=1}^{p} m_{ijk}\, \frac{x^0_{ijk}}{\rho}\right)}{\left(\sum_{i=1}^{m}\sum_{j=1}^{n}\sum_{k=1}^{p} n_{ijk}\, \frac{x^0_{ijk}}{\rho}\right)^2} =$$

$$\frac{\left(\sum_{i=1}^{m}\sum_{j=1}^{n}\sum_{k=1}^{p} l_{ijk}\,\tilde{z}_{ijk}\right)\left(\sum_{i=1}^{m}\sum_{j=1}^{n}\sum_{k=1}^{p} m_{ijk}\,\tilde{z}_{ijk}\right)}{\left(\sum_{i=1}^{m}\sum_{j=1}^{n}\sum_{k=1}^{p} n_{ijk}\,\tilde{z}_{ijk}\right)^{2}} =$$

$$= \left(\sum_{i=1}^{m}\sum_{j=1}^{n}\sum_{k=1}^{p} l_{ijk}\,\tilde{z}_{ijk}\right)\left(\sum_{i=1}^{m}\sum_{j=1}^{n}\sum_{k=1}^{p} m_{ijk}\,\tilde{z}_{ijk}\right).$$

From this expression and using (6.18), it results that

$$\left(\sum_{i=1}^{m}\sum_{j=1}^{n}\sum_{k=1}^{p} l_{ijk}\,\tilde{z}_{ijk}\right)\left(\sum_{i=1}^{m}\sum_{j=1}^{n}\sum_{k=1}^{p} m_{ijk}\,\tilde{z}_{ijk}\right) <$$

$$< \left(\sum_{i=1}^{m}\sum_{j=1}^{n}\sum_{k=1}^{p} l_{ijk}\,\bar{z}_{ijk}\right)\left(\sum_{i=1}^{m}\sum_{j=1}^{n}\sum_{k=1}^{p} m_{ijk}\,\bar{z}_{ijk}\right),$$

which contradicts (6.17). Hence the proof is complete now.

6.2. THE STOCHASTIC CASE

Let us revert now to problem (6.1)—(6.8) or its equivalent (6.10)—(6.11), and assume that the demands b_j (and hence B_{kj}) are random variables.

At various moments there will exist differences (plus or minus) between the transported quantities and those demanded. For such differences varying from failure in achieving to exceedance of demands one has to pay penalties.

Assume that the demands have a known distribution law. Then in calculating the efficiency function, account must be taken of the penalties.

The problem consists in determining a transportation programme such that the total transportation time (distance) plus the mean value of these penalties is minimized.

For ordinary transportation problems a linearization of such a function was attempted by Szwarc [152]. The mathematical model becomes:

$$\min Z = \left(\sum_{i=1}^{m}\sum_{j=1}^{n}\sum_{k=1}^{p} l_{ijk}\, x_{ijk}\right)\cdot\left(\sum_{i=1}^{m}\sum_{j=1}^{n}\sum_{k=1}^{p} m_{ijk}\, x_{ijk}\right) \Big/$$

$$\left(\sum_{i=1}^{m}\sum_{j=1}^{n}\sum_{k=1}^{p} n_{ijk}\, x_{ijk}\right)^2 +$$

$$+ \sum_{j=1}^{n}\sum_{k=1}^{p} \mathop{\mathbf{E}}_{B_{jk} \geqslant x_{jk}} [l_{jk}^1 (B_{jk} - x_{jk})] +$$

$$+ \sum_{j=1}^{n}\sum_{k=1}^{p} \mathop{\mathbf{E}}_{B_{jk} \leqslant x_{jk}} [l_{jk}^2 (x_{jk} - B_{jk})],$$

subject to

$$\sum_{j=1}^{n} x_{ijk} \leqslant A_{ik} \qquad (i = 1, \ldots, m;\ k = 1, \ldots, p),$$

$$\sum_{k=1}^{p} x_{ijk} \leqslant E_{ij}, \qquad (i = 1, \ldots, m;\ j = 1, \ldots, n),$$

$$x_{kj} = \sum_{i=1}^{m} x_{ijk}, \qquad (k = 1, \ldots, p;\ j = 1, \ldots, m),$$

$$x_{ijk} \geqslant 0, \qquad (i = 1, \ldots, m;\ j = 1, \ldots, n;\ k = 1, \ldots, p)$$

$$\underline{b}_{kj} \leqslant x_{kj} \leqslant \bar{b}_{kj}$$

$$\sum_{i=1}^{m} A_{ik} \geqslant \sum_{j=1}^{n} \underline{b}_{kj};\ \sum_{k=1}^{p} A_{ik} = \sum_{j=1}^{n} E_{ij}$$

$$\sum_{i=1}^{m} E_{ik} \geqslant \sum_{k=1}^{p} \underline{b}_{kj}$$

$$\sum_{i=1}^{m}\sum_{k=1}^{p} A_{ik} = \sum_{i=1}^{m}\sum_{j=1}^{n} E_{ij} \geqslant \sum_{j=1}^{n}\sum_{k=1}^{p} \underline{b}_{kj}$$

$$l_{jk}^1,\ l_{jk}^2 > 0,\ A_{ik},\ B_{kj},\ E_{ij} > 0,\ l_{ijk} \geqslant 0,\ m_{ijk} > 0,\ n_{ijk} > 0,$$

where $\underset{B_{jk}\geqslant x_{jk}}{\mathbf{E}} [l^1_{jk}(B_{jk}-x_{jk})]$ is the mean value of the random variable within brackets over the domain $B_{jk} \geqslant x_{jk}$, l^1_{jk} are the penalties paid for not meeting the consumer's demand and l^2_{jk} are the penalties paid for each unit transported in excess of the consumer's demand.

For a given k, assume that the variable B_{jk} is discrete and ranges over the interval $(\underline{b}_{kj}, \bar{b}_{kj}]$. Then the distribution function $f(b_{kj})$ is a stepwise function which is zero outside the given interval.

In order to linearize the last two terms of Z one can adopt a procedure similar to that used by Szwarc [152] and Corban [48] for ordinary transportation problems:

Using the change of variables $x_{ijk} = \dfrac{z_{ijk}}{t}$, the first term of the objective function becomes a second-degree term. This means that Z (which has a special form) can be transformed into a second-degree function.

Divide the interval $[\underline{b}_{jk}, \bar{b}_{jk}]$ into γ_j sub-intervals $\Delta_{\gamma_j,j}, \Delta_{\gamma_j-1,j}, \ldots, \Delta_{sj}, \ldots, \Delta_{2j}, \Delta_{1j}$ such that the function $f(b_{kj})$ is constant over the interval $\alpha^s_{kj}=[\bar{b}_{kj} - \Delta^s_{kj}, \bar{b}_{kj} - \Delta^{s-1}_{kj}]$, where, $\Delta^s_{kj}=\Delta_{1kj}+ + \ldots + \Delta_{skj}$, $\Delta^0_{kj} = 0$. Let $p_{skj} = P(\bar{b}_{kj}-\Delta^s_{kj} \leqslant B_{kj} \leqslant \bar{b} - \Delta^{s-1}_{kj})$. Denote:

$$H^s_{kj}(x_{kj}) = l^1_{kj} \underset{B_{jk}\geqslant x_{jk}}{\mathbf{E}} (B_{jk} - x_{jk}) + l^2_{jk} \underset{B_{jk}\leqslant x_{jk}}{\mathbf{E}} (x_{jk} - B_{jk}),$$

$$h^s_{kj}(x_{kj}) = \underset{B_{kj}\geqslant x_{kj}}{\mathbf{E}} (B_{kj}-x_{kj})\,; \quad \tilde{h}^s_{kj}(x_{kj}) = \underset{B_{kj}\leqslant x_{kj}}{\mathbf{E}} (x_{kj}-B_{kj}).$$

Hence,

$$H^s_{kj}(x_{kj}) = l^1_{kj}\, h^s_{kj}(x_{kj}) + l^2_{kj}\, \tilde{h}^s_{kj}(x_{kj}).$$

Using Szwarc's linearization procedure [152], one can approximate the decreasing convex function $h^s_{kj}(x_{kj})$ on each interval α^s_{kj} by the linear function $g_{kj}(x_{kj})$ and the increasing convex function $h^s_{kj}(x_{kj})$ by the linear function $\tilde{g}_{kj}(x_{kj})$, i.e. approximate $H^s_{kj}(x_{kj})$ by

$$G^s_{kj}(x_{kj}) = l^1_{kj}\, g^s_{kj}(x_{kj}) + l^2_{kj}\, \tilde{g}^s_{kj}(x_{kj}).$$

Hence, the last two terms of Z can be approximated by linear functions:

$$\sum_{j=1}^{n}\sum_{k=1}^{p}\mathop{\mathbf{E}}_{B_{jk}\geqslant x_{jk}}[(l_{jk}^{1}(B_{jk}-x_{jk})]+\sum_{j=1}^{n}\sum_{k=1}^{p}\mathop{\mathbf{E}}_{B_{jk}\leqslant x_{jk}}[l_{jk}^{2}(x_{jk}-B_{jk})]=$$

$$=\sum_{j=1}^{n}\sum_{k=1}^{p}l_{kj}^{1}\mathop{\mathbf{E}}_{B_{jk}\geqslant x_{jk}}(B_{jk}-x_{jk})+\sum_{j=1}^{n}\sum_{k=1}^{p}l_{jk}^{2}\mathop{\mathbf{E}}_{B_{jk}\leqslant x_{jk}}(x_{jk}-B_{jk})=$$

$$=\sum_{j=1}^{n}\sum_{k=1}^{p}H_{kj}^{s}(x_{kj})\approx\sum_{j=1}^{n}\sum_{k=1}^{p}G_{kj}^{s}(x_{kj})=\sum_{j=1}^{n}\sum_{k=1}^{p}l_{kj}^{1}\,g_{kj}^{s}(x_{kj})+$$

$$+\sum_{j=1}^{n}\sum_{k=1}^{p}l_{kj}^{2}\,g_{kj}^{s}(x_{kj}).$$

Chapter 7

Applications in economy

Apart from the specific mathematical interest, the methods presented so far may be directly applied in solving practical problems.

Some of the main types of problems which can be modelled by multiple objective functions and, hence, efficiently solved by such methods (either in the deterministic or in the stochastic case) are presented below.

7.1. PRODUCTION PLANNING WITH MULTIPLE EFFICIENCY CRITERIA [130]

This application was developed at a large Bucharest factory producing two main groups of goods : group A having 10 assortments and group B including 5 assortments. Specific production planning conditions and delivery schedules indicated a total value for the assortments under consideration and a global quantity of products and of each group. However, decision on the assortments in each group is not so strict (plus-minus 10% per assortment). The problem was to optimize the structure of the annual plan taking into account several efficiency criteria. The following criteria could be established by a detailed economic analysis.

1. minimization of the total cost ;
2. minimization of the imported material cost ;
3. minimization of the total material cost ;
4. minimization of the total man-hour expenditure ;
5. minimization of the man-hour expenditure on machinery groups I, II and III.

Table 7.1

No.	Product	Plan (thousands/year) minim	Plan (thousands/year) maxim	Selling price per piece
	Group A			
1	Type P_1	34.6	42.4	270
2	,, P_2	4.5	5.5	270
3	,, P_3	36	44	519
4	,, P_4	17.5	21.5	432
5	,, P_5	28.8	35.2	432
6	,, P_6	43.6	53.3	281
7	,, P_7	21.1	25.8	396
8	,, P_8	31.5	38.5	619
9	,, P_9	45	55	814
10	,, P_{10}	9	11	565
	Total mean plan	302.0		
	Group B			
11	Type P_{11}	29.7	36.3	1602
12	,, P_{12}	17.1	20.9	1602
13	,, P_{13}	7.2	8.8	1602
14	,, P_{14}	70.2	85.8	1813
15	,, P_{15}	18.9	23.1	1813
	Total mean plan	159.0		

Value of total annual production 421670.6

Table 7.2

Data for the optimization criteria

No.	Designation \ Criteria	C_1	C_2	C_3	C_4	C_5	C_6	C_7
1	Product P_1	40	213.75	236.69	44.6	12.91	5	290.67
2	Product P_2	18	212	272.33	51.22	8.35	6	290.8
3	Product P_3	103	385.32	459.70	85.13	12.03	10	541.3
4	Product P_4	77.3	302	575.78	130.52	23.19	16	462.56
5	Product P_5	51.5	495	621.28	111.74	27.32	18	692.3
6	Product P_6	53.9	314.3	419.74	110.74	26.32	5	466.95
7	Product P_7	54	387.93	613.27	147.13	28.36	16	573.25
8	Product P_8	93.4	651.31	886.74	206.04	67.4	30	937.40
9	Product P_9	280	980.63	917.86	212.65	68.62	30	1310.48
10	Product P_{10}	177.3	540	604.99	133.83	19.54	13	713.75
11	Product P_{11}	565.1	1592	1078.41	216.45	48.58	32	2122.9
12	Product P_{12}	565.1	1592	1078.41	216.45	48.58	32	1995.78
13	Product P_{13}	565.1	1570	1078.41	216.45	48.58	32	2058.72
14	Product P_{14}	581.8	1564	877.43	199.64	42.06	32	2183.9
15	Product P_{15}	581.8	1700	1040.4	211.4	175.23	40	2208.45

Using the data concerning the production plan (table 7.1) and the specific data concerning the optimization criteria (table 7.2.) we obtained the following mathematical model.

$$\min F_1 = 40\,x_1 + 18\,x_2 + 103\,x_3 + 77.3\,x_4 + 51.5\,x_5 + 53.9 x_6 + $$
$$+ 54\,x_7 + 93.4\,x_8 + 280\,x_9 + 177.3\,x_{10} + 565.1\,x_{11} + 565.1\,x_{12} +$$
$$+ 565.1\,x_{13} + 581.8\,x_{14} + 581.8\,x_{15}$$

$$\min F_2 = 213.75\,x_1 + 212\,x_2 + 385.43\,x_3 + 302\,x_4 + 495\,x_5 +$$
$$+ 314.3\,x_6 + 387.93\,x_7 + 615.31\,x_8 + 980.63\,x_9 + 540\,x_{10} +$$
$$+ 1592\,x_{11} + 1592\,x_{12} + 1570\,x_{13} + 1654\,x_{14} + 1700\,x_{15}$$

$$\min F_3 = 236.69\,x_1 + 272.33\,x_2 + 459.7\,x_3 + 575.78\,x_4 +$$
$$+ 621.28\,x_5 + 419.74\,x_6 + 613.27\,x_7 + 886.74\,x_8 + 917.86\,x_9 +$$
$$+ 604.99\,x_{10} + 1078.41\,x_{11} + 1078.41\,x_{12} + 1078.41\,x_{13} +$$
$$+ 877.43\,x_{14} + 1040.4\,x_{15}$$

$$\min F_4 = 44.63\,x_1 + 51.22\,x_2 + 85.13\,x_3 + 130.52\,x_4 +$$
$$+ 111.74\,x_5 + 110.74\,x_6 + 147.13\,x_7 + 206.04\,x_8 + 212.65\,x_9 +$$
$$+ 133.83\,x_{10} + 216.45\,x_{11} + 216.45\,x_{12} + 216.45\,x_{13} +$$
$$+ 199.64\,x_{14} + 211.4\,x_{15}$$

$$\min F_5 = 12.91\,x_1 + 8.35\,x_2 + 12.03\,x_3 + 23.19\,x_4 + 27.32\,x_5 +$$
$$+ 26.32\,x_6 + 28.36\,x_7 + 67.4\,x_8 + 68.62\,x_9 + 19.54\,x_{10} +$$
$$+ 48.58\,x_{11} + 48.58\,x_{12} + 48.58\,x_{13} + 42.06\,x_{14} + 175.23\,x_{15}$$

$$\min F_6 = 5\,x_1 + 6\,x_2 + 10\,x_3 + 16\,x_4 + 18\,x_5 + 5\,x_6 +$$
$$+ 16\,x_7 + 30\,x_8 + 30\,x_9 + 13\,x_{10} + 32\,x_{11} + 32\,x_{12} +$$
$$+ 32\,x_{13} + 32\,x_{14} + 40\,x_{15}$$

$$\min F_7 = 290.67\, x_1 + 290.8\, x_2 + 541.3\, x_3 + 462.56\, x_4 + \\ + 692.3\, x_5 + 466.95\, x_6 + 573.25\, x_7 + 937.4\, x_8 + \\ + 1310.48\, x_9 + 713.75\, x_{10} + 2122\, x_{11} + 1995.78\, x_{12} + \\ + 2058.72\, x_{13} + 2183.9\, x_{14} + 2208.45\, x_{15},$$

subject to

$$x_1 + x_2 + x_3 + x_4 + x_5 + x_6 + x_7 + x_8 + x_9 + x_{10} = 302$$

$$x_{11} + x_{12} + x_{13} + x_{14} + x_{15} = 159$$

$$270\, x_1 + 270\, x_2 + 519\, x_3 + 432\, x_4 + 432\, x_5 + 281\, x_6 + \\ + 396\, x_7 + 619\, x_8 + 814\, x_9 + 565\, x_{10} + 1602\, x_{11} + \\ + 1602\, x_{12} + 1602\, x_{13} + 1813\, x_{14} + 1813\, x_{15} = 421670.6$$

$$34.6 \leqslant x_1 \leqslant 42.4$$

$$4.5 \leqslant x_2 \leqslant 5.5$$

$$36 \leqslant x_3 \leqslant 44$$

$$17.5 \leqslant x_4 \leqslant 21.5$$

$$28.8 \leqslant x_5 \leqslant 35.2$$

$$43.6 \leqslant x_6 \leqslant 53.3$$

$$21.1 \leqslant x_7 \leqslant 25.8$$

$$31.5 \leqslant x_8 \leqslant 38.5$$

$$45 \leqslant x_9 \leqslant 55$$

$$9 \leqslant x_{10} \leqslant 11$$

$$29.7 \leqslant x_{11} \leqslant 36.3$$

$$17.1 \leqslant x_{12} \leqslant 20.9$$

$$7.2 \leqslant x_{13} \leqslant 8.8$$

$$70.2 \leqslant x_{14} \leqslant 85.8$$

$$18.9 \leqslant x_{15} \leqslant 23.1.$$

This mathematical model was solved on an IBM 360/40 F computer, using the maximum global utility method (Chapter 1, Section 1.4). According to this method, we solve the seven linear programming problems defined above and other seven problems in which min F_i was replaced by maxF_i ($i = 1, \ldots, 7$) (Table 7.3). In order to determine the synthesis function F^* we have to evaluate the weighting coefficients k_i as follows.

$$k_1 = 0.25 ; k_2 = 0.06 ; k_3 = 0.08 ; k_4 = 0.04 ;$$

$$k_5 = 0.04 ; k_6 = 0.02 ; k_7 = 0.51.$$

Hence, the synthesis function was found to be :

$$F^* = 0.04431\, x_1 + 0.033374\, x_2 + 0.067465\, x_3 +$$

$$+ 0.068\, x_4 + 0.078786\, x_5 + 0.060804\, x_6 +$$

$$+ 0.076363\, x_7 + 0.119748\, x_8 + 0.168226\, x_9 +$$

$$+ 0.097532\, x_{10} + 0.257827\, x_{11} + 0.249867\, x_{12} +$$

$$+ 0.253599\, x_{13} + 0.259591\, x_{14} + 0.26880\, x_{15} - 56.3933.$$

The minimum of this function has to be found. The optimal solution was found to be :

$$x_1 = 42.4 ; \quad x_2 = 5.5 ; \quad x_3 = 44 ;$$

$$x_4 = 21.5 ; \quad x_5 = 32.835 ; \quad x_6 = 43.6 ;$$

$$x_7 = 21.1 ; \quad x_8 = 35.264 ; \quad x_9 = 45 ; \quad x_{10} = 11 ;$$

$$x_{11} = 29.7 ; \quad x_{12} = 17.1 ; \quad x_{13} = 7.2 ;$$

$$x_{14} = 85.8 ; \quad x_{15} = 19.2.$$

Table 7.3

Criteria / Val.	min F_1	max F_1	min F_2	max F_2	min F_3	max F_3	min F_4	max F_4	min F_5	max F_5	min F_6	max F_6	min F_7	max F_7
x_1	40.13	37.01	42.40	34.60	42.40	34.60	42.40	34.60	41.84	35.34	40.74	36.43	42.40	34.60
x_2	5.50	4.50	5.50	4.50	5.50	4.50	5.50	4.50	5.50	4.50	4.50	5.50	5.50	4.50
x_3	37.77	42.28	44.00	36.00	44.00	36.00	44.00	36.00	44.00	36.00	43.99	36.00	44.00	36.00
x_4	21.50	17.50	21.50	17.50	17.50	21.50	20.85	18.39	21.50	17.50	17.50	21.50	21.50	17.50
x_5	35.20	28.80	28.80	35.20	28.80	35.19	35.20	28.80	35.20	28.80	28.80	35.20	28.80	35.29
x_6	43.60	53.30	43.60	53.30	50.58	46.48	43.60	53.30	43.60	53.30	53.30	43.60	43.60	53.30
x_7	25.80	21.10	24.31	22.83	21.10	25.80	21.10	25.79	21.10	25.79	21.10	25.80	24.31	22.83
x_8	38.50	31.50	35.88	34.06	31.50	38.50	31.50	38.50	31.50	38.50	31.50	38.50	35.88	34.06
x_9	45.00	55.00	45.00	55.00	49.61	50.41	46.84	53.10	46.76	53.25	49.55	50.46	45.00	55.00
x_{10}	9.00	11.00	11.00	9.00	11.00	9.00	11.00	9.00	11.00	9.00	10.99	9.00	11.00	9.00
x_{11}	29.70	36.29	29.70	36.29	29.70	36.30	29.70	36.30	30.00	36.29	29.70	36.30	29.70	36.29
x_{12}	17.10	20.90	17.10	20.90	17.10	20.90	17.10	20.90	17.10	20.90	17.10	20.90	17.10	20.90
x_{13}	7.20	8.80	7.20	8.20	7.20	8.79	7.20	8.80	7.20	8.50	7.20	8.80	7.20	8.80
x_{14}	85.80	74.10	85.80	70.20	85.80	70.20	85.80	70.20	85.80	70.20	85.80	70.20	85.80	70.20
x_{15}	19.20	18.90	19.20	19.19	19.20	22.80	19.20	22.80	18.50	23.10	19.20	22.80	19.20	22.79
F_i	122 208	124 461	404 871	411 271	327 520	337 894	71 307	73 706	19 427	20 871	101 178	10 523	544 164	552 279

This solution proved to be better than that considered by the factory management* and than that containing the optimum solutions calculated for each criterion.

One should also note that usually the coefficients of the objective functions are not constant but random variables. In such cases one deals with a stochastic programming problem with multiple objective functions.

7.2. ANOTHER EXAMPLE OF PRODUCTION PLANNING [149]

For a factory producing various house ware objects, the problem arised of optimising the plan structure for the production of pig-iron radiators taking into account several efficiency functions. The following data was considered :

— the pig-iron radiators are produced in 5 ranges

— these ranges of radiators differ in size, and so their mass and heating surface are likewise varied.

— the client has different annual demands for each range of radiators.

Using the above data one has to set up a production plan with due consideration of certain technological constraints, the maximum production capacity of the workshop, the value of the global production, contractual stipulations, etc.

Following consultation with the factory management it was established that the plan structure should meet the following demands :

1. Maximum value of commodity;
2. Minimum production costs;
3. Maximum earnings for the workers.

These objectives had to be met under the following types of constraints:

a. constraints concerning the annual number of radiators that can be produced in the workshop (say 4,900)

b. constraints concerning the total weight which cannot exceed 36,000 kg

---

* The factory management considered initially the solution : $x_1 = 38.5$ thousand ; $x_2 = 5$; $x_3 = 40$; $x_4 = 19.5$; $x_5 = 32$; $x_6 = 48.5$; $x_7 = 23.5$; $x_8 = 35$; $x_9 = 50$; $x_{10} = 10$; $x_{11} = 33$; $x_{12} = 19$; $x_{13} = 8$; $x_{14} = 78$; $x_{15} = 21$.

c. planning constraints. Certain radiators are in big demand and hence their number should exceed a planned limit.

The data necessary for establishing the mathematical model of the problem are given in Table 7.4 (actually this is hypothetical data !).

Table 7.4

No	Radiator type	Weight (kg/piece)	Cost (lei/piece)	Work force cost (lei/piece)	Selling price (lei/piece)
1	$300\times 250\times 3$	7.78	31.90	2.95	38.5
2	$500\times 150\times 2$	6.28	31.29	2.57	33.5
3	$600\times 135\times 2$	6.88	31.13	2.67	33.4
4	$600\times 150\times 2$	7.48	33.12	2.81	40.15
5	$600\times 200\times 3$	10.78	41.28	3.56	49.65

Denoting by i ($i = 1, \ldots, 5$) the types of radiators and by x_i (thousands) the quantity produced out of each type, one can set up the following model (F_1 = commodity, F_2 = cost, F_3 = workers' earnings) :

$$\max F_1 = 38.5\, x_1 + 33.5\, x_2 + 33.4\, x_3 + 40.15\, x_4 + 49.65\, x_5,$$

$$\min F_2 = 31.95\, x_1 + 31.29\, x_2 + 31.13\, x_3 + 33.12\, x_4 + $$
$$+ 41.28\, x_5,$$

$$\max F_3 = 2.95\, x_1 + 2.57\, x_2 + 2.67\, x_3 + 2.81\, x_4 + 3.65\, x_5,$$

subject to

$$7.78\, x_1 + 6.28\, x_2 + 6.88\, x_3 + 7.48\, x_4 + 10.78\, x_5 \leqslant 36\,000,$$

$$x_1 + x_2 + x_3 + x_4 + x_5 \leqslant 4\,990,$$

$$50 \leqslant x_1 \leqslant 160,$$

$$1620 \leqslant x_2 \leqslant 2000,$$

$$1300 \leqslant x_3 \leqslant 1600,$$

$$1350 \leqslant x_4 \leqslant 1650,$$

$$160 \leqslant x_5 \leqslant 200,$$

$$x_1, x_2, x_3, x_4, x_5 \geqslant 0.$$

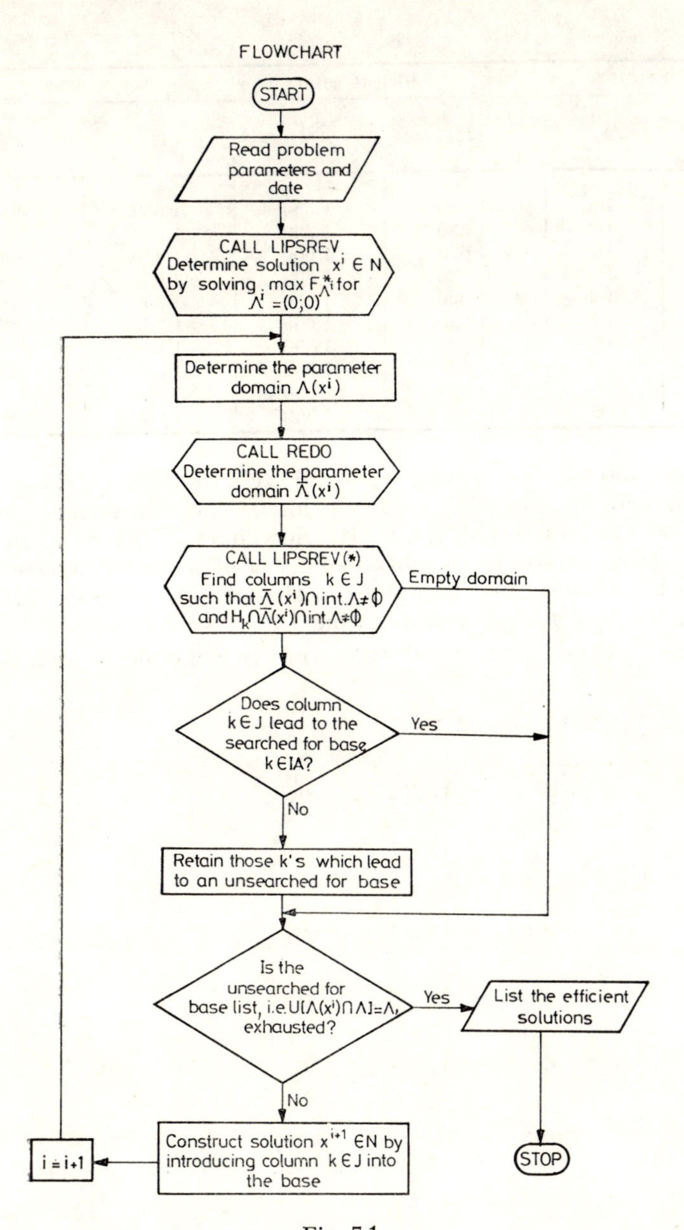
FLOWCHART
START
Read problem parameters and date
CALL LIPSREV. Determine solution x' ∈ N by solving max F*Λi for Λi = (0;0)
Determine the parameter domain Λ(xi)
CALL REDO Determine the parameter domain Λ̄(xi)
CALL LIPSREV (*) Find columns k ∈ J such that Λ̄(xi)∩ int.Λ≠ϕ and Hk∩Λ̄(xi)∩int.Λ≠ϕ
Empty domain
Does column k ∈ J lead to the searched for base k ∈ IA?
Yes
No
Retain those k's which lead to an unsearched for base
Is the unsearched for base list, i.e. U[Λ(xi)∩Λ]=Λ, exhausted?
Yes
List the efficient solutions
No
Construct solution xi+1 ∈ N by introducing column k ∈ J into the base
i = i+1
STOP

Fig. 7.1

Table

	Efficient solutions x^i				
	x_1	x_2	x_3	x_4	x_5
x^1	160	1 620	1 360	1 650	200
x^2	160	1 680	1 300	1 650	200
x^3	160	1 620	1 600	1 410	200
x^4	160	1 620	1 300	1 650	200
x^5	160	1 620	1 300	1 350	200
x^6	160	1 620	1 600	1 350	200
x^7	160	1 620	1 300	1 650	160
x^8	50	1 620	1 300	1 650	160
x^9	50	1 620	1 300	1 350	160
x^{10}	160	1 620	1 300	1 350	160

This model was solved with Zeleny's method [168] in order to determine the efficient solutions. The computer program was written in FORTRAN. Its flowchart is given in fig. 7.1. The efficient solutions obtained as well as the corresponding function values and the values of the straight lines defining the domains $\Lambda(X^i)$ are given in Table 7.5.

The straight lines defining the parametric domains are:

(A) $2.43\ \lambda_1 - 1.86\ \lambda_2 = 0,$

(B) $-47.21\ \lambda_1 + 15.29\ \lambda_2 = 0,$

(C) $-11.51\ \lambda_1 + 12.65\ \lambda_2 = 0,$

(D) $-12.05\ \lambda_1 + 11.81\ \lambda_2 = 0,$

(E) $-13.29\ \lambda_1 + 12.78\ \lambda_2 = 0,$

(F) $-12.60\ \lambda_1 + 12.31\ \lambda_2 = 0.$

These domains are sketched in fig. 7.2. Any of the efficient solutions $x^1, \ldots, x^{10}$ can be used to elaborate the production plan.

7.5

Function value			Parametric range
F_1	F_2	F_3	
178 031.5	161 042.6	18 889.1	$A \leqslant 1$; $B \leqslant 1$; $C \leqslant 1$
178 039.5	161 042.2	18 883.1	$A \geqslant 1$
176 411.5	162 432.4	18 885.5	$B \geqslant 1$; $C \leqslant 1$; $F \leqslant 1$
176 027.5	159 174.8	18 728.9	$C \geqslant 1$; $E \leqslant 1$; $F \leqslant 1$
163 982.5	149 238.8	17 885.9	$C \geqslant 1$; $E \geqslant 1$; $F \leqslant 1$
174 002.5	158 577.8	18 686.9	$C \leqslant 1$; $E \geqslant 1$
174 041.5	158 523.6	18 582.9	$D \leqslant 1$; $E \leqslant 1$; $F \geqslant 1$
169 806.5	155 009.1	18 258.4	$D \geqslant 1$; $E \leqslant 1$
157 761.5	145 075.1	17 415.4	$D \geqslant 1$; $E \geqslant 1$
161 996.5	148 589.6	17 739.9	$D \leqslant 1$; $E \geqslant 1$; $F \geqslant 1$

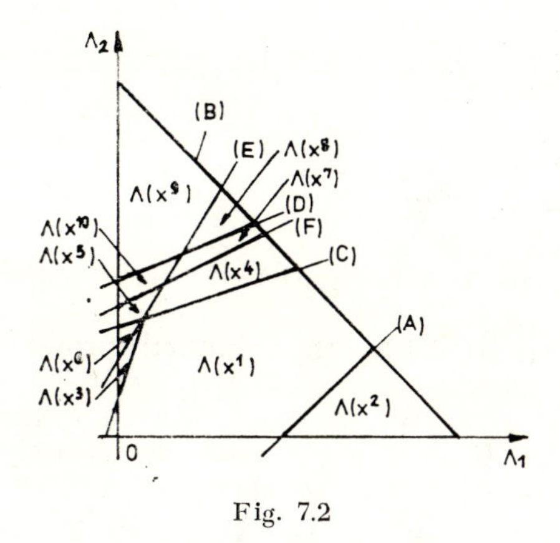

Fig. 7.2

7.3. A PROBLEM OF MACHINERY LOADING

In production planning one is often confronted with the problem of efficient machinery loading. We give below an example of such situations.

In a production workshop one has to machine n types of items, $r_1, \ldots, r_n$, each in quantities $N_1, \ldots, N_n$ using m machine tools $U_1, \ldots, U_m$, each available for periods $D_1, \ldots, D_m$ over the time period taken to draw up the schedule.

Let t_{ij} be the unit machining time of machine U_j for the item r_i, let e_{ij} be the unit cost of machining this item on that machine, and let p_j be the unit idling cost of the same machine U_j.

From the observation of the actual production process, one could derive that the unit machining times t_{ij} have random variable characteristics.

Denoting by x_{ij} the quantity machined on U_j out of item r_i, one can set up the following stochastic programming problem with multiple objective functions:

$$(7.1) \qquad \min \sum_{i=1}^{n} \sum_{j=1}^{m} c_{ij} x_{ij}$$

$$(7.2) \qquad \min \sum_{j=1}^{m} p_j \left(D_j - \sum_{i=1}^{n} t_{ij} x_{ij} \right)$$

subject to

$$(7.3) \qquad \sum_{j=1}^{m} x_{ij} = N_i, \quad (i = 1, \ldots, n)$$

$$(7.4) \qquad P\left(\sum_{i=1}^{n} \sum_{j=1}^{m} t_{ij} x_{ij} \leqslant D_j \right) \geqslant \beta_j, \quad (j = 1, \ldots, m)$$

$$(7.5) \qquad x_{ij} \geqslant 0, \ (i = 1, \ldots, n; \ j = 1, \ldots, m).$$

Under constraint (7.3) the planned production demands must be fully met.

Constraint (7.4) refers to the probability that the service time of machine U_j does not exceed the available capacity D_j and is at least equal to a given threshold β_j.

The objective function (7.1) represents the total machining cost which must be minimized.

The objective function (7.2) represents the total penalty for machinery idling time which must be minimized. This function is random since it contains t_{ij}.

7.4. A MIXING PROBLEM [41], [133], [134]

It is common knowledge that the chemical processes of an oil refinery have become exceedingly complicated and hence the economic analysis and production planning in such environments requires sophisticated mathematical tools. Several inconveniences may arise in setting up models that should be as faithful as possible to the processes involved, e.g. a mixing process.

Thus, though the constraints referring to the quantities involved may be easily expressed mathematically, the constraints referring to the quality of the end product are substantially more involved. Most authors use models for which the quality constraints are made under the assumption that the important features have additive behaviour. Hence the characteristic of the mixture is the weighted sum of the characteristics of its constituents :

$$c = \frac{\sum_{i=1}^{n} c_i x_i}{\sum_{i=1}^{n} x_i},$$

where c is the characteristic of the mixture, c_i is that of component i, and x_i denotes the quantity of component i which is involved in the mixture. However, in applications the above law does not always comply with reality and it is not infrequent that the characteristics of the mixture obtained in this manner exceed by far the admissible limit given in STAS code requirements. This, for instance, was the case at a refinery which was producing two petrol types (CO 75 and 90) out of 8 constituents. Then it was decided to determine the actual relations between the mixture characteristics and those of the constituents, such that the quality specifications may be met.

The main shortcoming of the quality relationships used in the literature is that too much emphasis is placed on the economic efficiency of the model containing these relationships, but little is thought of determining the domain over which these relationships hold, their accuracy as well as the probability with which the characteristics determined in such a fashion will meet the specifications, either in laboratory or in the factory.

An analysis of the technical characteristics of the constituents and that of the resulting mixtures revealed that most importance can be attributed to the octane number and to the distillation index to the gradations of 10, 50, 90 and final.

In order that the final model should be as simple as possible, linear relations were sought, i.e.

$$c = \frac{\sum_{i=1}^{n} k_i c_i x_i}{\sum_{i=1}^{n} x_i} = \sum_{i=1}^{n} k_i c_i \left(\frac{x_i}{\sum_{i=1}^{n} x_i} \right) = \sum_{i=1}^{n} k_i c_i x'_i = \sum_{i=1}^{n} x_i y_i,$$

where c is the mixture characteristic and k_i are the regression coefficients for that type. k_i can be determined by use the correlation method.

The products $c_i x_i$ were assumed to be independent or factorial variables.

Since some of the characteristics of a certain component, as for instance the octane number, vary substantially from one time period to another, it was assumed that the constraints should be given a random interpretation. For instance, the constraint for the octane number is

$$P\left[\sum_{j=1}^{n} a_{ij} x_j \geqslant b_i\right] \geqslant \delta,$$

where b_i is the lower limit prescribed in code requirements and δ is a value obtained by laboratory tests.

The following efficiency functions may exist :

— the minimum cost : $F_1(x_1, \ldots, x_n) = \sum_{i=1}^{n} c_i x_i$

— the maximum profit : $F_2(x_1, \ldots, x_n) = p \sum_{i=1}^{n} x_i - \sum_{i=1}^{n} c_i^* x_i - c'$,

where c_i^* is the cost of component i, p is the selling price of the resulting product and c' denotes other costs.

One can likewise choose other efficiency functions, such that, for example, the petrol of a certain quality is produced in the maximum quantity.

7.5. OPTIMISATION OF FLOUR TRANSPORTATION [132]

The optimisation of goods transportation is of national interest. Basically, it consists in shorter transportation distances, reduced transportation time, minimum transportation costs and thus maximum economic efficiency of the whole process.

An efficient scheme for goods transportation can be obtained by influencing the social and economic environment, e.g. the geographical distribution of the production and consumption centers, which must be as close as possible to one another.

The problem presented below refers to the optimisation of flour transportation on the national scale. The numerical values refer to 1971 and are given for 3-month periods [132].

The following types of flour were considered:
— white flour transported in bulk;
— white flour ready-packed;
— half-white flour;
— whole-grain flour.

Thus 20 classical transportation models resulted:
— model 1: white bulk flour, throughout 1971;
— models 2—5: white bulk flour, periods I, II, III and IV of the year 1971;
— model 6: white ready-packed flour, throughout the year 1971;
— models 7—10: white ready-packed flour, periods I—IV of the year 1971;
— model 11: half-white flour, throughout the year 1971;
— models 12—15: half-white flour, periods I—IV of the year 1971;
— model 16: whole-grain flour, throughout the year 1971;
— models 17—20: whole-grain flour, periods I—IV of the year 1971.

To models 1—5 we associated 33 producers, and 90 consumers. For each model the available quantity was less than required.

The objective functions were three in number and aimed at:
— minimization of the total transportation cost;
— minimization of the total transportation time;
— maximization of the transported quantity.

One of the mathematical models is

$$\min \sum_{i=1}^{33} \sum_{j=1}^{90} c_{ij} x_{ij}$$

$$\min \sum_{i=1}^{33} \sum_{j=1}^{90} t_{ij} x_{ij}$$

$$\max \sum_{i=1}^{33} \sum_{j=1}^{90} x_{ij}$$

$$\sum_{j=1}^{90} x_{ij} \leqslant a_i, \qquad (i = 1,2, \ldots, 33)$$

$$\sum_{i=1}^{33} x_{ij} \geqslant b_j, \qquad (j = 1,2, \ldots, 90)$$

$$\sum_{i=1}^{33} x_{ij} > \sum_{j=1}^{90} x_{ij},$$

$$x_{ij} \geqslant 0, \ (i = 1, \ldots, 33; \ j = 1, \ldots, 90),$$

where c_{ij} and t_{ij} were the transportation costs and times, respectively. The model mimics in form the model presented in Chapter 6 and was solved on computer.

A comparison of the optimised solution with the one recorded for that year (based on empirical data) showed that an economic efficiency of 10.7% would have been possible, i.e. more than one million lei for the whole year.

7.6. EFFICIENT USE OF PRODUCTION CAPACITY IN A METALLURGICAL WORKSHOP

The problem analysed here is that of splitting the monthly production plan of a given workshop into plans for each production line. The application is made for a net welding workshop from a metallurgical plant. We denote by

P the set of net types to be produced,

M the set of machine tool existing in the workshop.

$T_i(i \in P)$ the time taken to produce the net type i,

$N_i(i \in P)$ the requirement of net type i during that month,

$C_j(j \in M)$ the hourly capacity of machine j,

$\bar{s}_j(j \in M)$ the mean loading time of machine j.

We introduce the following variables :
x_{ij} — quantity of net type i produced on machine j,

$$a_{ij} = \begin{cases} 1 \text{ if net type } i \text{ can be welded on machine } j \\ 0 \text{ otherwise.} \end{cases}$$

The mathematical model for this problem was :

$$(7.6) \qquad \max F_1 = \sum_{i \in P} \sum_{j \in M} T_i a_{ij} x_{ij}$$

$$(7.7) \qquad \min F_2 = \sum_{j \in M} \sum_{i \in P} (T_i a_{ij} x_{ij} - \bar{s}_j)^2,$$

subject to

$$(7.8) \qquad \sum_{j \in M} a_{ij} x_{ij} \geqslant N_i, \qquad i \in P$$

$$(7.9) \qquad \sum_{i \in P} T_i a_{ij} x_j \leqslant C_j, \quad j \in M$$

$$(7.10) \quad x_{ij} \geqslant 0, \qquad i \in P, \; j \in M.$$

Constraint (7.8) is that of producing a minimum necessary quantity of each net type. This constraint appears with " $\geqslant$ " since some of the nets may not be accepted by the client and, hence, one has to produce more than is required in order to overcome rejects. Moreover, if the available capacity of the machine can produce more than required, then constraints (7.8) allow exactly this.

Constraints (7.9) are concerned with not exceeding the time budget available for each machine. The time T_i is the same for all the machine on which net type i can be produced.

The objective function F_1 represents the total service time of the machines and F_2 refers to a uniform loading of all the machines.

In our problem T_i are random variables. Just as well C_i are random too. Due to some specific conditions, idle times vary from one period to the next.

One can restate the above problem taking two limit values $\bar{F}_1$ and $\bar{F}_2$ for the objective functions and then we will have to maximize the function :

$$\max \; P\{\sum_{i \in P} \sum_{j \in M} T_i a_{ij} x_{ij} \geqslant \bar{F}_1 ; \; \sum_{j \in M} \; (\sum_{i \in P} T_i a_{ij} x_{ij} - \bar{s}_j)^2 \leqslant \bar{F}_2\}$$

subject to (7.8)—(7.10).

7.7. THE MATHEMATICAL MODEL FOR PIG-IRON PRODUCTION

The problem is to find the optimum recipe for obtaining pig-iron out of m raw materials (e.g.: special primary pig-iron, scrap pig-iron, scrap steel, rejects, etc.).

The pig-iron mixture should contain n chemical elements in different percentages (e.g., C, Si, Mn, P, C, etc.). The quality of the mixture depends on the charge composition but also on the proportion of various chemical elements in the raw materials.

Let:

— x_i be the quantity (in kg) of raw material i in a ton of mixture;
— c_i the cost of component i ($i = 1, \ldots, m$);
— a_{ij} the percentage of the chemical element j in the raw material i;
— a_i the maximum quantity (in kg) of raw material i which may come into a ton of mixture;
— p_j, q_j the maximum and minimum values, respectively, of the percentage in which element j must come into the pig-iron mixture;
— r_j the variation coefficient for element j due to the technological process in the furnace.

Due to such factors as the construction of the furnace, the time required for various chemical reactions, the melting temperatures, for type of fuel, the air draught and others, the element content of the charge is different from that of the molted mass, the variation facing around several percent.

The mathematical model for obtaining a ton of mixture is

$$(7.11) \quad \min F_1 = \sum_{i=1}^{m} c_i x_i$$

$$(7.12) \quad \min F_2 = x_k,$$

subject to

$$(7.13) \quad \sum_{i=1}^{m} x_i = 1000$$

$$(7.14) \quad 10q_i \leqslant \left(1 + \frac{r_i}{100}\right) \sum_{j=1}^{n} \frac{a_{ji}}{100} x_j \leqslant 10p_i, \quad i = 1, \ldots, m$$

$$(7.15) \quad 0 \leqslant x_i \leqslant a_i, \; i = 1, \ldots, m.$$

The objective function F_2 represents the condition for the minimum consumption of the k raw material.

Due to the technological process, the coefficients r_i are not constant. Instead, they are random variables and hence we consider that constraints (7.14) are satisfied with a certain probability:

$$P\left[10q_i \leqslant \left(1 + \frac{r_i}{100}\right)\sum_{j=1}^{n} \frac{a_{ji}}{100} x_j \leqslant 10p_i\right] \geqslant \delta,$$

where the value of δ can be estimated in practice.

The costs c_i are not constant. They may vary over time periods and are taken as random variables of given distribution. Thus one obtains a stochastic programming problem with multiple objective functions.

7.8. OPERATION SCHEDULING

Suppose that in a given workshop one considers the items $r_1, \ldots, r_p$ which are machined on the machine-tools $U_1, \ldots, U_m$. The items go through machining in various orders of operation. It is also possible that one item should be worked upon by machine U_j, go on to other machines and then come back to U_j. Denote by O_{ijk} the operation of rank k ($k = 1, \ldots, m_j$) done upon the item r_i on the machine U_j, and by t_{ijk} the time taken to do this. t_{ijk} may include various technological idle times, or one can consider these as dummy operations. Let x_{ijk} be the moment when operation O_{ijk} starts and τ_{ijk} the waiting time on machine U_j before the operation of rank k is started on item r_i (the wait is due to lack of items to be machined). The waiting time can be calculated as

$$\tau_{jik} = x_{ijk} - (x_{hjl} + t_{hjl}),$$

where $x_{hjl} < x_{ijk}$ and $x_{ijk} = \min_{\lambda,\mu} x_{\lambda j\mu}$; $x_{\lambda j\mu} \geqslant x_{ijk}$.

There is also another waiting time θ_{ijk}, as item r_i waits for the machine U_j to finish work k on other items. This time is calculated as:

$$\theta_{ijk} = x_{ijk} - (x_{ihl} + t_{ihl}).$$

The technological schedule indicates that an operation of rank l upon item i on machine U_h is followed by an operation of rank k on machine U_j. Denote by σ_j the unit cost of idle time on machine U_j and by Γ_{ijk} the unit cost of penalties for delaying the item r_i before starting operation O_{ijk}. If item r_i is machined on U_j during an operation of rank k in baches of q_{ijk} items, denote by γ_i the cost of penalties per item per unit time of delay. Then, $\Gamma_{ijk} = q_{ijk}\, \gamma_i$.

A possible executive scheme for the operations given on the given items and corresponding machines is represented below in a time-graph of Gantt type.

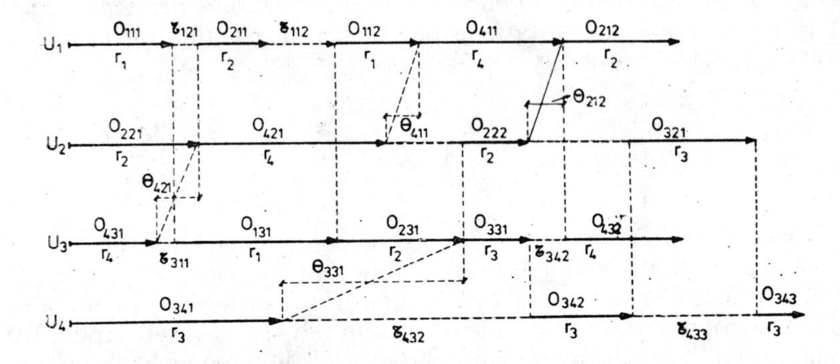

It is required that the starting moments of each operation be chosen in such a manner that the total waiting cost be minimized both for the machines and for the items. The constraints to be introduced in the mathematical model are :

a) The technological and organizational sequence :

$$x_{ijk} - x_{ihl} \geqslant t_{ihl} \quad i = 1, \ldots, p\,;\; j = 1, \ldots, m$$

$$l, k = 1, \ldots, m_j$$

b) The noninterference conditions between operations on the same machine :

$$(x_{ujv} - x_{hjl} \geqslant t_{hjl}) V (x_{hjl} - x_{ujv} \geqslant t_{ujv})$$

$$u, h = 1, \ldots, p\,;\; j = 1, \ldots, m\,;\; l, v = 1, \ldots, m_j$$

c) the non-negativeness condition for the starting times:

$$x_{ijk} \geqslant 0,\ i = 1, \ldots, p;\ j = 1, \ldots, m;\ k = 1, \ldots, m_j.$$

The following two cost functions are considered:

$$\min F_1 = \sum_{i=1}^{p} \sum_{j=1}^{m} \sum_{k=1}^{m_j} \tau_{jik}\ \sigma_j$$

$$\min F_2 = \sum_{i=1}^{p} \sum_{j=1}^{m} \sum_{k=1}^{m_j} \theta_{ijk}\ \Gamma_{ijk}.$$

The above problem is rather a stochastic programming problem since not all the variables in the model are precisely defined. For example, σ_j is a random variable. Its value cannot be precisely defined since one needs to know the duration the machine U_j is under service, and this duration is a random variable itself, as well as the down-time which is also a random variable. Some considerations apply for Γ_{ijk} too.

7.9. THE ASSIGNMENT PROBLEM

In a workshop, n workers have to perform n jobs. Define a boolean variable x_{ij} indicating whether worker M_i is given the job L_j or not:

$$x_{ij} = \begin{cases} 1 & \text{if worker } M_i \text{ is given job } L_j \\ 0 & \text{otherwise.} \end{cases}$$

Denote by t_{ij} the time needed by worker M_i to do the job L_j. Since each worker has different qualifications and performance, two workers do not require the same time to do the same job. Let c_{ij} be the cost assigned to worker M_i doing job L_j. This cost includes material consumption, general costs, etc.

Let p_{ij} be the productivity of worker M_i doing job L_j. The productivity is expressed as items per unit time. If cs_{ij} denotes the specific consumption of scarce materials when worker M_i does the job L_j and if csn_j is the normed specific consumption of scarce materials, then cs_{ij}/csn_j is the index of specific consumption.

Let p^*_{ij} be the preference of worker M_i to do work L_j.

The problem is to assign one and only one job to each worker such that the total time taken to do the jobs by the workers is minimum, the total cost is also minimum and the total productivity is maximum while the total index of specific consumption is minimum and the workers' preferences are maximally satisfied. The corresponding mathematical model is :

$$\min \sum_{i=1}^{n} \sum_{j=1}^{n} t_{ij} x_{ij}$$

$$\min \sum_{i=1}^{n} \sum_{j=1}^{n} c_{ij} x_{ij}$$

$$\max \sum_{i=1}^{n} \sum_{j=1}^{n} p_{ij} x_{ij}$$

$$\min \sum_{i=1}^{n} \sum_{j=1}^{n} \frac{cs_{ij}}{csn_j} x_{ij}$$

$$\max \sum_{i=1}^{n} \sum_{j=1}^{n} p^*_{ij} x_{ij}$$

subject to

$$\sum_{i=1}^{n} x_{ij} = 1 \qquad j = 1, \ldots, n$$

$$\sum_{j=1}^{n} x_{ij} = 1 \qquad i = 1, \ldots, n$$

$$x_{ij} \in \{0, 1\} \qquad i, j = 1, \ldots, n.$$

The above mentioned functions are random since t_{ij}, c_{ij}, p_{ij}, p_{ij}^* and cs_{ij} (csn_j) are random variables. For example, the time taken to do the job varies from worker to worker depending on his training, motivation and other psychological factors. It can also vary for the same worker from a time period to the next. Since t_{ij} is random, c_{ij} (and p_{ij}) are also random. Moreover, p_{ij} can be affected by the machine condition. Just as well, c_{ij} depends on workers' training . Workers' preferences for certain jobs differ and may vary from time to time.

7.10. THE INPUT-OUTPUT MODEL FOR RESOURCE ALLOCATION

The Input-Output Model was first formulated by Leontieff [215]. In order to define the general structure of this model, consider a re-production system (national economy, regional economy, etc.) with n branches (industry, agriculture, constructions, transports and telecommunications, etc.) consisting of the following elements :

$$\{X_i,\ x_{ij},\ y_i\ \ i, j = 1, \dots, n\},$$

where

X_i is the production of branch i

x_{ij} is the production of branch i consumed in branch j (inter-branch flux)

y_i is the part of branch i doing the consumption (for example, for accumulation, reserves, export, etc.)

Between these elements one can set up the expression

$$(7.16)\quad X_i = \sum_{j=1}^{n} x_{ij} + y_i,\ i = 1, \dots, n,$$

called the system equilibrium equation.

Due to the concrete significance of elements X_i, x_{ij}, y_j one has :

$$X_i \geqslant 0,\ x_{ij} \geqslant 0,\ y_i \geqslant 0,\ i, j = 1, \dots, n.$$

Using the direct consumptions (a_{ij}) of the producing branches,

$$(7.17) \quad X_i = \sum_{j=1}^{n} a_{ij} X_j + y_i, \qquad i = 1, \ldots, n,$$

or in matrix notation,

$$(7.18) \quad AX + Y = X.$$

The technology coefficients a_{ij} represent the quantity produced by branch i and absorbed by branch j per unit of its total output X_i.

The balance equations (7.18) are constructed under the assumption that the fluxes between branches (x_{ij}) are proportional to the production in the consuming branches (X_i) and the proportionality factor is the technological consumption, i.e.

$$x_{ij} = a_{ij} X_i.$$

Solving the matrix equation (7.18) (if Leontieff matrix $I - A$ is nonsingular) yields :

$$(7.19) \quad X = (I - A)^{-1} Y.$$

The Input-Output Models can be formulated as mathematical programming models with multiple objective functions. Many objectives are considered to reflect the needs and aspirations : maximization of export ($z_1(x)$), income ($z_2(X)$), employment ($z_3(X)$) or minimization of pollution ($Z_4(X)$), of materials consumption and of energy usage ($x_5(x)$) [228].

Thus, the multiobjective input-ouput mathematical programming model is :

(7.20) maximize $(z_1(X), z_2(X), z_3(X), -z_4(X), -z_5(X))$,

subject to

$$X(I - A) \geqslant Y$$

$$(7.21) \quad g_i(X) \leqslant 0, \; i = 1, \ldots, n,$$

$$X \geqslant 0$$

where to the constraints already defined one has also added other constraints of type $g_i(X) \leqslant 0$, $i = 1, \ldots, n$.

In the classical input-output models, the coefficients are assumed to be exactly known. This assumption is seldom satisfied in practice. Thus, productions X can fluctuate from a period to the next, the inter-branch fluxes x_{ij} and the direct consumptions a_{ij} can vary due to changes and improvements in technology, the change in the relationship between two branches and the price dynamics, whereas the final consumptions y_i can themselves vary due to changes in investment plans, perturbations in the international economical conditions and other factors.

Accordingly, these quantities can be treated as random variables, which means that the model (7.20—7.21) represents a problem of stochastic programming with multiple objective functions (for a stochastic-informational approach to input-output models, see ref. [223]).

References

1. Aggarwal, S. P., *Indefinite quadratic fractional programming.* Ekonomiko-Matematicky Obzor **8**, *2*, 1972, 191–199.
2. Aggarwal, S. P., *Indefinite quadratic fractional programming with a quadratic constraint.* Cahiers du Centre d'Etudes de Recherche Opérationnelle, **15**(**4**), 1973, 405–410.
3. Aggarwal, S. P., Saxena, P. C., *Duality theorems for nonlinear fractional programs.* Z. Agnew. Math. Mech., **55**, 1975, 523–524.
4. Balintfy, J. L., *Nonlinear programming for models with joint chance constraints.* In : Integer and Nonlinear Programming, 337–352, J. Abadie (ed.) North-Holland, Amsterdam, 1970.
5. Beale, E. M. L., *On minimizing a convex function subject to linear inequalities.* J. Roy. Statist. Soc., **B17**, 1955, 173–184.
6. Beale, E. M. L., *The use of quadratic programming in stochastic linear programming.* Tech. Report, Rand Corporation, Santa Monica, 1961.
7. Bector, C. R.,*Some aspects of non-linear indefinite-fractional functional programming.* Cahiers du Centre d'Etudes de Recherche Opérationnelle, **12**, *1*, 1970, 24–34.
8. Bector, C. R., *Indefinite quadratic fractional functional programming.* Metrika, **18**, *1*, 1971, 21–30.
9. Bector, C. R., *On convexity, pseudo-convexity and quasi-convexity of composite functions.* Cahiers du Centre d'Etudes de Recherche Opérationnelle, **15**, *4*, 1973, 411–428.
10. Belenson, S. M., Kapur, K. C., *An algorithm for solving multicriterion linear programming problems with examples.* Operational Research Quarterly, **24**, *1*, 1973, 65–77.
11. Benayoun, R., J. de Montgolfier, Tergny, J., Larichev, O. I., *Linear programming with multiple objective functions* : STEP Method (STEM), Mathematical Programming, **1**, *3*, 1971, 366–375.
12. Benayoun, R., Tergny, T., *Critères multiples en programmation mathématique: une solution dans le cas linéaire,* Revue Française d'Informatique et de Recherche Opérationnelle, **3**, *5–2*, 1969, 31–56.

13. Ben-Israel, A., Charnes, A., Kirby, M. J. L., *On stochastic linear approximation problems*. Operations Research, **18**, 1970, 555—558.

14. Bereanu, B., *On stochastic linear programming*. I : *Distribution problems. A single random variable*. Rev. Roumaine Math. Pures Appl., **8**, *4*, 1963, 683—697.

15. Bereanu, B., *O proprietate a funcţiilor convexe cu aplicaţii la programarea neliniară*. Com. Acad. R.P.R., **13** *9*, 1963, 783—788.

16. Bereanu, B., *Soluţii cu risc minim în programarea liniară*. Studii de Statistică. Lucrările celei de a 3-a Consfătuiri Ştiinţifice de Statistică 26—30 (Dec. 1963). Also in An. Univ. Bucureşti, Mat. Mec., **13**, 1964, 121—140.

17. Bereanu, B., *Programme de risque minimal en programmation linéaire stochastique*. C.R. Acad. Sci. Paris, **259**, 1964, 981—983.

18. Bereanu, B. *Distribution problems and minimum risk solutions in stochastic programming*. In : Colloquium on Applications of Mathematics to Economics, Budapest, 1963, 37—42, A. Prékopa (ed.), Publishing House of the Hungarian Academy of Sciences, Budapest, 1965.

19. Bereanu, B. *On stochastic linear programming* II. *Distribution problems* : *Nonstochastic technological matrix*. Rev. Roumaine Math. Pures Appl., **11**, *6*, 1966, 713—725.

20. Bereanu, B., *On stochastic linear programming. The Laplace transform of the distribution of the optimum and applications*. J. Math. Anal. Appl., **15**, 1966, 280—294.

21. Bereanu, B., *On stochastic linear programming. Distributions problems, stochastic technology matrix*. Z. Wahrscheinlichkeitstheorie Verw. Gebiete, **8**, 1967, 148—152.

22. Bereanu, B., *Some numerical methods in stochastic linear programming and their computer implementation*. Tech. Report. Institut für Ökonometrie und Operations Research, University of Bonn, 1972. Published under the title "On stochastic linear programming IV. Some numerical methods and their computer implementation" in B. Bereanu, M. Iosifescu, T. Postelnicu, P. Tăutu (Eds.), Proc. Fourth Conf. on Probability Theory, Braşov, 1971, 13—36, Bucharest, Ed. Academiei Republicii Socialiste România, 1973.

23. Bereanu, B., *Large group decision making with multiple criteria*. The Conference on Multiple Criteria Decision-Making, may 21—23, 1975, Jouy-en-Josas, France.

24. Bereanu, B., *The generalized distribution problem of stochastic linear programming*. In : Symposia Matematica, Academic Press, New York, 1976.

25. Bergthaller, C., *A quadratic equivalent for the minimum risk problem*. Rev. Roumaine Math. Pures Appl., **15**, *1*, 1970, 17—23.

26. Bhatia, H. L., Puri, M. C., *Time-cost trade-off in a solid transportation problem*. Z. Agnew. Math. Mech., **52**, 1972, 371—373.

27. Blau, R., *Decomposition techniques for the Chebyshev problem*. Operations Research, **21**, *5*, 1973, 1157—1163.

28. Bod, P., *Lineáris programozás több, egyidejüleg adott célfüggvény szerint.* Publication of the Mathematical Institute of the Hungarian Academy of Sciences, Series B, **8**, *4*, 1963, 541–554.
29. Bod, P., *Lineare optimierung mittels simultan gegebener zielfunktionen.* In : A. Prékopa ed., *Colloquium on Applications of Mathematics to Economics*, Akadémiai Kiadó, Budapest, 55–60, 1965.
30. Boldur Gh., Stancu-Minasian, I. M., *Méthodes de résolution de certains problèmes de programmation linéaire multidimensionnelle.* Rev. Roumaine Math. Pures Appl., **16**, *3*, 1971, 313–327.
31. Boldur Gh., Stancu-Minasian, I. M., *Programarea liniară cu mai multe funcţii obiectiv. Privire de ansamblu.* Stud. cerc. mat., **24**, *8*, 1972, 1169–1191.
32. Brucker, P., *Diskrete parametrische optimierungs probleme und wesentliche effiziente Punkte.* Zeitschrift für Operations Research, **16**, 1972, 189–197.
33. Charnes, A., Cooper, W. W., Symonds, G. H., *Cost horizons and certainty equivalents: an approach to stochastic programming of heating oil.* Management Sci., 1958, **4**, 235–263.
34. Charnes, A., Cooper, W. W., *Chance-constrained programming.* Management Sci., **6**, 1959, 73–79.
35. Charnes, A., Cooper, W. W., *Management Models and Industrial Applications of Linear Programming*, vol. 1, John Wiley, New York, 1961.
36. Charnes, A., Cooper, W. W., *Programming with linear fractional functionals.* Nav. Res. Log. Quart., **9**, 1962, 181–186.
37. Charnes, A., Cooper, W. W., Thompson, G. L., *Characterizations by chance-constrained programming.* In : Recent Advances in Mathematical Programming (R. L. Graves and P. Wolfe, eds. 1963), Symposium for Mathematical Programming held at the University of Chicago, 1962.
38. Charnes, A., Cooper, W. W., *Deterministic equivalents for optimizing and satisfying under chance constraints.* Operations Research, **11**, 1963, 18–39.
39. Charnes, A., Kirby, M., *Optimal decision rules for the E-Model of chance-constrained programming.* Cahiers du Centre d'Etudes de Recherche Opérationnelle, **8**, 1966, 5–44.
40. Charnes, A., Kirby, M. J. L., Raike, W. M., *An acceptance region theory for chance-constrained programming.* J. Math. Anal. Appl., **32**, 1970, 38–61.
41. Chelaru, H., Stancu-Minasian, I. M., *Folosirea metodelor statistico-matematice la elaborarea funcţiilor de calitate în probleme de amestec.* Studii şi Cercetări de Calcul Economic şi Cibernetică Economică, *4*, 1971, 17–25.
42. Ciobanu, Gh., *Applying digital simulation in solving problems of linear programming with several objective functions.* Economic Computation and Economic Cybernetics Studies and Research, *1*, 1976, 63–68.
43. Cochrane, J. L., Zeleny, M., eds. *Multiple Criteria Decision Making.* University of South Carolina Press, Columbia, S.c., 1973.

44. Contini, B., *A stochastic approach to goal programming*, Operations Research., **16**, *3*, 1968, 576—586.

45. Corban, A., *A multidimensional transportation problem.* Rev. Roumaine Math. Pures Appl., **9**, *8*, 1964, 721—735.

46. Corban, A., *Un model multidimensional de transport.* Stud. cerc. mat., **24**, *7*, 1972, 1019—1082.

47. Corban, A., *Programming with fractional linear objective function.* Rev. Roum. Math. Pures Appl., **18**, *5*, 1973, 633—637.

48. Corban, A., *Un model de transport multidimensional II.* Stud. cerc. mat., **27**, *4*, 1975, 413—424.

49. Dantzig, G. B., *Linear programming under uncertainty.* Management Sci., *1*, 1955, 197—206.

50. Dantzig, G. B., Madansky, A., *On the solution of two-stage linear programms under uncertainty.* In : Fourth Berkeley Symposium on Mathematical Statistics and Probability, 165—176, University of California Press, Berkeley, 1961.

51. Dempster, M. A. H., *On stochastic programming* I. *Static linear programming under risk.* J. Math. Anal., Appl., **21**, *2*, 1968, 304—343.

52. Dinkelbach, W., *Unternehmerische Entscheidungen bei mehrfacher Zielsetzung*, Zeitschrift für Betriebswirtschaft, **32**, *12*, 1962, 739—747.

53. Dinkelbach, W., *On nonlinear fractional programming.* Management Science, **13**, *7*, 1967, 492—498.

54. Dinkelbach, W., *Über einen Lösungsansatz zum Vektormaximumproblem.* In : M. Beckman and H.P. Künzi, eds., *Unternehmensforschung-Heute*, Springer-Verlag, Berlin, 1971.

55. Dragomirescu, M., *On sensitiveness in linear programming.* Stud. cerc. mat., **20**, 1968, 965—967.

56. Dragomirescu, M., *An algorithm for the minimum risk problem of stochastic programming.* Operations Research., **20**, 1972, 154—164.

57. Egglestone, H. G., *Convexity.* Cambridge University Press, Cambridge, 1963.

58. Eisenberg, E., *A note on semidefinite matrices.* Research Roport 9, Operations Research Center, Univ. of California, Berkeley, 1961.

59. Eisenberg, E., *Suports of a convex function.* Bull. Amer. Math. Soc., **68**, *3*, 1962.

60. Elmaghraby, S. E., *An approach to linear programming under uncertainty.* Operations Research, **7**, 1959, 208—216.

61. Ewbank, J. B., Foote, B. L., Kumin, H. J., *A method for the solution of the distribution problem of stochastic linear programming.* SIAM J. Appl. Math., **26**, 1974, 225—238.

62. Fandel, G., *Optimale Entscheidung bei Mehrfacher Zielsetzung.* Springer-Verlag, New York, 1972.

63. Fenchel, W., *Convex cones, sets and functions.* Princeton University Press, Princeton, 1953.

64. Fiacco, A. V., McCormick G. P., *Nonlinear programming: sequential unconstrained minimization techniques.* Wiley, New York, 1968.

65. Fishburn, P. C., *Additive utilities with incomplete product sets : Applications to priorities and assignments.* Operations Research, **15**, *3*, 1967, 537–542.

66. Fishburn, P. C., *Utility theory for decision making,* Wiley, New York, 1970.

67. Fishburn, P. C., Keeney, R. L., *Generalized utility independence and some applications.* Operations Research, **23**, *5*, 1975, 928–940.

68. Freund, R. J., *The introduction of risk into a programming model.* Econometrica, **24**, 1956, 253–263.

69. Focke, J. *Vektormaximumprobleme und parametrische Optimierung.* Mathematische Operationsforschung und Statistik, **4**, *5*, 1973, 365–369.

70. Fortet, R., *Programmes linéaires stochastiques.* In : *Mathématique des programmes économiques,* 109–126, Monographies de Recherche Opérationnelle, 1, Dunod, Paris, 1964.

71. Gale, D., *The theory of linear economic models.* McGraw-Hill, New York, 1960.

72. Geoffrion, A. M., *A parametric programming solution to the vector maximum problem, with applications to decisions under uncertainty.* Technical Report No. 11, Operations Research Programs, Stanford University, February, 1965.

73. Geoffrion, A. M., *Stochastic programming with aspiration and fractile criteria.* Management Sci., **13**, 1967, 672–679.

74. Geoffrion, A. M., *Solving bicriterion mathematical programs.* Operations Research **15**, *1*, 1967, 39–54.

75. Geoffrion, A. M., *Proper efficiency and the theory of vector maximization,* Journal of Mathematical Analysis and Applications, **22**, *3*, 1968, 618–630.

76. Goicoechea, A., Duckstein, L., Bulfin, R. L., *Multi-objective, stochastic programming : the PROTRADE Method.* Presented at Operations Research Society of America, Miami Beach, Florida, november 3–5, 1976.

77. Gupta, Reeta, *Time-cost transportation problem.* Ekonomicko-Matematicky Obzor, *4*, 1977, 431–443.

78. Hillier, F. S., *Chance-constrained programming with 0–1 on bounded continuous decision variables.* Management Sci., **14**, 1967, 34–57.

79. Huangy, H. Y., *Unified approach to quadratically convergent algorithms for function minimization.* Journal of Optimization Theory and Applications, **5**, *6*, 1970, 405–423.

80. Huang, S. C., *Note on the mean-square strategy of vector-valued objective function.* Journal of Optimization Theory and Applications, **9**, *5*, 1972, 364–366.

81. Ijiri, Y., *Management goals and accouting for control.* North-Holland, Amsterdam, 1965.

82. Iosifescu, M., Theodorescu, R., *Asupra programării liniare stochastice.* Comunicările Acad. R.S.R., **12**, 1962, 299–302.

83. Iosifescu, M., Theodorescu, R., *Sur la programmation linéaire.* C. R. Acad. Sci. Paris, **256**, 1963, 4831—4833.
84. Iosifescu, M., Theodorescu, R., *Statistical decisions and linear programming.* C.R. Bulgare Acad., Sci., **17**, 1964, 223—226.
85. Iosifescu, M., Theodorescu, R., *Teoria jocurilor și programarea liniară stochastică.* In: *Calcul Economic*, Ed. Acad. R.P.R., București, 1964.
86. Iosifescu, M., Theodorescu, R., *Linear programming under uncertainty.* In: Colloquium on Applications of Mathematics to Economics, Budapest, 1963, 133—140, A. Prékopa (ed.) Publishing House of the Hungarian Academy of Sciences, Budapest, 1965.
87. Izerman, H., *An algorithm for solving the transportation problem with multiple linear objective functions.* 9-th International Symposium on Mathematical Programming. Budapest-Hungary, August 23—27, 1976.
88. Jagannathan, R., *Chance-constrained programming with joint constraints.* Operations Research, **22**, 1974, 359—372.
89. Kall, P., *Some remarks on the distribution problem of stochastic linear programming.* Operations Research Verfahren, **16**, 1973, 189—196.
90. Karlin, S., *Mathematical Methods and Theory in Games, Programming and Economics*, vol. 1, Addison-Wesley, Reading-Mass., 1959, 216—217.
91. Kataoka, Shinji, *A stochastic programming model.* Econometrica, **31**, 1963, 181—196.
92. Kataoka, Shinji., *Stochastic programming maximum probability model.* Hitotsubashi J. Arts. Sci., **8**, 1967, 51—59.
93. Keeney, R. L., *Multiplicative utility functions.* Operations Research, **22**, *1*, 1974, 23—34.
94. Kelley, J. E., *The cutting-plane method for solving convex programs.* J. Soc. Indus. Appl. Math., **8**, 1960, 703—712.
95. Klinger, A., *Improper solutions of the vector maximum problem.* Operations Research, **15**, *3*, 1967, 570—572.
96. Kuhn, H. W., Tucker, A. W., *Non-linear programming.* Proc. IInd Berkeley Symposium Math. Statist. Probab., 1951, 481—492.
97. Lebedev, B. D., Podinovskii, V. V., Styrikovic, R. S., *An optimization problem with respect to the order totality of criteria* (in Russian), Ekon. Mat. Metody, SSSR, **7**, 1971, 612—616.
98. Lee, S. M., *Goal programming for decision analysis.* Auerbach Publishers, Inc., Philadelphia, Pennsylvania, 1972.
99. Lee, Sang M., Moore L. J., *Optimizing transportation problems with multiple objectives.* AIIE Transactions, **5**, 1973, 333—336.
100. Luce, R. D., Raiffa, H., *Games and decisions.* New York, Wiley, 1958.
101. Lukacs, E., *Characteristic Functions.* Charles Griffin and Co., London, 1960.
102. Madansky, A., *Some results and problems in stochastic linear programming.* Tech. Report, p. 1596, RAND Corporation, 1959.

103. Mangasarian, O., *Pseudo-convex functions*. SIAM J. Control, **8**, 281—290, 1965.

104. Markowitz, H., *Portfolio selection : Efficient Diversification of Investment*, New York, John Wiley, 1959.

105. Martos, B., *The direct power of adjacent vertex programming methods*. Management Science, **12**, *3*, 1965, 241—252.

106. Maruşciac I., Rădulescu, M., *Un problème général de la programmation linéaire à plusieurs fonctions économiques*. Studia Universitatis Babeş-Bolyai, 2, 1970, 55—65.

107. Maruşciac I., Rădulescu, M., *Un problème de la programmation quadratique à plusieurs functions*. Studia Universitatis Babeş-Bolyai, **11**, 1970, 81—89.

108. Maruşciac, I., Rădulescu, M., *Sur l'ensemble des points efficients d'un problème de la programmation mathématique*. Analele Ştiinţifice ale Universităţii "Al. I. Cuza" Iassy, **18**, *1*, 1972, 210—226.

109. Maruşciac, I., *Metode de rezolvare a problemelor de programare neliniară*. Editura Dacia, Cluj-Napoca, 1973.

110. Mihoc, Gh., *Unele precizări în legătură cu aplicarea programării lineare*. Revista de Statistică, **12**, 1959, 13—18.

111. Miller, L. B., Wagner, H.. *Chance-constrained programming with joint constraints*. Operations Research, **13**, 1965, 930—945.

112. Naslund, B., *Decisions under risk*, Tech. Report, Stockholm School of Economics, 1967.

113. Nykowski, I., *Problem pogodzenia kilku kryteriów w jednym programie liniowym*, Przeglad Statystyczny, **13**, *4*, 1966, 367—374.

114. Nykowski, I., *Multi-criteria linear models*. Ekonomistka, **4**, 1970, 721—733.

115. Onicescu, O., Iosifescu, M., *Cîteva consideraţii asupra programării liniare stochastice*. Studii şi Cercetări de Calcul Economic şi Cibernetică Economică, **6**, 1967, 69—72.

116. Philip, J., *An algorithm for combined quadratic and multiobjective programming*. Department of Mathematics, TRIRAR-MAT-1973—5, Royal Institute of Technology, Stockholm, 1973.

117. Philip, J., *Algorithms for the vector maximization problem*. Mathematical Programming, **2**, *2*, 1972, 207—229.

118. Prékopa, A., *On the probability distribution of the optimum of a random linear program*. SIAM J. Control, **4**, 1966, 211—222.

119. Rosen, J. B., *The gradient projection method for nonlinear programming*. Part I, *Linear constraints*, SIAM J, **8**, 1960, 181—217; Part II, *Nonlinear constraints*, SIAM, J., **9**, 1961, 514—532.

120. Roy, B., *Pourquoi des approches multi-critères et comment?* Note de travail 108, SEMA, novembre, 1969.

121. Roy, B., *Problems and Methods with Multiple Objective Functions.* Note de travail 60, SEMA, september, 1970.

122. Saska, J., *Lineárni Multiprogramování.* Ekonomicko-Mathematicky Obzor, **4**, *3*, 1968, 359—373.

123. Schaible, S., *Quasi-concave, strictly quasi-concave and pseudoconcave functions.* Operations Research Verfahren, **17**, *2*, 1972, 308—316.

124. Sengupta, J. K., *Methods of linear programming under risk.* Metroeconomica, **21**, 1969, 195—213.

125. Sengupta, S. S., Podrebarac, M. L., Fernando, T. D. H., *Probabilities of Optima in Multi-Objective Linear Programmes.* In : *Multiple Criteria Decision Making,* J. L. Cochrane ; M. Zeleny(eds.) USC, Columbia, 1973, 217—236.

126. Sharma, I. C., *Transportation technique in non-linear fractional programming.* Trabajos de Estadistica, **24**, *1—2*, 1973, 131—139.

127. Simon, H. A., *Dynamic programming under uncertainty with a quadratic criterion functions.* Econometrica, **24**, 1956, 74—81.

128. Sinha, S. M., *Stochastic programming.* ORC 63—82 (RR), 1963.

129. Sinha, S. M., *A duality theorem for non-linear programming.* Management Sci., **12**, 1966, 385—390.

130. Stancu-Minasian, I. M., Boldur, Gh., *Elaborarea planului de producţie al unei întreprinderi ţinînd sema de mai multe criterii de eficienţă.* Studii şi Cercetări de Calcul Economic şi Cibernetică Economică, **1**, 1970, 13—21.

131. Stancu-Minasian, I. M., *The solution of transportation network in the case of multiple criteria.* 7-th Mathematical Programming Symposium. The Hague, September 1970.

132. Stancu-Minasian, I. M., C. N. Sima, *Optimizarea transporturilor făinii destinată fondului pieţii.* Studii şi Cercetări de Calcul Economic şi Cibernetică Economică, **4**, 1972, 83—88.

133. Stancu-Minasian, I. M., Chelariu, H., Spircu, L., *Modelarea statistico-matematică a funcţiilor de calitate în probleme de amestec.* Studii şi Cercetări de Calcul Economic şi Cibernetică Economică, **6**, 1972, 25—37.

134. Stancu-Minasian, I. M., Spircu, L., Vatcu, M., *Modelarea matematică a procesului de amestec al produselor petroliere.* Studii şi Cercetări de Calcul Economic şi Cibernetică Economică, **5**, 1973, 35—45.

135. Stancu-Minasian, I. M., *A three-dimensional transportation problem with a specially structured objective function.* Bull. Math. Soc. Sci. Math., R. S. Roumanie, **18**, **(66)**, *3—4*, 1974, 385—397.

136. Stancu-Minasian, I. M., *Stochastic programming with multiple objective functions.* Econom. Comp. Econom. Cybernet. Stud. Res., **1**, 1974, 46—67.

137. Stancu-Minasian, I.M., *A selected bibliography of works related to stochastic programming.* Preprint nr. 1, Chair of Economic Cybernetics, ASE, 1974.

138. Stancu-Minasian, I. M., *Notă asupra programării stocastice cu mai multe funcţii obiectiv, avînd vectorul c aleator.* Stud. cerc. mat., **27**, *4*, 1975, 453—459.

139. Stancu-Minasian, I. M., *A selected bibliography of works related to the multiple criteria decision making.* Preprint no. 2, Chair of Economic Cybernetics, ASE, 1975.

140. Stancu-Minasian, I. M., *Asupra problemei de risc minim multiplu* I: *cazul a două funcţii obiectiv.* Stud. cerc. mat., **28**, *5*, 1976, 617—623.

141. Stancu-Minasian, I. M., *Asupra problemei de risc minim multiplu* II: *cazul a r* ($r > 2$) *funcţii obiectiv.* Stud. cerc. mat., **28**, *6*, 1976, 723—734.

142. Stancu-Minasian, I. M., *Criterii multiple în programarea stohastică.* Doctoral thesis. Centrul de statistică matematică, Bucureşti, 1976.

143. Stancu-Minasian, I. M., *Analiza unor puncte de vedere în programarea matematică cu mai multe funcţii obiectiv.* Studii şi Cercetări de Calcul Economic şi Cibernetică Economică, **1**, 1976, 37—48.

144. Stancu-Minasian, I. M., Wets, M. J., *A research bibliography in stochastic programming*, Operations Research, **24**, *6*, 1976, 1078—1119.

145. Stancu-Minasian, I. M., *Asupra problemei lui Kataoka*, Stud. cerc. mat., **28**, *1*, 1976, 95—111.

146. Stancu-Minasian, I. M., *Bibliography of Fractional Programming; 1960 — 1976*, Preprint no. 3, Chair of Economic Cybernetics, ASE, 1977, also in Pure Appl. Math. Sci., **13**, *1—2*, 35—69, 1981.

147. Stancu-Minasian, I. M., *On stochastic programming with multiple objective functions.* In: *Proceedings of the Fifth Conference on Probability Theory.* September 1—6, 1974, Braşov, Romania, Editura Academiei R.S.R., Bucureşti, 1977, 429—436.

148. Stancu-Minasian, I. M., *Problema Cebîşev stohastică. Funcţia de repartiţie a optimului.* Stud. cerc. mat., **30**, *5*, 1978, 567—577.

149. Stancu-Minasian, I. M., Orzan, Gh., *Stabilirea structurii optime a planului la UREMOAS prin programarea multidimensională.* Proceedings of the 2nd Symposium of Information and Management Sciences, Cluj-Napoca, 1976, Editura Dacia, Cluj-Napoca, 1977, 23—29.

150. Swarup, K., *Quadratic programming.* Cahiers de Centre d'Etudes de Recherche Opérationnelle, **8**, *4*, 1966, 223—243.

151. Swarup, K., Aggarwal, S. P., Gupta, R. K., *Stochastic indefinite quadratic programming.* Z. Agnew. Math. Mech., **52**, 1972, 371—373.

152. Szwarc, W., *The transportation problem with stochastic demand.* Management Sci., **11**, 1964, 33—50.

153. Tamm, M. I., *Linear programming problem solution with several objective functions* (in Russian), Ekon. Mat. Metody SSSR, **9**, *2*, 1973, 328—329.

154. Teodorescu, N., Boldur, Gh., Stoica, M., Stancu-Minasian, I. M., Băncilă, I., *Metode ale cercetării operaţionale în gestiunea întreprinderilor*, Editura tehnică, Bucureşti, 1972.

155. Tintner, G., *Stochastic linear programming with applications to agricultural economics*. In : *Proceedings of the Second Symposium in Linear Programming*, pp. 197—228, H. A. Antonosiewicz (ed.), National Bureau Standard, Washington, 1955.

156. Tintner, G., *A note on stochastic linear programming*. Econometrica, **28**, 1960, 490—495.

157. Van de Panne, C., Popp, W., *Minimum-cost cattle feed under probabilistic protein constraints*. Management Sci., **9**, 1963, 405—430.

158. Walkup, D., Wets, R., *Stochastic programs with recourse: Special forms*. In : *Proceedings of the Princeton Symposium on Mathematical Programming, 139—162*, H. Kuhn (ed.), Princeton University Press, Princeton, 1970.

159. Wets, R., *Programming under uncertainty: the complete problem*. Z. Wahrscheinlichkeitstheorie Verw. Gebiete, **4**, 1966, 316—339.

160. Wets, R., *Programming under uncertainty: the equivalent convex program*. SIAM J. Appl. Math., **14**, 1966, 89—105.

161. Wets, R., *Programming under uncertainty: the solution set*, SIAM J. Appl. Math., **14**, 1966, 1143—1151.

162. White, D. J., *Dynamic programming and probabilistic constraints*. Operations Research, **22**, 1974, 654—664.

163. Williams, A. C., *On stochastic linear programming*. SIAM J. Appl. Math., **13**, 1965, 927—940.

164. Wolfe, P., *The simplex method for quadratic programming*. Econometrica, **27**, *3*, July 1959, 382—398.

165. Yu, P. L., Zeleny, M., *The set of all nondominated solutions in linear cases and multicriteria simplex method*. J. Math. Anal. Appl., **49**, *2*, 1975, 430—468.

166. Zacková, J., *Minimax Stochastic Programs with Nonconvex Nonseparable Penalty Functions*. In : *Progress in Operations Research*, vol. I, II, Proceedings of the Sixth Hungarian Conference, which was also the second International Conference held in Eger, 1974. Edited by A. Prékopa. Colloquia Mathematica Scietatis János Bolyai, vol. 12. North-Holland Publishing Co., Amsterdam-London ; Bolyai János Matematikai Társulat, Budapest, 1976, 303—316.

167. Zak, Y. A., *Models and methods of constructing compromise plans in problems of mathematical programming with several objective functions* (in Russian). Kibernetika, **4**, 1972, 102—107.

168. Zeleny, M., *Linear Multiobjective Programming*. Lecture Notes in Economics and Mathematical Systems. Springer Verlag, Berlin, Heidelberg, New York, 1974.

169. Zionts, S., *Programming with linear fractional functionals*. Naval Research Logistics Quarterly, *3*, 1968, 494—451.
170. Zoutendijk, G., *Nonlinear programming*. SIAM Journal on Control, **4**, 1960.
171. Zoutendijk, G., *Method of feasible directions*. Elsevier, Amsterdam, 1960.
172. Zuhovitskii, S. I., Avdeeva, L. I., *Lineinoe i vîpulkoe programmirovanie*. Second edition, Moscow, Nauka, 1967.

Bibliography added for the English edition

173. Achiles, A., Elster, K. -H., Nehse, R., *Bibliographie zur Vector optimierung (Theorie und Anwendungen)*. Math. Operationsforsh. Statist., Ser. Optimization, **10**, *2*, 1979, 277—321.
174. Aggarwal, S. P., *Parametric linear fractional functionals programming*. Metrika, **12**, *2—3*, 1968, 106—114.
175. Aggarwal, S. P., Swarup, K., *Fractional functional programming with a quadratic constraint*. Cahiers Centre Etudes Recherche Opér., **14**, *5*, 1966, 950—956.
176. Avenhaus, R., Beedgen, R., Chernavsky, S., Schrattenholzer L. (with Annex A by A. Hölzl), *Handling uncertainties in linear programming models*. Working Paper, WP-80-170, November 1980, IIASA-Laxenburg, Austria.
177. Banker, R. L., Guignard Monique, Gupta K. Shiv, *An algorithm for generating efficient solutions to the multiple objective integer programming problem*. Technical Report. University of Pennsylvania, Pennsylvania, December 1978.
178. Basar, T., "Stochastic multicriteria decision problems with multilevels of hierarchy". *Proceedings of the 19th IEEE Conference on Decision & Control Including the Symposium on Adaptive Processes*, Albuquerque, NM, USA, 10—12, Dec. 1980 (New York, USA : IEEE 1980), 1171—1174. Also in : IEEE Trans. Autom. Control (USA), Vol. **AC—26**, *2*, April 1981, 549—553.
179. Bereanu, B., *The distribution problem in stochastic linear programming and minimum risk solutions* (in Romanian). Dissertation paper, University of Bucharest, 1963.
180. Bereanu, B., "Some numerical methods in stochastic linear programming under risk and uncertainty". In : *Stochastic Programming*, edited by M.A.H. Dempster (Proc. Internat. Conf. Univ. Oxford, Oxford, 1974), 196—205, Academic Press, London, 1980.
181. Bernardo, J. J., *A stochastic multiattribute heuristic model of investor choice*. Decis. Sci. (USA), **11**, *3*, 1980, 425—438.
182. Bitran, G. R., *Linear multiple objective problems with interval coefficients*. Technical Report No. 167, Operations Research Center, MIT, Massachusetts, August 1979.

183. Bitran, G. R., *Theory and algorithms for linear multiple objective programs with zero-one variables*. Math. Programming, **17**, 1979, 362—390.

184. Bitran, G. R., Rivera, J. M., *An efficient point-utility theory approach to solve binary multicriteria problems*. Technical Report No. 181, Massachusetts Institute of Technology, September 1980.

185. Bowman, V. J. Jr., "On the relationship of the Tchebycheff norm and the efficient frontier of multiple criteria objectives". *Multiple Criteria Decision Making*, edited by H. Thiriez and S. Zionts (Jouy-en-Josas, France, 1975), Springer-Verlag, Berlin, 1976.

186. Chai, K. C., *The analysis of practical mathematical programming problems using stochastic programming and multiobjective linear programming*, Ph. D. Thesis, Brunel University, Sept. 1977.

187. Charnes, A., Stedry, A. C., *Search-theoretic models of organization control by budgeted multiple goals*. Management Sci., **12**, *5*, 1966, 457—482.

188. Cojocaru, I., Dragomirescu, M., *The continuity of the optimal value of a general linear program*. Rev. Roumaine Math. Pures Appl., **27**, *6*, 1982, 663—676.

189. Dubov, Ju. A., *Necessary and sufficient conditions for Pareto optimality in mean*. Izv. Akad. Nauk SSSR, Tehn. Kibernet. 1979, *6*, 137—141, 200 (in Russian); translated as Engrg. Cybernetics, *17* (1979), *6* (1980), 109—114.

190. Duca, D. I., *On vectorial programming problem in complex space*. Studia Univ. Babeş-Bolyai, Mathematica, *1*, 1979, 51—56.

191. Ecker, J. G., Kouada, I. A., *Finding efficient points for linear multiple objective programs*. Math. Programming, **8**, *3*, 1975, 375—377.

192. Ecker, J. G., Shoemaker, Nancy E., "Multiple objective linear programming and the trade-off compromise set. Multiple criteria decision making theory and application " (Proc. Third Conf., Hagen/Königswinter, 1979), pp. 60—73, *Lecture Notes in Econom. and Math. Systems*, 177, Springer, Berlin, 1980.

193. Fandel, G., Gal, T., "Multiple Criteria Decision Making: Theory and Applications". Proceedings of the third conference held at Hagen/Königswinter, August 20—24, 1979. *Lecture Notes in Economics and Mathematical Systems*, 177, Springer-Verlag, Berlin, New York, 1980, XVI + 570 pp.

194. Fishburn, P. C., *Methods of estimating additive utilities*. Management Sci., **13**, *7*, 1967, 435—453.

195. Geoffrion, A. M., Dyer, J. S., Feinberg, A., *An interactive approach for multi-criterion optimization with an application to the operation of an academic department*. Management Sci., **19**, *4*, 1972, B357—B368.

196. Goicoechea, A., *A multi-objective, stochastic programming model in watershed management*. Ph. D. dissertation, Dept. of Systems & Industrial Engineering, Univ. of Arizona, Tucson, Arizona, 1977.

197. Goicoechea, A., Hansen, D. R., Duckstein, L., *Multiobjective decision analysis with engineering and bussiness applications*. John Wiley & Sons, 1982.

198. Gupta, S. N., Swarup, K., *Note on stochastic programming for minimization of variance*. New Zealand Oper. Res. **8**, *2*, 1980, 185—187.

199. Hendrix, G. C., Stedry, A., *The elementary redundancy-optimization problem: a case study in probabilistic multiple-goal programming*. Operations Res., *22*, 1974, 639—653.

200. Hiljuk, L. F., *A stochastic multicriterial problem in controlling a complex of users of water resources in river basins. Probabilistic criteria for choosing a solution in the Pareto Region*. Avtomatika *4*, 1978, 60—70 (in Russian); translated as Soviet Automat. Control **11**, *4*, 1978, 49—57.

201. Huckert, K., Rhode, R., Roglin, O., Weber, R., *On the interactive solution to a multicriteria scheduling problem*. Zeitschrift für Operations Research., *24*, 1980, 47—60.

202. Iserman, H., *Proper efficiency and the linear vector maximum problem*. Operations Res., **22**, *1*, 1974, 189—191.

203. Kall, P., *Stochastic programming*, European J. Oper. Res., **10**, *2*, 1982, 125—130.

204. Kaplan, R., Soden, J., *On the objective function for the sequential P-model of chance-constrained programming*. Operations Res., **19**, *1*, 1971, 105—114.

205. Kaul, R. N., Gupta, B., *Multi-objective programming in complex space*. Z. Angew. Math. Mech., **61**, *11*, 1981, 599—601.

206. Koopmans, T. C., *Analysis of production as an efficient combination of activities. Activity analysis of production and allocation*, edited by T. C. Koopmans, Cowles Commis Monogr. 13, New York, 1951, 33—97.

207. Kornbluth, J. S. H., *A survey of goal programming*. Omega **1**, *2*, 1973, 193—205.

208. Lau, H. S., *The Newsboy problem under alternative optimization objectives*. J. Oper. Res. Soc. Vol. 31, 1980, 525—535.

209. Lavrinenko, E. P., *Pareto-optimal solutions in problems of vector stochastic optimization*; pp. 40—51, 55, In: Questions in the study of multicriteria optimization problems (Preprint 81, 22 (in Russian)). Acad. Nauk Ukrain. SSR, Inst. Kibernet., Kiev, 1981, 55 pp.

210. Leclercq, J.-P., *Résolution de programmes linéaires stochastiques par des techniques multicritères*. Thèse de doctorat. Faculté des Sciences de l'Université de Namur, 1979.

211. Leclercq, J.-P., *La programmation linéaire stochastique: une approche multicritère*. Partie I: Formulation. Cahiers Centre Etudes Recherche Opér., **23**, *1*, 1981, 31—41.

212. Leclercq, J.-P., *La programmation linéaire stochastique: une approche multicritère. Partie II: Un algorithme interactif adapté aux distributions multinormales*. Cahiers Centre Etudes Recherche Opér., **23**, *2*, 1981, 121—132.

213. Leclercq, J.-P., *Stochastic programming: An interactive multicriteria approach*. European J. Oper. Res. **10**, *1*, 1982, 33—41.

214. Lee, S. M., *Interactive integer goal programming: methods and application,* Paper presented at Conference on Multiple Criteria Problem Solving: *Theory, Methodology, and Practice,* Buffalo, N. Y. August, 1977.

215. Leontieff, W., *Input-Output Economics,* Oxford University Press, New York, 1966.

216. Lupşa, L., *Asupra unei probleme de programare vectorială în variabile întregi.* Seminarul itinerant de ecuaţii funcţionale, aproximare şi convexitate, 17—19 May 1979, Cluj-Napoca.

217. Marquez, J. D. C., *On the equivalence between the safety first and min-variance criterion for portfolio selection.* European J. Oper. Res., **10**, *2*, 144—150, 1982.

218. Martin, D.H., *On the conti nuity of the maximum in parametric linear programming.* J. Optim. Theory Appl., **17**, *3—4*, 1975, 205—210.

219. Mitra, G., Chai, K. C., *A multicriterion stochastic programming model for manpower planning.* Operations Research-Verfahren **35**, 1979, 291—312.

220. Nehse, R., *Bibliographie zur Vektoroptimierung. Theorie und Anwendungen* (1. Fortsetzung). Math. Operationsforsch. Statist., Ser. Optimization, **13**, *4*, 1982, 593—625.

221. J. von Neumann, Morgenstern, O., *Theory of Games and Economic Behaviour,* 2nd ed., Princeton University Press, Princeton, New Jersey, 1947.

222. Nijkamp, P., Spronk, J., *Interactive multidimensional programming models for locational decisions.* European J. Oper. Res., **6**, *2*, 1981, 220—223.

223. Onicescu, O., Botez, M. C., *A stochastic informational approach to input-output analysis.* Econom. Comput. Econom. Cybernet. Stud. Res., **12**, *3*, 1978, 19—27.

224. Osyczka, A., *Multicriterion network optimization problem.* Computing **25**, 4, 1980, 363—368.

225. Pareto, V., *Cours d'économie politique.* Lausanne, Switzerland, Rouge, 1896.

226. Patkar, V., Stancu-Minasian, I. M., *Approaches for solving a class of nondifferentiable nonlinear fractional programming problems.* Nat. Acad. Sci. Letters (India), **4**, *12*, 1981, 477—480.

227. Payne, Anthony, N., Polak, Eliyah, *An interactive rectangle elimination method for biobjective decision making.* IEEE Trans. Automat. Control, **25**, *3*, 1980, 421—432.

228. Reitano, B., Hendricks, D., *Input-output modeling for facility level water planning.* Water Supply Management **4**, 1980, 379—396.

229. Sakawa, Masatoshi, Seo, Fumiko, *Interactive multiobjective decision making for large-scale systems and its application to environmental systems.* IEEE Trans. Systems Man. Cybernet., **10**, *12*, 1980, 796—806.

230. Slowinski, Roman, *Multiobjective network scheduling with efficient use of renewable and nonrenewable resources.* European J. Oper. Res., **7**, *3*, 1981, 265—273.

231. Stadler, W., *A survey of multicriteria optimization of the vector maximùm problem*, Part I: 1776—1960. J. Optim. Theory Appl. **29**, *1*, 1979, 1—52.

232. Stancu-Minasian, I. M., *Recent Results in Stochastic Programming with multiple objective functions*. Task Force Meeting on Multiobjective and Stochastic Optimization. 30 November—4 December 1981, IIASA Laxemburg, Austria.

233. Stancu-Minasian, I. M., "A survey of methods used for solving the linear fractional programming problems with several objective functions". Symposium on Operations Research. Methods of Operations Research, edited by R. E. Burkard and T. Ellinger, vol. 40, Hain Meisenheim, Königstein/Ts. 1981, 159—162.

234. Stancu-Minasian, I. M., *Multiple minimum risk solutions in stochastic programming*. Fifth International Conference on Multiple Criteria Decision Making, Mons, 9—13 August 1982.

235. Stancu-Minasian, I. M., *A second bibliography of fractional programming: 1977—1981*. Pure Appl. Math. Sci. (India), **XVII**, *1—2*, 1983, 87—102.

236. Stancu-Minasian, I. M., Ţigan , Şt., *The minimum risk approach to special problems of mathematical programming. The distribution function of the optimal value*. Revue d'Analyse Numérique et de Théorie de l'Approximation (in print).

237. Starr, M. K., Zeleny. M. (eds.), *Multiple criteria decision making*. North Holland, N. Y., 1977.

238. Takeda, E., Nishida, T., *Multiple criteria decision problems with fuzzy domination structures*. Fuzzy Sets and Systems, *3*, 1980, 123—136.

239. Tamm, Ebu, *Minimization of a probability function* (in Russian), Eesti NSV Tead. Akad. Toimetised Füüs. -Mat. **28**, *1*, 1979, 17—24, 93.

240. Tammer, K., *Relations between stochastic and parametric programming for decision problems with a random objective function*. Math. Operationsforsch. Statist. Ser. Optim., **9**, *4*, 1978, 523—535.

241. Tammer, K., *Beiträge zur Theorie der parametrischen Optimierung, zu den mathematischen Grundlagen ihrer Anwendung und zu Lösungsverfahren*, Diss. (B), Humboldt-Universität, Berlin, 1979.

242. Tammer, K., *Behandlung stochastischer Optimierungsprobleme unter dem Gesichtspunkt des Strategie der Vektoroptimierung*, Wiss. Z. TH Leipzig, *4*, 1980, 295—302.

243. Tamura, K., *A method for constructing the polar cone of a polyhedral cone, with applications to linear multicriteria decision problems*, J. Optim. Theory Appl. **19**, *4*, 1976, 547—564.

244. Thiriez, H., Zionts, S. (eds.), *Multiple criteria decision making*, Jony-en-Josas, France, Springer-Verlag, New York, 1976.

245. Ţigan, St., *Sur une méthode pour la résolution d'un problème d'optimization fractional par segments.* Mathematica — Revue d'analyse numérique et de théorie de l'approximation, **4**, *1*, 1975, 87—97.
246. Ţigan, Şt., Stancu-Minasian, I. M., "Criteriul riscului minim pentru problema Cebîşev". *Lucrările celui de al IV-lea Simpozion "Modelarea cibernetică a proceselor de producţie", 26—28 mai 1983,* ASE—Bucureşti, vol. I, 338—342.
247. Vajda, S., *Probabilistic programming.* Academic Press, New York, 1972.
248. Veličko, D. A. *Random search for variants of solutions according to a vector criterion* (in Russian). *Mathematical methods in operations research and reliability theory* (in Russian), pp. 8—13, Acad. Nauk Ukrain. SSR, Inst. Kibernet, Kiev, 1978.
249. Villarreal, B., Karwan, M. H., *Multicriteria integer programming: A (Hybrid) dynamic programming recursive approach.* Mathematical Programming, **21**, *2*, 1981, 204—223.
250. Yu, P. L., *Convexity, cone extreme points, and nondominated solutions in decision problem with multiobjectives.* J. Optim. Theory, Appl., *14,* 1974, 319— 377.
251. Zangwill, W., *Nonlinear programming. A unified approach.* Prentice Hall International Series in Management, N. Y., 1969.
252. Zeleny, M. (ed.), *Multiple Criteria Decision Making,* Kyoto, 1975, Springer-Verlag, New York, 1976.
253. Zeleny, M., *MCDM bibliography 1975, Multiple criteria decision making,* edited by Zeleny, M. Kyoto, 1975. *Lect. Notes in Econom. and Math. Systems,* Vol. 123, Springer, Berlin—Heidelberg—New York, 1976, 291—321.
254. Zimmermann, H. -J., *Fuzzy programming and linear programming with several objective functions.* Fuzzy Sets and Systems, *1*, 1978, 45—55.
255. Zionts, S., Wallenius, J., *An interactive programming method for solving the multiple criteria problem.* Management Sci., *22*, 1976, B652—B663.
256. Zionts, S., "Integer linear programming with multiple objectives". *Annals of Discrete Mathematics,* North-Holland, Amsterdam, vol. 1, 1977, 551—562.
257. Zionts, S. (ed.), *Multiple Criteria Problem Solving,* Springer-Verlag, New York, 1978.
258. Zionts, S., *A survey of multiple criteria integer programming methods. Discrete Optimisation,* vol. 2, 1979, 389—398, Amsterdam, North-Holland.

Index

Academic planning 28
Active approach 102
Aggarwal's method 235
σ-algebra 72
Algorithms of metric variable 211
Aspiration criterion 167
Assignment problem 307
Assortments 287
Asymptotically normal 90
Auxiliary function 174
Auxiliary problem 194
Average solution (value) 55

Bases 75
Basic feasible solution 24
Basic variables 24
Best compromise solution 8
Black Mesa region 172
Block diagram 245
Boolean variable 307
Borel set 72
Boundary hyperplane (point) 22
Branch-and-bound approach 69

Cartesian integration method 73
Cartesian product 35
Cauchy distribution 264
Cauchy random variables 264
Central processing unit time 81
Certain event 82
Certainty conditions 184
Certainty equivalent 114
Chance-constrained programming 113
Chance constraints 114
Characteristic (solutions) values 79, 137
Chebyshev's inequality 91
Chebyshev (stochastic) problem 9
Chemical processes 298
Chi-square distribution 91
Column vectors 75
Commodity 293
Compact set 63
Complementary variable 13
Complete description method 82
Complete problem 99
Complete recourse problem 103
Compromise solution 8
Computational phase 266
Computer program STOPRO 81
Concave function 15
Conditional probability 77
Cone 211
Confidence region 143
Constrained median type 100
Constraints 11
Consumption centers 270
Continuous function 63
Contradictory criteria 269
Conventional unit 81

Convex function 13
Convex polyhedral cone 22
Convex polyhedral set 24
Convex polyhedron 20
Correlation matrix 116
Correlation method 300
Cost of penalties 306
Covariance matrix 94
Critical regions 137
Critical values 79
Cumulated distribution 115
Cumulative distribution function 172
Cutting-plane approach 69

Decision-maker 34
Decision phase 266
Decision problem 11
Decision region 73
Degree of freedom 200
Determinant 76
Deterministic constraints 115
Deterministic equivalent 71
Deterministic problem 71
Diagonal matrix 144
Dietitian's problem 37
Dinkelbach's algorithm 225
Dinkelbach's theorem 210
Dirichlet distribution 155
Discrete distributions 183
Discrete independent random variable 84
Discrete values 183
Discretization method 86
Distance function 9
Distillation index 299
Distribution density 142
Distribution-free minimum-risk solution 157
Distribution-free optimal basis 157
Distribution-free optimal solution 157
Distribution function 71
Distribution problem 72
Domination structures 70
Down-time 307
Dual cutting plane method 69
Duality principle 220
Duality theory 21
Dual problem 100
Dual variables 276
Dummy operations 305

Efficiency function 33
Efficient-base solution 271
Efficient integer solutions 70
Efficient solution (point) 8, 11
ε-Efficient solution 166
Efficient with probability 1, 165
Ellipsoid 143
E-model 229
Equally probable criterion 185
Essential efficiency 70
Euclidean distance 29
Euclidean space 23
Event 76
Expected win per play 43
Explicitly quasi-concave function 16
Explicitly quasi-convex function 16
Extreme solutions 56
Extreme values 220
Factory management 293
Farcas' lemma 23
Feasible direction 266
Feasible solution 7
Financial decisions 28
Fixed recourse 102
Flowchart 57
Flow time 70
Fractile criterion 167
Fractional programming problem 9
Free minimum-risk solution 157
Fuzzy approach 70

Fuzzy convex cones 70
Fuzzy polar cones 70

Gale's theorem 75
Gamma density 97
Gamma-distributions 117
Game-type simulation method 156
Generalized distribution problem 73
Generalized inverse 32
Generalized inverse method 32
Generalized Tchebycheff norm 69
Geoffrion-Dyer-Feinberg algorithm 68
Geoffrion's method 95
Global maximum 209
Goal programming 28
Goal vector 28
Group decision-making 153

Half-plane 22
Half-space 25
Here-and-now approach 98
Heuristically efficient solutions 70
Hölder's inequality 29
Hölder's norm 29
Hurwicz's criterion 186
Hyperplane 12
Hyperbolic programming problem 189
Hypersphere 124

Identity matrix 99
Idle times 303
Idling cost 298
Implicit enumeration method 69
Incommensurable objectives 33
Incomplete description method 86
Incomplete Gamma function 117
Indefinite programming 222
Indefinite quadratic fractional function 235
Index value 248
Input-Output model 309
Integer and mixed integer programming 69
Interactive approach 261
Interactive multiple-objective methods 68
Interactive rectangle elimination method 68
Inter-branch flux 309
Items 297

Joint probability 230

Kataoka's model 168
Kataoka's problem 92
Kronecker-Capelli's theorem 82
Kronecker's symbol 193
Kuhn-Tucker conditions 221
Kuhn-Tucker saddlepoint theorem 101

Lagrange-multiplier method 169
Lagrange multipliers 170
Lagrange's function 169
Lagrangian approach 169
Land reclamation 172
Laplace function 91
Laplace's criterion 185
Laplace transform 73
Lebesque measure 160
Leontieff matrix 310
Levels 28
Linear stochastic program 72
Loading time 302
Local maximum 209
Location problems 69
Lower bound 92

Machine-tools 305
Man-hour expenditure 287
Mapping 192
Marginal conditioning method 60
Marginal solution 53
Marketing 28
Matrix inequality 75
Matrix of regrets 186
Maximum global utility method 33
Maximum utility solution 40
Maximum verisimility function 143
Mean profit 115
Mean-square strategy 17
Mean value 62
Measurable bounded function 75
Medical care 28
Melting temperatures 304
Minimal degradation method 60
Minimax approach 103
Minimum-risk model 167
Minimum-risk problem 92
Minimum-risk solution 92
Mixed strategies 43
Mixing problem 298
Mixt criteria 56
Mixt probability density 77
Mixt solution 53
Moment-generating function 97
Monotonous norm 152
Multicriterial simplex method 272
Multicriteria optimization 68
Multi-dimensional decision problem 33
Multi-dimensional location model 68
Multiple criteria problem 8
Multiple minimum-risk problem 230
Multiple minimum-risk solution 232
Multiplicative (multidimensional) utility function 174
Mutually independent criteria 35

Natural states 183
Neighbourhood 88
Nets welding workshop 302
Nonanalytical techniques 185
Non-degenerated solution 27
Non-degenerate n-dimensional normal distribution 228
Nondominated extreme points 159
Nondominated solution 8
Non-improvable solution 265
Noninferior solution 8
Nonlinear constraints 258
Non-negative solutions 75
Nonzero-order rule 114
Norm 28
Normal distribution 94
Normalized pay-off matrix 45
Normal random variables 90
North-West corner rules 276
Null space 32

Octane number 299
Operation scheduling 305
Optimal base 88
Optimal solution 8
Optimum value 28
Ordering relation 71
Ordering schemes 12

Parameter space 23
Parametric fractional problem 137
Parametric programming 20
Pareto optimality 8
Pareto's optimum solution 8
Partial derivatives 278
Passive stochastic programming 73
Pay-off matrix 43
Penalties 98

Penalty functions 103
Perturbation factor 142
Pessimistic values 36
Piecewise linear convex function 123
Pig-iron production 304
Planning constraints 294
P-model 92
Polar cone 68
Polyhedral cone 68
Polyhedron 53
POP method 12
Positive (semi-)definite matrix 95
Positive stochastic linear programs 74
Probability convergence 85
Probability density 77
Probability distribution 23
Probability distribution function 92
Probability laws 73
Probability mixture 34
Probability space 71
Production centers 270
Production costs 293
Production planning 28
Productivity 308
Programming in complex space 69
Properly efficient solutions 12
PROTRADE method 172
Prudent's law 185
Pseudoconcave/Pseudoconvex function 16
Psychological factors 309

Quadratic constraints 230
Quadratic form 142
Quadratic programming problem 40
Quantile of order α 92
Quasi-convex (concave) function 16

Radiators 293
Random matrices 72
Random value 71
Random variables 71
Rank of matrix 82
Raw materials 304
Recourse 98
Recursive dynamic programming approach 69
Redundant constraints 125
Regions of integration 79
Regression coefficients 300
Regrets criterion 186
Regularity condition 211
Relaxation factors 256
Relaxation values 232
Risk conditions 184
Running time 81

Satisfactory solution 47
Satisfactory value 176
Savage's criterion 186
Scale factors 99
Scheduling problem 70
Search procedure 12
Second-order moment 167
Second-stage program 105
Selling price 294
Separable function 106
Separation hyperplane 27
Sequential decision 115
Sequential unconstrained minimization technique (SUMT) 210
Service time 303
Shipping cost 269
Shipping time 269
Simple recourse program 99
Simplex method 30
Simplex multipliers 276
Simulation method 85
Slack variable 240
Specific consumption 308
Square matrices 75
Stable distribution 118
Stable optimal basis 157

Stable probability laws 189
Standard deviation 202
Standard normal distribution 92
STEM method 12
Stepwise function 289
Stochastic efficient solutions at level α 172
Stochastic fractional programming 135
Stochastic linear programs with simple randomization 78
Stochastic programming problem 71
Stochastic vector 93
Strategies 33
Strictly concave/convex function 16
Strong feasibility 103
Symmetric distributions 263
Synthesis function 8
System equilibrium equation 309
Szwarc's linearization procedure 284
Swarup's method 235

Table of consequence 53
Taylor series 95
Technological coefficients 310
Technological constraints 293
Technological consumption 310
Theory of games 43
Three-dimensional transportation problem 269
Threshold values 55
Time-graph of Gantt type 306
Total cost 287
Total probability formula 77
Trade-off compromise set 68
Transformation method 95
Transportation cost 272
Transportation problem 269
Two-stage programming under uncertainty 72
Typical values 72

Unbounded set 212
Uncertainty conditions 185
Uncomparable criteria 11
Uniform distribution 102
Uniform domination 52
Unimodal function 67
Unit prices 37
Upper bound 92
Utility function 33
Utility theory 271

Variable transformation 216
Variance 62
Vector function 22
Vector space 8
V-model 229
Von Neumann-Morgerstern's method 36

Wait-and-see approach 73
Waiting cost 306
Waiting time 305
Wald's criterion 185
Weak feasibility 103
Weighted mean 23
Weighting coefficients 30
Weighting factors 145
Welfare function 154
Wets' algorithm 109
Win of player 43
Wolfe's algorithm 41
Workshop 293

Zero-order rule 114
Zero-sum two-person game 43
Zionts-Wallenius algorithm 69

PRINTED IN ROMANIA